Mathematics for Industry

Volume 11

Aims & Scope

The meaning of “Mathematics for Industry” (sometimes abbreviated as MI or MfI) is different from that of “Mathematics in Industry” (or of “Industrial Mathematics”). The latter is restrictive: it tends to be identified with the actual mathematics that specifically arises in the daily management and operation of manufacturing. The former, however, denotes a new research field in mathematics that may serve as a foundation for creating future technologies. This concept was born from the integration and reorganization of pure and applied mathematics in the present day into a fluid and versatile form capable of stimulating awareness of the importance of mathematics in industry, as well as responding to the needs of industrial technologies. The history of this integration and reorganization indicates that this basic idea will someday find increasing utility. Mathematics can be a key technology in modern society.

The series aims to promote this trend by (1) providing comprehensive content on applications of mathematics, especially to industry technologies via various types of scientific research, (2) introducing basic, useful, necessary and crucial knowledge for several applications through concrete subjects, and (3) introducing new research results and developments for applications of mathematics in the real world. These points may provide the basis for opening a new mathematics-oriented technological world and even new research fields of mathematics.

More information about this series at http://www.springer.com/series/13254

Robert S. Anderssen · Philip Broadbridge
Yasuhide Fukumoto · Kenji Kajiwara
Tsuyoshi Takagi · Evgeny Verbitskiy
Masato Wakayama
Editors

Applications + Practical Conceptualization + Mathematics = fruitful Innovation

Proceedings of the Forum of Mathematics for Industry 2014

Springer

Editors
Robert S. Anderssen
CSIRO Computational Informatics
Canberra, ACT
Australia

Philip Broadbridge
La Trobe University
Melbourne, VIC
Australia

Yasuhide Fukumoto
Institute of Mathematics for Industry
Kyushu University
Fukuoka
Japan

Kenji Kajiwara
Institute of Mathematics for Industry
Kyushu University
Fukuoka
Japan

Tsuyoshi Takagi
Kyushu University
Fukuoka
Japan

Evgeny Verbitskiy
Leiden University
Leiden
The Netherlands

Masato Wakayama
Institute of Mathematics for Industry
Kyushu University
Fukuoka
Japan

ISSN 2198-350X ISSN 2198-3518 (electronic)
Mathematics for Industry
ISBN 978-4-431-55341-0 ISBN 978-4-431-55342-7 (eBook)
DOI 10.1007/978-4-431-55342-7

Library of Congress Control Number: 2015946073

Springer Tokyo Heidelberg New York Dordrecht London

Printed on acid-free paper

Springer Japan KK is part of Springer Science+Business Media (www.springer.com)

Preface

This book is the proceedings of the conference "Forum Math-for-Industry 2014," for which the unifying theme was "Applications + Practical Conceptualization + Mathematics = fruitful Innovation." This epigram encapsulates the dynamics of the process that takes an application through to an innovation. Industrial mathematics can be viewed as the causal engine that implements the epigram by taking an Application such as input and convolving it with a mixture of Practical Conceptualization and Mathematics to generate a fruitful Innovation as output.

In industrial mathematics, the questions spawned by real-world applications are what drive the resulting two-way interaction between a particular application and the associated mathematics that is utilized and developed, and that sometimes involves, quite unexpectedly, deeper aspects and new areas of mathematics than initially anticipated.

The plan for the talks at the forum was to illustrate various aspects of this two-way interaction between applications and the associated highlighting of how the practical conceptualization assists with the linking of the question that encapsulates the current application to the relevant mathematics. The organizers believe the plan was quite successful. Readers will find in this proceedings that the forum can actually be viewed as a way for unifying the two-way interaction between applications and mathematics.

In a mathematics-for-industry situation, although the application context and the desired innovation are notionally relatively clear, it takes time to identify the questions to be resolved. It is at this stage that conceptualization plays a key role through the generation of a plethora of possibilities of how to link various questions to the mathematics that will generate reliable and useful answers. In a way, one is reminded of this quotation from Edward David: "The importance of mathematics is not self-evident."

It acknowledges the fact that the role of mathematics plays in solving real-world problems is often taken for granted. Circumspectly, using mathematics to solve real-world problems is similar to a sculptor working with mechanical devices to chisel out one of the possible forms hidden in the block of wood or stone being carved. The one chosen is the result of the current subconscious conceptualization of the artist. That sculpturing could be viewed from this perspective can be found in Soseki Natsume's series of short stories "Ten Nights of Dreams" (*Yume-Juya*) in the Meiji Period (1868–1912). On the sixth night, the dreamer subconsciously visualizes Unkei, the famous Japanese twelfth-century sculptor (1150–1223), in the act of carving the two forbidding guardians of the Buddha Nio guarding the main gate of the Gokoku-ji Temple. Unkei is so absorbed in the carving, he is unaware of the noisy crowd gathered around him. The dreamer then visualizes that some onlookers theorize that the sculpture of the guardians is already hidden in the wood which Unkei is discovering rather than creating. The dreamer then concludes that he should also be able to find Nio in the wood and heads home to attempt it. Unfortunately, he is not able to find one. The dreamer thereby concludes that Nio is no longer in the wood of the Meiji Period.

The first Japanese Field Medallist Kunihiko Kodaira, in 1954, echoed this analogy in his remark that his theory of elliptic surfaces was not invented by him but was just sculptured, using paper and pencil, from the wood of mathematics where it was waiting to be discovered. The corresponding analogy for industrial mathematics, independent of the dynamics outside the context of the application being examined, is: "Answers to the questions that arise in an application are sculptured into reality using the tools of mathematics, which are thereby developed and sharpened by this process."

We would like to thank the participants of the forum, especially the members of the Scientific Board of the Forum. Without their cooperation and support, we would never have experienced the great excitement and success of the forum. Moreover, we would like to express our deep appreciation for the great help of the conference secretaries, especially Tsubura Imabayashi during the preparation and organization of the forum, and Chiemi Furutani for the proceedings.

Fukuoka, Japan
April 2015

Masato Wakayama
On behalf of the Organizing Committee
of the Forum Math-for-Industry 2014
and the Editorial Committee of the Proceedings

FMI2009	FMI2009	FMI2010	FMI2011	FMI2012	FMI2013	FMfI2014
Tokyo	Fukuoka	Fukuoka	Honolulu	Fukuoka	Fukuoka	Fukuoka
Sep. 19-17	Nov. 9-13	Oct. 21-23	Oct. 24-28	Oct. 22-26	Nov. 4-8	Dec.27-31
The 1st Forum:Consortiu m Math For Industry	Casimir Force, Casimir Operators and the Riemann Hypothesis	Information Security, Visualization, and Inverse Problems, on the basis of Optimization Techniques	TSUNAMI - Mathematical Modelling- Using Mathematics for Natural Disaster Prediction, Recovery and Provision for the Future -	Information Recovery and Discovery	-The Impact of Applications on Mathematics-	-Applications + Practical Conceptualiza tion + Mathematics = fruitful Innovation-

Forum "Math-for-Industry" 2014
Applications + Practical Conceptualization + Mathematics = fruitful Innovation
October 27(Mon) - 31(Fri), 2014
Venue: Nishijin Plaza, Kyushu University
2-16-23 Nishijin, Fukuoka City
Invited Speakers
Gary Froyland University of New South Wales
Masahito Hasegawa RIMS, Kyoto University
Hans-Christian Hege Zuse-Institute Berlin (ZIB)
Thorsten Koch TU Berlin / Zuse-Institute Berlin (ZIB)
Kerry Landman The University of Melbourne
Vladimir Lorman CNRS & Université Montpellier 2
Reinout Quispel La Trobe University
Ernesto G. Birgin University of São Paulo
Daniel Braak University of Augsburg
Íñigo L. Egusquiza University of the Basque Country, Bilbao
Farid Melgani University of Trento
Robert Norman RMIT University
Konrad Polthier Freie Universität Berlin
Enrique Solano University of the Basque Country, Bilbao
Akira Takada Asahi Glass Co., Ltd.
Roger C. E. Tan National University of Singapore
Kenichi Arai NTT Communication Science Laboratories
Zainal Abdul Aziz Universiti Teknologi Malaysia
Troy Farrell Queensland University of Technology
Luke Fullard Massey University
Arnab Ganguly University of Louisville
Sachiko Ishida Meiji University
Shinsaku Kiyomoto KDDI R&D Laboratories, Inc.
Akira Ohata Toyota Motor Corporation
Takashi Sasaki Yokogawa Electric Corporation
Toshinao Yoshiba Bank of Japan
Alexandra Hogan Australian National University
Keita Iida Tohoku University
Yasuaki Kobayashi Hokkaido University
Lucas Lamata University of the Basque Country, Bilbao
Mikel Sanz University of the Basque Country, Bilbao
Eriko Shinkawa Graduate School of Mathematics, Kyushu University
Supported by
Institute of Mathematics for Industry
Kyushu University
Co-Sponsored by
CESS
Organizing Committee
Bob Anderssen CSIRO, Australia
Philip Broadbridge La Trobe University
Yasuhide Fukumoto IMI
Kenji Kajiwara IMI
Tsuyoshi Takagi IMI
Evgeny Verbitskiy Leiden University / University of Groningen
Masato Wakayama IMI, Chair
Supported by
Institute of Mathematics for Industry, Kyushu University
FMfI2014 Organizing Office
IMI: Institute of Mathematics for Industry, Kyushu University
fmi2014@imi.kyushu-u.ac.jp http://fmi2014.imi.kyushu-u.ac.jp
Tel: 81 92 802 4401/4404 Fax: 81 92 802 4405
Scientific Board
Hirokazu Anai FUJITSU LABORATORIES LTD.
Shin-Ichiro Ei Hokkaido University
Tim Hoffmann Technische Universität München
Miyuki Koiso IMI
Ryuei Nishii IMI
Jill Pipher Brown University
Hisayoshi Sato Yokohama Research Laboratory, Hitachi, Ltd.
Tomoyuki Shirai IMI
Katsuyuki Takashima Mitsubishi Electric Corporation
Hayato Waki IMI
Takeshi Yamada NTT Communication Science Laboratories
Asia Pacific Consortium of Mathematics for Industry
九州大学
KYUSHU UNIVERSITY

Forum "Math-for-Industry" 2014

-Applications + Practical Conceptualization + Mathematics = fruitful Innovation-

October 27 to 31, 2014, Nishijin Plaza, Fukuoka, Japan

	October 27, 2014		October 28, 2014		October 29, 2014		October 30, 2014		October 31, 2014
	Registration		Registration		Registration		Registration		Registration
9:50	Opening Ceremony								
10:00 - 10:50	Thorsten Koch TU Berlin / Zuse-Institute Berlin (ZIB) Mixed-Integer Nonlinear Programs for Gas Network Optimization	9:45 - 10:35	Kerry Landman University of Melbourne An Interaction with Biologists: Insights into Development and Disease	9:45 - 10:35	Gary Froyland University of New South Wales Dynamical Systems Approaches to Analysing Transport in Geophysical Fluid Flow	9:45 - 10:35	Hans-Christian Hege Zuse-Institute Berlin (ZIB) Visual Analysis of Molecular Dynamics Data Using Geometric and Topological Methods	9:45 - 10:35	Masahito Hasegawa RIMS, Kyoto University Programming Languages, Tensor Categories, and Quantum Topology
10:55 - 11:30	Kenichi Arai NTT Communication Science Laboratories Synchronization of Semiconductor Lasers for Secret Key Distribution	10:40 - 11:15	Arnab Ganguly University of Louisville Modeling Biochemical Reaction Systems with Markov Chains	10:40 - 11:15	Akira Takada Asahi Glass Co., Ltd. Mathematical Modeling of Inorganic Glass Materials	10:40 - 11:15	Shinsaku Kiyomoto KDDI R&D Laboratories Inc. Towards Privacy-Preserving Services, -Risk Analysis and Solution-	10:40 - 11:15	Takashi Sasaki Yokogawa Electric Corporation Modeling and Control in Paper-Making Machines
11:35 - 12:10	Zainal Abdul Aziz Universiti Teknologi Malaysia UTM-CIAM: Transformation and Beyond Malaysian Mathematics for Industry	11:20 - 12:05	Daniel Braak University of Augsburg Analytical Solutions of Basic Models in Quantum Optics	11:20 - 12:05	Akira Ohata Toyota Motor Corporation Boundary Model Identification for Automotive Engine Controls	11:20 - 12:05	Robert Norman RMIT University Platform Technologies for Space Atmosphere and Climate: An Australian Space Research Program Project	11:20 - 12:05	Farid Melgani University of Trento Recent Methods for Reconstructing Missing Data in Multispectral Satellite Imagery
	LUNCH		LUNCH		LUNCH		LUNCH		LUNCH
13:40 - 14:30	Vladimir Lorman CNRS & Université Montpellier 2 Group Theory Methods in Virology: Landau Density Wave Approach	13:35 - 14:20	Enrique Solano University of the Basque Country, Bilbao Quantum Simulations and Quantum Technologies	13:35 - 14:25	Reinout Quispel La Trobe University Geometric Numerical Integration of Differential Equations and its Applications	13:35 - 14:10	Luke Fullard Massey University Modelling the Discharge of Particulates from a Hopper-Silo System– How a 6 week Industrial Contract Turned into a Research Career	13:35 - 14:10	Roger C. E. Tan National University of Singapore Sensitivity Analysis and its Numerical Methods for Quadratic Eigenvalue Problems
14:35 - 15:20	Ernesto G. Birgin University of São Paulo Applications of Nonlinear Programming to Packing Problems	14:25 - 15:35	Young Reseacher Session Alexandra Hogan A Model for Respiratory Syncytial Virus (RSV) Transmission Keita Iida Research to Construct a Mathematical Theory that can Compute Stochastic Cellular Processes Yasuaki Kobayashi Mathematical Modeling for Epidermal Structure	14:30 - 15:05	Íñigo L. Egusquiza University of the Basque Country, Bilbao Beyond Adiabatic: Effective Hamiltonians and Singular Perturbation	14:15 - 15:25	Young Reseacher Session Lucas Lamata Quantum Simulations with Trapped Ions and Superconducting Circuits Mikel Sanz Quantum Biomimetics and Bio-Inspired Quantum Devices Eriko Shinkawa Uniqueness and Stability for Double Crystals in the Plane	14:15 - 14:50	Toshinao Yoshiba Bank of Japan Risk Aggregation with Copula for Banking Industry
	COFFEE BREAK		COFFEE BREAK		COFFEE BREAK		COFFEE BREAK		COFFEE BREAK
15:40 - 16:15	Sachiko Ishida Meiji University Origami Evolution from Mathematics to Engineering			15:30 - 16:00	APCMfI Launch Ceremony -"The Importance and Future of APCMfI"-			15:10 - 15:55	Konrad Polthier Freie Universität Berlin Reliable Computing in Geometry Processing and CAGD
16:20 - 16:55	Troy Farrell Queensland University of Technology Mathematics for Industry: A Mining and Resources Case Study	15:55 - 18:00	Poster Session	16:05 - 17:00	Poster Session Voting	15:45 - 18:00	Poster Session Award Ceremony	16:00	Closing
	Welcome Party (Snack & Beverages)		PHOTO	18:00	Banquet Hilton Fukuoka Sea Hawk		PHOTO		Excursion

Organizing Committee

Bob Anderssen (CSIRO, Australia)
Philip Broadbridge (La Trobe University)
Yasuhide Fukumoto (IMI)
Kenji Kajiwara (IMI)
Tsuyoshi Takagi (IMI)
Evgeny Verbitskiy (Leiden University, University of Groningen)
Masato Wakayama (IMI), Chair

Scientific Board

Hirokazu Anai (FUJITSU LABS. LTD.)
Shin-Ichiro Ei (Hokkaido University)
Tim Hoffmann (Technische Universität München)
Miyuki Koiso (IMI)
Ryuei Nishii (IMI)
Jill Pipher (Brown University)
Hisayoshi Sato (Yokohama Res. Lab., Hitachi.)
Tomoyuki Shirai (IMI)
Katsuyuki Takashima (Mitsubishi Electric.)
Hayato Waki (IMI)
Takeshi Yamada (NTT Comm. Sci. Labs.)

Contents

UTM-CIAM: Transformation and Beyond Malaysian Mathematics for Industry

Zainal Abdul Aziz and Arifah Bahar

Abstract This article deliberates on how the Malaysian industries and mathematicians have come to revive the synergy of mathematics and industry through the Malaysian Mathematics in Industry Study Groups (MISG 2011, 2014). The Malaysian setting of pre MISG was a disengaged connection between industries and mathematics. Post MISG 2011 and 2014 have seen intensified partnership between local industries and mathematical community beginning to crystallize. The founding of UTM Centre for Industrial and Applied Mathematics (UTM-CIAM) with seven permanent staff at the end 2012 was a follow-up effect of the MISG 2011. It is a positive transformation for the Malaysian Mathematics for Industry scene. The Malaysian MISG is organized in cooperation with Oxford Centre for Industrial and Applied Mathematics (OCIAM), University of Oxford. These are collaborative problem-solving workshops where more than seventy mathematicians, operational researchers and statisticians deal with real life problems brought up by six private and public companies. These workshops assist to find out promptly the key scientific issues and mathematical challenges to be confronted. The meeting provides opportunities for bridging the gap between academics and scientists from Malaysian industry, and encourages innovative knowledge and technology transfer. This work also summarizes the successful collaboration formed between academics and industry practitioners in solving specific problems from the national high revenue industries during the Malaysian MISG 2011 and 2014.

Keywords Industrial mathematics · UTM-CIAM · MISG · OCIAM · Malaysian industry · STEM · Mathematical modelling

---

Z.A. Aziz (✉) · A. Bahar
Faculty of Science, Department of Mathematical Sciences,
UTM Centre for Industrial and Applied Mathematics,
Universiti Teknologi Malaysia,
81310 UTM Johor Bahru, Johor, Malaysia
e-mail: zainalaz@utm.my

A. Bahar
e-mail: arifah@utm.my

R.S. Anderssen et al. (eds.), *Applications + Practical Conceptualization + Mathematics = fruitful Innovation*, Mathematics for Industry 11,
DOI 10.1007/978-4-431-55342-7_1

1 Introduction

Scientific research methods and outcomes are essential to the progress of industrial innovation. These related researches are strengthened and driven basically by industrial mathematics. However, due to the nature of mathematics as an enabling technology for the innovative industry, its input is rarely visible in the final product that industry delivers. Current manufacturing and technology, engineering and service industries have progressed dramatically in modern times due to the upsurge in the knowledge-based economy. Speedy and inexpensive computing, the growth and exploitation of huge databases to cope with this sudden requirement, certainly have necessitated advanced methods to meet new demands. Industrial mathematics is principally one of the enabling factors in carrying out and implementing these methods. One of the important fundamental factors in this respect is the use of theoretical mathematics as the primary driving feature for innovation. Only through mathematics can the complex processes and products in current major technologies and innovation cycles be administered in a cost-effective, strong and sustainable way. In this new millennium, realistically it is most useful to assume a thorough outlook of industrial mathematics research. In particular, this can be based primarily on the precise idea of "construction and analysis of models", as portrayed in Fig. 1, [1, 2]. This treatment becomes more effective if the underlying fundamental knowledge is based on theoretical mathematics and the closely related fields as in STEM (Science, Technology, Engineering and Mathematics). This work demonstrates that industrial mathematics will be a major enabling technology for the Malaysian industries.

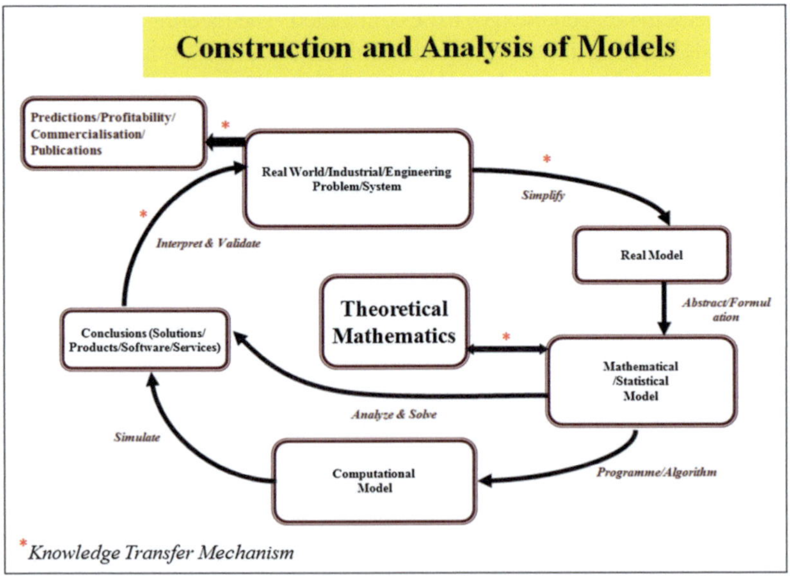

Fig. 1 Construction and analysis of models

This work elucidates on how the Malaysian industries have come to stimulate the synergy of mathematics and industry through the Malaysian Mathematics in Industry Study Groups (MISG 2011 and MISG 2014). The Malaysian setting before MISG experienced a disengaged relation between industries and mathematics. There was no earnest relationship for the local industries to provide industrial problems and equally no takers from the Malaysian mathematical community to offer decision support and solutions. Post MISG 2011 and 2014 have seen intensified partnership between local industries and mathematical community beginning to crystallize. Letters of Intention which lead to Memorandums of Understanding were signed and further collaborative problem solving efforts ensued. These involve companies and agencies like J-Biotech Environment Sdn. Bhd, National Hydraulics Research Institute Malaysia (NAHRIM), PROTON (Automotive) Berhad, KPJ Healthcare, PETRONAS (Oil & Gas) Melaka, Hospital Universiti Sains Malaysia (HUSM). The founding of Universiti Teknologi Malaysia Centre for Industrial and Applied Mathematics (UTM-CIAM) at the end of 2012 was a strong follow-up of the MISG 2011. It is a crucial transformation for the Malaysian mathematics for industry scene. This establishment creates a custom-built programme for active linkup particularly via the centre's flagship event in study group.

The Malaysian Mathematics in Industry Study Group is the centre's flagship event in study group and is organized in cooperation with Oxford Centre for Industrial and Applied Mathematics (OCIAM), University of Oxford. These are collaborative problem-solving workshops where more than seventy applied mathematicians, operational researchers and statisticians deal with real life problems brought by six private and public companies, and presented by the respective industrial representatives [3, 4]. These intense problem-solving sessions help to find out rapidly the main scientific and engineering issues, and mathematical challenges to be confronted. These MISGs have been a key tool for delivering the underpinning discipline of mathematics, operational research and statistics to difficult industrial problems. The meeting provides opportunities for linking the gap between academics and scientists from industry, encourages practical innovative knowledge and technology transfer and certainly occasions for young STEM graduates and postdoctoral fellows to explore possibilities of future employment. This article also summarizes the successful collaboration formed between academics and industry practitioners in solving specific problems from our national high revenue industries during the Malaysian MISG 2011 and 2014.

2 UTM Centre for Industrial and Applied Mathematics (UTM-CIAM)

UTM Centre for Industrial and Applied Mathematics (UTM-CIAM) is a research centre of excellence in Universiti Teknologi Malaysia. It is established with the strategic cooperation of Oxford Centre for Industrial and Applied Mathematics (OCIAM).

It was launched via a collaborative agreement which was signed on 12 December 2012 in the presence of the then Malaysian Minister for Higher Education at St. Anne College, University of Oxford. The establishment of UTM-CIAM is to facilitate the research cooperation with Malaysian industry and other disciplines including science, engineering and technology. This research would notably embed modelling and prepare to generate an innovative and optimal solution to be shared with the industry. In parallel with UTM-CIAM's vision to become a global renowned centre and to pioneer mathematics in the nation's industry, the main agendas of UTM-CIAM are to advance multidisciplinary research, to promote collaborative mathematical research with various disciplines, and to encourage practical applications of industrial and applied mathematics.

Mirroring the setting up of OCIAM, the objectives of UTM-CIAM are specifically laid down to ensure our capacity building in advanced research areas of industrial & applied mathematics, intensification of research & development (R & D) on joint industrial & applied mathematics with fields related to science, engineering and technology, advance multidisciplinary R & D by strengthening the collaborative research, modelling and formulation of optimal & innovative mathematical modelling solutions as the crucial enabling technology for the local industry (use of mathematical techniques: differential equations, operational research (OR) methods & statistical analysis) and last but not least in promoting practical & innovative applications of industrial and applied mathematics in industry to generate high quality applied research output.

Given this development (refer to Fig. 2), UTM-CIAM has launched many initiatives to persuade the government (policy makers), industry, academia & community user/civil society that the time is ready for change. Besides MISGs, the initiatives include the organization of International Seminar on Mathematics in Industry (ISMI)

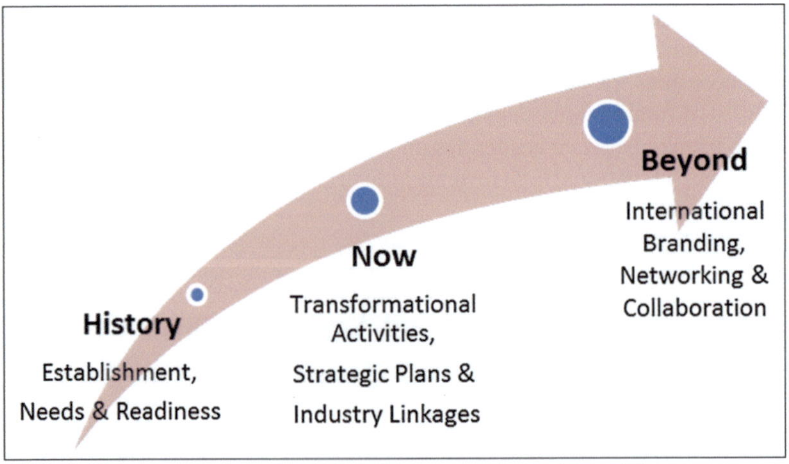

Fig. 2 UTM-CIAM pioneering Malaysian mathematics for industry

2013, National Seminar on Mathematics in Industry 2014, procurement of STEM research grants in the 2015 Public Private Research Network (PPRN) Malaysia, and paperwork to the Malaysian Ministry of Education on the national STEM graduate employability activities. The transformational activities, strategic plans and industry linkages are planned to bring about the important insight that industrial mathematics work is an increasingly central and necessary part of significant areas of investigation in medicine & healthcare, advanced design, environment, manufacturing, oil & gas, finance, agriculture and many more relevantly related to the Malaysian industries.

The targets for industrial linkages are chosen mainly from Small and Medium Enterprises (SMEs), Government Linked Companies (GLCs) types of industries and national research institutes with strong link with such industries. These industries contribute significantly to the national GDP. For example, in 2011 45 % of the government's budget was dependent on PETRONAS' dividend [5], in 2012 PROTON provides a fair share of the automotive sector contribution to the nation's GDP of 3.2 % [6], and similarly as for 2011 the KPJ Healthcare imparts a large share in the total healthcare spending in Malaysia amounted to 4.7 % of the country's GDP [7]. These industrial relationships can be further exploited with international linkages, networking and branding that UTM-CIAM is having with OCIAM UK, Centre for Mathematics in Industry (CMI), New Zealand, Institute of Mathematics for Industry (IMI), Kyushu University and Centre for Mathematical Modelling and Simulation, Institut Teknologi Bandung, Indonesia. The objective is to promote and foster the effective use of industrial mathematics and closely related knowledge and expertise (particularly related to STEM) in industrial, science & technology and management settings. More specifically, to perceive what is required by industry and business, to consider what is available and to discuss various plans to fill the gaps. The areas of mathematics are chosen based on the industrial problems being proposed in the MISGs which are predominantly in the areas of oil & gas and environment. These include differential equations, advanced mechanics (classical and fluid mechanics), mathematical modelling & simulations, OR methods and statistical analysis.

3 Transformation: Pioneering Malaysian Mathematics for Industry

In the efforts to increase research alliances with the industry, UTM-CIAM's researchers regularly visit potential industries to discuss the associated research projects and in the long run to strengthen the respective industrial partnership. Some of the industries that have been identified to currently retain active working relationship are NAHRIM, KPJ Healthcare, PETRONAS Melaka, J-Biotech Environment Sdn. Bhd, PROTON Berhad, Department of Civil Aviation (DCA), HUSM, Malaysian Agriculture Research & Development Institute (MARDI), and PROSPECT (Chemical processes). As a consequence, a Memorandum of Understanding (MOU) has been agreed between UTM-CIAM and HUSM, and UTM-

CIAM and J-Biotech Environment. Further MOUs are in the pipeline and soon will be activated.

As mentioned above, one of the main flagship activities being organized by UTM-CIAM is Mathematics in Industry Study Group (MISG). In 2014, UTM-CIAM collaborated with OCIAM, University of Oxford in holding the 2nd MISG Malaysia. Initially the industrial problems provided by the local industries are prepared and identified, and thus finally endorsed during an earlier discussion session between UTM-CIAM and OCIAM. Subsequently it is then determined that for MISG 2014, there are six industrial problems which would involve rigorous mathematical modelling. These include oxidation pond problem (J-Biotech Environment Sdn. Bhd), riverbank filtration problem (NAHRIM), electromagnetic shock absorber problem (PROTON Berhad), paddy field pest population problem (MARDI), blood flow at bifurcated artery (KPJ Healthcare), and water complex and heat integration in industrial process (PROSPECT). The study group is a rigorous workshop for problem solving where mathematician's expertise is being employed to overcome these real life problems proposed by the industry. In relation to that, many STEM academics with relevant backgrounds from OCIAM international network and local institutions are invited. Together with the industrial representatives, they would quickly help to determine the important scientific-industrial concerns and challenges of the study group during the problem solving sessions of the industrial problems.

As a result, MISG has given us the opportunity to end gradually the previous rift between academics, and industrial practitioners and scientists from the industries. It is shown practically that MISG has become a pragmatic platform to disseminate methods and emphases on the field of industrial and applied mathematics in numerous difficult industrial problems. We would like to emulate our European counterparts [8, 9] with excellent examples of European industry and mathematical sciences coming together to overcome various complex challenges of European industries face. Moreover in 2015, the new scheme is to organize the first Malaysia Mathematical Modelling Camp 2015 (MMMC 2015) before the annual MISG 2015. The aim of MMMC 2015 is to train and expose our young postgraduates and postdoctoral fellows to hands-on experiences on a broad range of real industrial problem-solving skills such as mathematical modelling & analysis, scientific computation and critical assessment of solutions. The problems to be considered are inspired by real problems that have arisen in industry.

In addition, we contemplate that our current and proposed programmes (e.g. MISG, MMMC, planned internships in industry) will have certain impact on the increase in our STEM graduate employability [10]. Undergoing these activities would facilitate our STEM graduates to be ready made for various sorts of industrial purposes and thus opening up a vast opportunity for young mathematical scientists to opt for other more satisfying careers in industry. Experience shows that successful and rewarding industrial problems require an interdisciplinary team and mathematical scientists can be indispensable contributors to this set up.

4 Action Plan and Roadmap

Based on Fig. 3 and the lessons learned from the first Malaysian MISG 2011, the action plan and roadmap for UTM-CIAM are strategized to realize the potential of industrial mathematics for breakthroughs and innovations in industrial and other societal problems. Our next steps are to continuously implement the flagship events including ISMI, MMMC and MISG. These are rooted in a natural and constant growth of activities which steps up on their strength over the six year period (2012–2018). All these developments are vital for the long term growth and competitive power, and demand an in-depth review of the overall progression of UTM-CIAM towards becoming a national higher institution centre of excellence (HiCOE) or simply a national referral point in industrial mathematics. This carries implications on the nature and scale of research and operational funding. The total amount of various funding secured so far by UTM-CIAM sums up to more than USD500K or RM1.65M (until early 2015). Its ambitious aim is to become the national dedicated one-stop centre to coordinate and facilitate the needed interactions in the domain of application-driven mathematical research and its development for innovations in industry, science and society. A national centre implies that other researchers in the field of industrial mathematics from other Malaysian universities will be able to link-up with the local industries via UTM-CIAM's activities, particularly ISMI, MMMC and MISG. For ISMI 2013, MISG 2011 & 2014, almost a third of the academics came from other Malaysian universities including Universiti Sains Malaysia (USM), Universiti Malaya (UM), Universiti Malaysia Perlis (UNIMAP), Universiti Putra

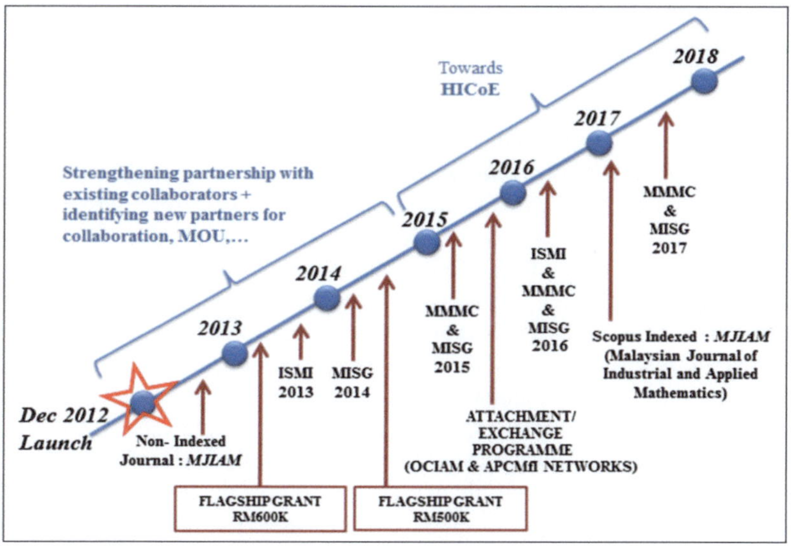

Fig. 3 Roadmap (2013–2018)

Malaysia (UPM) and Universiti Malaysia Terengganu (UMT). The MISG 2015 will be co-organized by UPM. The local researchers also will be able to take part in the planned attachment and exchange programmes, which will be made available via networks of OCIAM and APCMfI (Asia Pacific Consortium of Mathematics for Industry; refer to Sect. 6).

Undoubtedly, the continuous multidisciplinary research and novel mathematical, operational, statistical and computational methods are required to provide necessary tools and methods of solution for Malaysian industrial innovation and competitive edge. These will be documented in our Malaysian Journal of Industrial and Applied Mathematics (MJIAM). Although the mathematics-industry interaction has reached an acceptable level in some countries of the European Union and Australia-New Zealand, it is far from being equally developed in Malaysia and South East Asia. Thus the action plan and roadmap of UTM-CIAM are strategized to address these issues directly.

5 Industry Collaboration/Linkages

In recent years, university-industry linkages in Malaysia have enlarged given the rapid pace in knowledge generation as well as the escalating costs associated with R & D activities. UTM-CIAM has recognized that collaborations with industry and other external parties bring with them benefits to teaching and skills development; access to funding and empirical data from industry; reputation enhancement; application of knowledge; opportunities for talent development by both students and staff; and promotion of entrepreneurship. Concurrently with the Malaysian Ministry of Education's initiatives in promoting linkages and knowledge transfer, UTM-CIAM efforts in creating partnerships and collaborations with industry need to be streamlined and enhanced.

In order to strengthen the industrial linkages, UTM-CIAM plans to embrace the "*Quadruple Helix* (QH) *innovation model*", which refers to a model or framework describing the interaction or innovation cooperation between four clusters, namely government, industry, academia & community user/civil society. A general definition of the QH innovation model [11] refers to an innovation cooperation model or innovation environment in which users, private firms, universities and public authorities cooperate in order to produce innovations. These innovations can be anything that is considered useful for the partners in innovation cooperation, for example, technological, social, product, service, commercial, non-commercial, private-sector and public-sector innovations. Applying this QH innovation model implies that UTM-CIAM's linkage between the four clusters would involve interaction in the forms of *relation* (involving contract research, consultancy, etc.), *mobility* (involving research training, modelling workshop, study group, etc.), *transfer* (involving copyright, commercialization, etc.,) and *formality* (involving MOU, MOA, LOI, etc.). This framework has been used to describe the inner workings of regional innovation systems. According to [12], this framework already encourages the perspective of the knowledge society and of knowledge democracy for knowledge generation and innovation; and the sustainable development of a knowledge economy requires a co-evolution with the knowledge society. Thus this description suits well with UTM-CIAM's action plan and roadmap in strengthening, sustaining and identifying further interactive partnerships with industries.

The following notes are given as solvable samples of brief equation-free reports on two industrial problems in the MISG 2014 Malaysia.

5.1 Case Study on Oxidation Pond

The oxidation pond that becomes the pilot scale study is Taman Timor Oxidation Pond, Tampoi, Johor, Malaysia estimated about 1,909 square meters and has a depth about 1.5 meters and total volume of water about 2,864.13 cubic meters or 2,864,125.13 litres. To improve the efficiency of the water treatment, beneficial microorganism based product (called mPHO, a bio-product used to improve water quality) has been added to the pond to provide efficient removal and break down of solid content of influent and pathogens, which are generally not sufficiently removed by oxidation pond. The product mPHO is made from the selected species of phototrophic bacteria (PSB). About 1,375 litres of mPHO have been applied to the Oxidation Pond, throughout 3 months of treatments. Samples were collected at two points of the pond which are CP1 (influent and application of mPHO) and CP2 (effluent). The microbial approach can help in purifying heavily polluted water especially in the area that is exposed to sunlight. The ability of mPHO to photosynthesize and photometabolize many organic substances may help to reduce BOD (biological oxygen demand), COD (chemical oxygen demand), TSS (total suspended solids), NH3-N (ammoniacal nitrogen) and bacteria (E. coli and Coliform) content in wastewater and increases the amount of DO (dissolved oxygen) in the pond. Therefore, the right amount of mPHO must be determined to ensure that sufficient dose is added, thus maintaining its efficiency and prevent excessive use.

However, in this study the industrial partner wants to investigate the ability of mPHO in reducing the population of E. coli and Coliform in the pond. This is because that microorganism (E. coli and Coliform) is the most harmful living organism in the pond that may affect human's health. To solve this problem, we have built a model that includes the interaction of these three microorganisms (PSB, E. coli and Coliform).

Oxidation pond techniques have practically proved to be effective for wastewater treatment process (WWTP) because of their low construction and operating cost. Cumbersome sampling is required to monitor the dynamics of the WWTP which also involves enormous costly work. Stochastic model accommodating the correlation between the amount of phototrophic bacteria in mPHO and pollutant (bacteria E. coli and Coliform) existing in oxidation pond is developed to facilitate the analysis of this process. This study presents a stochastic model for an oxidation pond to investigate the effect of mPHO on the degradation of pollutant. The model consists of a system of stochastic differential equations with coupled reaction equations for the pollutant and phototrophic bacteria. The parameters of the model is estimated using the real data collected from the oxidation pond located in Taman Timor, Johor, Malaysia to illustrate a real life application of this model. The simulation results provide a better understanding of the ecological system in the oxidation pond.

5.2 Case Study on Groundwater Modelling

In tropical countries like Malaysia where the rainfalls continuously recharge river flow, the main source of dependable water supply is essentially river water. However, as development and economic activities spread, the management of water resources can be very critical due to increased demand as well as environmental degradation. Pollution of rivers has made surface water unsuitable for sources of raw water for treatment and in certain cases has caused the treatment costs to rise unexpectedly. One of the alternative ways to improve dependable clean potable water supply that is being considered is through Riverbank filtration (RBF) technology as a second source of water supply to guarantee clean and dependable watersupply solution without neglecting the polluted surface water.

Riverbank filtration (RBF) is a natural technique for surface water treatment, based on the natural removal of pollutants from water during its transfer through the aquifer to pumping well. This technology is applied in USA and several European countries. Recently, RBF technique has been applied by our industrial partner for the first time in Malaysia in a pilot project conducted in Jenderam Hilir, located in Langat Basin, Selangor, Malaysia. This approach had been approved to be a very effective technology that reduces pollutants concentration and it has potential benefits for drinking water supply in Malaysia. To manage and operate RBF system efficiently, it is extremely important to evaluate the potential for contamination of drinking water wells by river pollution.

To manage and protect the water supply, transport processes need to be predicted by using the RBF method. This study specifically simulates the groundwater flow and contaminants transport induced by pumping well and investigates the evolution of the water chemistry during bank-filtrations systems.

Riverbank filtration (RBF) [13] is a natural technique for surface water treatment, based on the natural removal of pollutants from water during its transfer through the aquifer to pumping well. In this study a mathematical model is used to simulate three major processes related to RBF which are: (1) groundwater flow and contaminants transport from river towards the pumping well (2) arrival time needed by contaminants to reach the well and (3) evolution of water chemistry due to microbial activity. Whiles the first and second problems were solved numerically by using MODFLOW, the third problem was solved analytically. MODFLOW simulation highlighted the capture zone influenced by each pumping well. Also it was found that the contaminant needs 75 days to reach the pumping well. Furthermore, the analytical model for the third problem showed that NO_3 and O_2 were consumed within 10 cm of the aquifer whiles SO_4 consumed within 30 cm.

The impact of these MISG efforts is to put forward workable and practical solutions to the respective industries in order for the companies to adhere to certain best practices, particularly in implementing their standard operating procedures, cost effective and optimal measures on the specific problems. Post MISG collaborative endeavour requires ongoing follow-ups on the specific problems, ensuring a more developed solution will be finally obtained. Besides, the ongoing venture will assure a long-term university-industry relationship.

6 Beyond Malaysian Mathematics for Industry

In order to progress beyond the Malaysian landscape and to strengthen the partnership between UTM-CIAM and the industries globally, UTM-CIAM has been selected as the sole Malaysian representative at a special meeting in Canberra, Australia on 31 Mac–2 April 2014. This encounter discussed the formation and launching of Asia Pacific Consortium of Mathematics for Industry (APCMfI), where the major interest is on mathematics for industry. The centre of APCMfI is agreed to be located at the Institute of Mathematics for Industry, Kyushu University, Fukuoka, Japan. The meeting and discussion basically prepared UTM-CIAM for the next level of engagement. The formation of APCMfI has obtained the strong support and encouragement from China, Korea, Malaysia, Singapore, Australia, New Zealand and also Japan. UTM-CIAM strongly hopes that this affiliation would bring us to another level of commitment which goes beyond the Malaysian mathematics for industry.

In the face of challenges as one of the high-tiered national research universities and to increase her capacity and capability, Universiti Teknologi Malaysia (UTM) is compelled to reassess her current R & D ecosystem. Strengthening the competitive advantage and increasing the benchmark performance in R & D locally and globally requires a rethinking process of research, innovation and commercialization. The strategy adopted and implemented is the consolidation and reorganization of UTM centres of excellence (COE) into a Research Institute. It is now acknowledged that UTM-CIAM will be consolidated and reorganized as an entity within a research institute, namely the Ibnu Sina Institute for Scientific and Industrial Research (Ibnu Sina ISIR). The Ibnu Sina ISIR consists of six COEs and forty permanent research staff with various backgrounds and research inclinations, as portrayed in Fig. 4. UTM-CIAM deeply foresees this as another forward look initiative and level of engagement to strengthen the current collaboration and consolidation beyond Malaysian mathematics for industry.

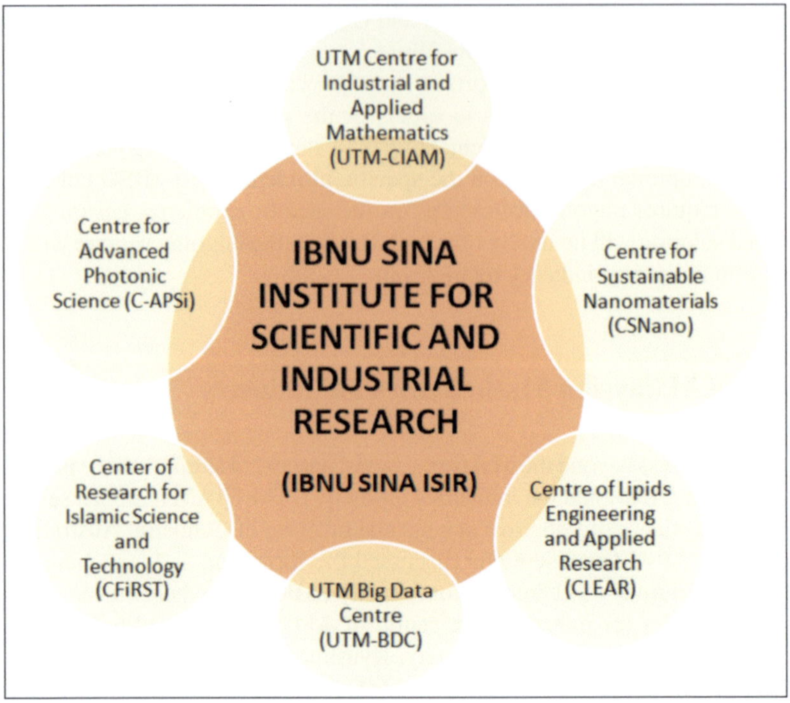

Fig. 4 Strengthening collaboration & consolidation beyond mathematics, where Islamic science is the concept as discussed in Aziz [14]

7 Conclusion

From the Malaysian perspective, the forward look at mathematics and industry arose from the firm belief that UTM-CIAM has the potential to become an important economic and STEM skilled human capital resources for the Malaysian industry, helping its innovation and hence its capacity and capability of competing on the global market. The creation of UTM-CIAM is indeed to steadily close the gap between the industrial mathematicians and the industrialists, government, and civil society. Thus based on the above discussions, the streamlined programmes and initiatives set out by UTM-CIAM, the future plan is to generate more affirmative impact and response via the quadruple helix innovation model mechanism. Besides mathematics, UTM-CIAM is increasingly reliant on science, technology and engineering to help boost innovation and economic growth and to improve our quality of life [15]. With this dependence comes a growing need for improved education and training in STEM subjects including mathematical modelling, operational research and statistics, both for the scientific and technical workforce and for the general public in a gradually more technological and digital world.

Furthermore we hope that from these programmes and initiatives, the hundreds of the Malaysian SMEs, multinational GLCs and the private ones will become progressively more aware of the significance of the applications of mathematics for their commercial survival. These Malaysian industries should now come to terms and appreciate that industrial mathematics is one of the keys to expand and sustain their economies and profitability. Moreover, the professional and academic associations, including the Malaysian Mathematical Science Society (PERSAMA) and the Malaysian Academy of Mathematical Scientists, ought to be proactive in order to play a major role as catalysts in influencing the culture and development of industrial mathematics and related mathematics-industry interface. These can be done by actively supporting high-profiled activities such as conferences, MISG workshops, and weekly industrial problem solving gatherings in the field of industrial mathematics.

Acknowledgments The authors are thankful to the reviewers for the constructive comments and to University Teknologi Malaysia for providing financial support through 01G22 and 02G00.

References

1. Aziz, Z.A.: An Overview of the importance of industrial mathematics. J. Sci. Math. **2**(2), 9–17 (2010). ISSN: 1985-7918
2. Bahar, A., Aziz, M.K.B.M., Salleh, M.M.: Boundedness of the solution of stochastic power law logistic model. In: Aziz, Z.A. (ed.) Research in Applied Mathematics, pp. 151–160. Publisher UTM, Skudai (2008). ISBN: 978-983-52-0606-1
3. The First Mathematics in Industry Study Group (MISG): Malaysia 2011 Short Reports, Universiti Teknologi Malaysia, Johor Bahru, Malaysia (14–18 March 2011). Retrieved from http://www.utm.my/partners/utm-ciam/files/2015/03/MISG2011_Short-Reports.pdf
4. The Second Mathematics in Industry Study Group (MISG): Malaysia 2014 Short Reports, Universiti Teknologi Malaysia, Johor Bahru, Malaysia (17–21 March 2014). Retrieved from http://www.utm.my/partners/utm-ciam/files/2015/03/report-full.pdf
5. The Daily Star (3 July 2012). PETRONAS chafes at its role as Malaysia's piggy bank. Retrieved from http://archive.thedailystar.net/newDesign/news-details.php?nid=240660
6. Raja Zainab Raja Sabaradin (5 December 2013). Industry brief: Malaysia's automotive industry (The national carmaker). Retrieved from http://www.exim.com.my/sites/default/files/industry_brief_-_automotive_in_malaysia_28nov_201329.pdf
7. Inside Malaysia (July 2012). Sound growth projected for KPJ Healthcare. Retrieved from http://etp.pemandu.gov.my/upload/Inside%20Investor%20-%20Healthcare.pdf
8. Lery, T. et al. (eds.): European Mathematical Society: European Success Stories in Industrial Mathematics. Springer, Berlin (2012). doi:10.1007/978-3-642-23848-2, ISBN: 978-3-642-23847-5
9. Schilders, W.: Industrial mathematics in Europe. In: Wakayama, M. (ed.) Proceedings of the Forum of Mathematics for Industry 2013. Springer, Tokyo (2014). ISBN: 9784-431-54906-2
10. Cameron, M.A.: Developing mathematicians for industry research teams. In: Wakayama, M., et al. (eds.) Proceedings of the Forum of Mathematics for Industry 2013, pp. 157–166. Springer, Tokyo (2014). ISBN: 978-4-431-54906-2
11. Arnkil, R., Jarvensivu, A., Koski, P., Piirainen, T.: Exploring Quadruple Helix—Outlining user-oriented innovation models, Work Research Centre, Institute for Social Research, University of Tampere, Tampere, Finland (2010). ISBN: 978-951-44-8208-3

12. Carayannis, E.L., Barth, T.D., Campbell, D.F.J.: The Quintuple Helix innovation model: global warming as a challenge and driver for innovation. J. Innov. Entrep. **1**(2) (2012). doi:10.1186/2192-5372-1-2
13. Mustafa, S., Bahar, A., Aziz, Z.A., Suratman, S.: Review of the role of analytical modelling methods in riverbank filtration system. Jurnal Teknologi (Sci. Eng.) **71**(1), 59–69 (2014)
14. Aziz, Z.A.: Islamic science and some physico-philosophical thoughts from a mathematical physics point of view. Ulum Islamiyyah J. **7**, 65–80 (2011). ISSN: 1675 5936
15. Chen, G.: Interdisciplinary research—mathematical interactions viewed from four portrayals. Not. AMS **59**(9), 1246–1252 (2012)

Group Theory Methods in Virology: Landau Density Wave Approach

Vladimir L. Lorman and Sergei B. Rochal

Abstract Viruses are organized biological nanosystems which display high level of spatial organization. In the present work we focus on the group theory methods application to the problems of virus self-assembly and resulting viral structure formation. The approach is based on the successive application of methods of representation theory for continuous and discrete groups and invariant theory for the groups not-generated by reflections. It generalizes the Landau density wave theory of crystallization to the case of compact crystal-like manifold assembly. To compare the predictions of the theory with the available cryoelectronic microscopy data we use the calculated density distribution functions which generate the protein positions on a spherical surface of the cage protecting viral genome. We also discuss the relation between density distribution functions and viral infectivity.

Keywords Group theory · Virology · Density waves · Self-assembly · Icosahedral viruses · Crystallization

1 Introduction

Viruses occupy the "gray area" between living and non-living matter. In contrast with other forms of living matter viruses cannot replicate independently, they need a host cell and its biosynthetic machinery to reproduce new viral particles. Their genetic material is vulnerable to degradation before the infection into the cell can occur. Therefore, it is protected by the viral protein shell (capsid) which encloses the genome. Though virus capsid formation involves biologically specific events, some

---

V.L. Lorman (✉)
Laboratoire Charles Coulomb, UMR 5221, CNRS-Université de Montepllier,
PI. E. Bataillon, 34095 Montepiller Cedex 5, France
e-mail: vladimir.lorman@univ-montp2.fr

S.B. Rochal
Southern Federal University, ul. Zorge 5, 344090 Rostov-on-Don, Russia
e-mail: rochal_s@yahoo.fr

R.S. Anderssen et al. (eds.), *Applications + Practical Conceptualization + Mathematics = fruitful Innovation*, Mathematics for Industry 11,
DOI 10.1007/978-4-431-55342-7_2

steps of the self-assembly are similar to passive physical processes and show universal features. All viruses follow some common scenario: they deliver viral genomic material into host cell, subvert cell's biosynthetic mechanisms into producing viral genome and proteins, then, new viral particles self-assemble in the infected cell, and finally, new generation of virus leaves the cell [1]. Genomic material of viruses is strongly diversified: it can be DNA or RNA (both "+" and "-" type), single- or double-stranded, linear or circular, one or several copies. On the contrary, capsid structure is quite universal. Viral capsids are made of many copies of identical subunits (one or few proteins) self-assembling in a two-dimensional (2D) shell. The positions and orientations of subunits display high level of spatial organization. With a typical diameter of the order of 50 nm and the regular protein organization, capsids represent nano-systems well-suited to modern structural methods of study like synchrotron radiation diffraction or cryoelectron tomography [2]. Recent structural data obtained due to the progress of cryoelectronic technique rise a whole number of questions concerning unconventional positional order of subunits in the shell, thermodynamics and physical mechanisms of the self-assembly.

From the point of view of geometry, viral capsids are divided into two wide classes: (i) open cylindrical structures with helical rod-like protein arrangement; (ii) spherical ones. Several viral families constitute notable exception to this classification. For example, capsids of many retro-viruses including HIV are conical shells with continuously varying curvature. In the present work we focus on viruses with compact spherical topology, on successive steps of the spherical virus assembly process, and on the underlying physical mechanisms and mathematical formalism. We are interested mainly in the first step of the process. At this step a solid spherical protein shell self-assembles from the aqueous solution of individual proteins.

In their pioneer work Crick and Watson (CW) [3] stated that viruses with spherical topology should have the symmetry (but not necessarily the shape) of one of regular polyhedra. Using more detailed X-ray diffraction data Caspar and Klug (CK) precised that spherical capsids adopt icosahedral point symmetry [4]. Decades of experimental studies showed that lateral, strongly orientation-dependent type of capsid protein interaction, together with intrinsic curvature and, especially, asymmetry of capsid proteins define particular docking preferences. Consequently, protein interaction results in specific geometric arrangements and influences the capsid assembly phenomenon. Both CW and CK insisted on the fact that typical viruses have very small genome (with respect to any other biological system, i.e. bacteria). Thus, it can code only for a few proteins, and among them there is typically only one "coating" protein. Viral capsid is then constituted with multiple copies of the same coating protein. They also stressed that the interaction of identical proteins should lead to identical local environments, including local orientational and positional order and local chemical bonding, and proposed to construct the icosahedral shells possessing these properties.

Due to these advances the main problems in the field of physical structural virology could be formulated: (i) *how to construct a regular 2D shell with the icosahedral symmetry formed by multiple copies of identical asymmetric proteins in identical local environments*, including local orientational and positional order and local chemical bonding; (ii) *how to relate the symmetry of individual proteins and the symmetry of the assembled shell to the free energy* of the assembly process.

2 Geometric Restrictions

A whole number of physical and geometric restrictions arise in this way and lead to the selection rules for the viral capsid structures. To understand the implementation of these rules it is necessary to take into account the fact that they are combined with the evolutionary optimization of viral genome. This remarkable feature of biological systems distinguishes them from the non-living ones. Evolutionary pressure pushes the virus to increase its genome, and, consequently, to increase the volume of the protective capsid shell. Taking into account WC and CK ideas about "identical subunits in identical environments" one can easily formulate the simplest group theory problem: *what is the 3D irreducible symmetry group of the protein distribution forming a spherical shell of maximal volume and composed with a finite number of 2D proteins of fixed finite area*?

2.1 Protein Asymmetry and Rotational Icosahedral Symmetry of Viral Capsids

Since the proteins are asymmetric they can form the structures with *rotational* symmetry elements only, excluding inversion and mirror planes. Indeed, because of the asymmetry at least several first moments of their mass and charge distribution are different from zero. Let us take into account the simplest one, usually called chirality. Corresponding moment has the symmetry of a pseudo-scalar ε of the fixed sign. The sum of pseudo-scalars of the same sign over any distribution being non-zero $\sum_i \varepsilon_i \neq 0$, the average capsid chirality is also non-zero. Thus the capsid structure is chiral, and its symmetry group contains no spatial inversion nor mirror planes.

Identical asymmetric proteins can be put in identical environments if they are located in the positions which form one *regular orbit* of the 3D rotational point group G only. To occupy positions on a n-fold axis the molecules should have corresponding symmetry themselves. The number of positions of identical asymmetric proteins is equal to the dimension $dim\,[Orb\,G]$ of a regular orbit of a discrete symmetry group G which is equal to the group order $|G|$. For the tetrahedral rotational group $G = T$ the number of proteins constituting the shell is $dim\,[Orb\,T] = 12$, for the octahedral group $G = O$ this number is $dim\,[Orb\,O] = 24$ while for the icosahedral rotational group $G = I$ it is $dim\,[Orb\,I] = 60$. It is evident that the icosahedral symmetry of the protein distribution leads to the shell constituted with 60 copies of the same

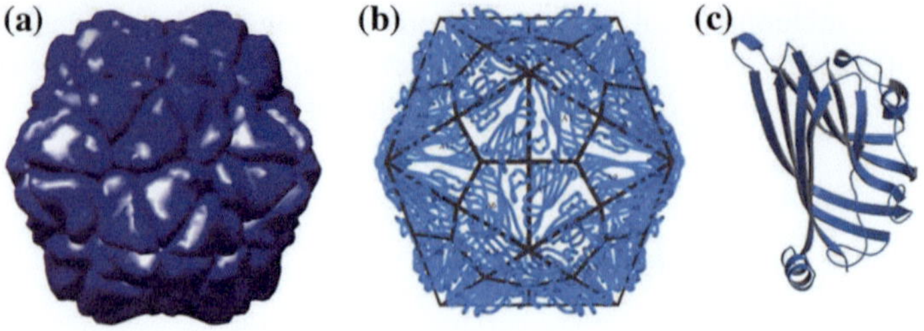

Fig. 1 Typical small spherical virus (Satellite Tobacco Necrosis Virus) with the icosahedral symmetry of protein organization: **a** Shaded Surface Diagram of the cryoelectronic density; **b** Ribbon Diagram with schematic protein presentation; **c** Asymmetric Protein

protein. Corresponding icosahedral shell has the volume much bigger than the shell volume in tetrahedral and octahedral cases.

The majority of spherical viruses are indeed icosahedral and the ordered distribution of proteins in their capsids has the symmetry group I (Fig. 1). The tessellation of a sphere by the action of its elements (called Möbius tessellation) is presented in Fig. 2 [5]. Note that the fundamental domain of the group I (shown in Fig. 2 the yellow triangle with one 5-fold and two 3-fold axes in the vertices) is usually called "asymmetric protein unit" in structural virology (though this group theory notion is not used explicitly in this field). Without taking into account evolution arguments, 60 would be the maximum possible number of proteins with identical local environments in an icosahedral shell. However, evolutionary pressure compels virus to increase more and more its genome size and the capsid volume. Experimentally, many viruses contain more than 60 identical proteins in the shell. In general, their number is equal to $60N$, where N is a positive integer. Their positions belong to different regular orbits of the group I and cannot be equivalent.

Fig. 2 Möbius tessellation of a sphere by the action of icosahedral symmetry elements [5]. Fundamental domain of the rotational icoshedral group I is shown in *yellow*, symmetry axes are given by corresponding polygons

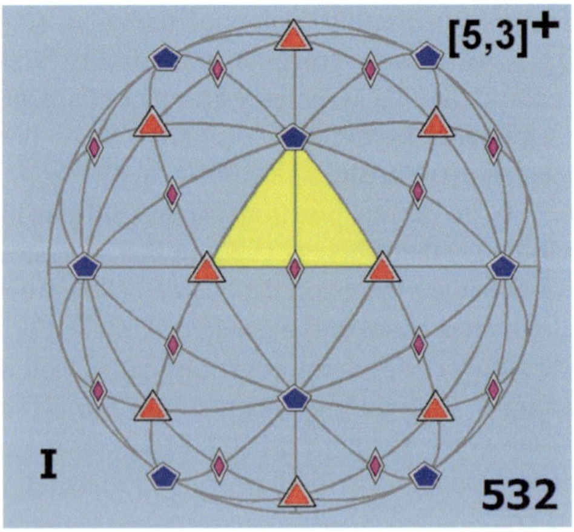

2.2 Selection Rules: Caspar and Klug Geometric Model

A partial explanation of this discrepancy was given by CK [4]. The task of CK was to find a way to put identical proteins in different but nearly equivalent positions and to explain the origin of this "quasi-equivalence". For that aim CK proposed to use the properties of plane periodic structures formed by the same type of asymmetric particles, and then to transfer these properties onto a sphere submitted to the action of the icosahedral group. Indeed, the above restriction on the group orbit size is proper to point groups. By contrast, the translational symmetry of a 2D plane lattice makes infinite the dimension of the regular orbit of the 2D space group $dim[OrbG] = \infty$. Consequently, the number of asymmetric proteins which can be put in equivalent environments becomes infinite in a 2D planar crystal. Then CK were looking for an almost regular mapping of the periodic planar structure on the icosahedron surface. Mapping splits one regular orbit of the plane structure symmetry group into several different regular orbits of the icosahedron group I but maintains some "traces" of their former "equivalence" in the plane structure.

Specific properties of the purely geometric CK model impose the selection rules for the value of N, i.e. for the number of different regular orbits of the I group. Consequently, they select the number $60N$ of proteins constituting the shell. Only the values which satisfy the relation $N = h^2 + k^2 + hk$, where h and k are non-negative integers are allowed by the CK selection rules. Constructively, CK proposed to map a net of an icosahedron slitted in the 2D hexagonal lattice to the icosahedron surface (Fig. 3). The mapping is chosen in such a way that the vertices of regular triangular faces of the icosahedron coincide with the 6-fold axes of the plane hexagonal struc-

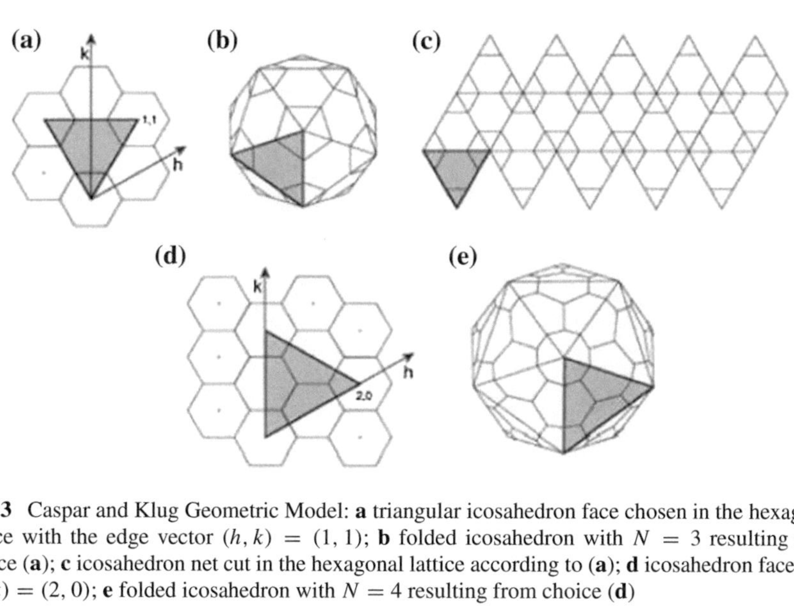

Fig. 3 Caspar and Klug Geometric Model: **a** triangular icosahedron face chosen in the hexagonal lattice with the edge vector $(h, k) = (1, 1)$; **b** folded icosahedron with $N = 3$ resulting from choice (**a**); **c** icosahedron net cut in the hexagonal lattice according to (**a**); **d** icosahedron face with $(h, k) = (2, 0)$; **e** folded icosahedron with $N = 4$ resulting from choice (**d**)

ture. By cutting out 60°-sector one can transform the 6-fold axis into the 5-fold one and then join the sector edges on the icosahedron surface. The edge length of the icosahedron face is determined by the vector joining vertices of the net triangles. Vertices being situated in the nodes of the hexagonal lattice, the vector length square is given by $N = h^2 + k^2 + hk$, where h and k are non-negative integers. This number is evidently equal to the number of the lattice nodes contained in two net triangles. Figure 3 shows first two non-trivial icosahedron nets with $N = 3$ and $N = 4$, respectively. In the hexagonal coordinates (h, k) the first net is indexed as (1, 1), the index of the second one is (2, 0). The procedure imposes also specific local arrangement of protein positions induced by the CK mapping. They are forced to form "pentamers" in the vicinity of 5-fold axes and "hexamers" elsewhere.

From the theorist's point of view the procedure proposed by CK has several problems of arbitrariness. It concerns the mapping choice, which is not unique, and the interpretation of geometric properties of manifolds. However, they do not constitute the main problem of the CK geometric model application to the virus structures and self-assembly process. The main problem, as it is often the case, comes from the refined experimental data. Though a big number of virus capsid structures are in a good agreement with the CK scheme, there is a growing number of experimentally

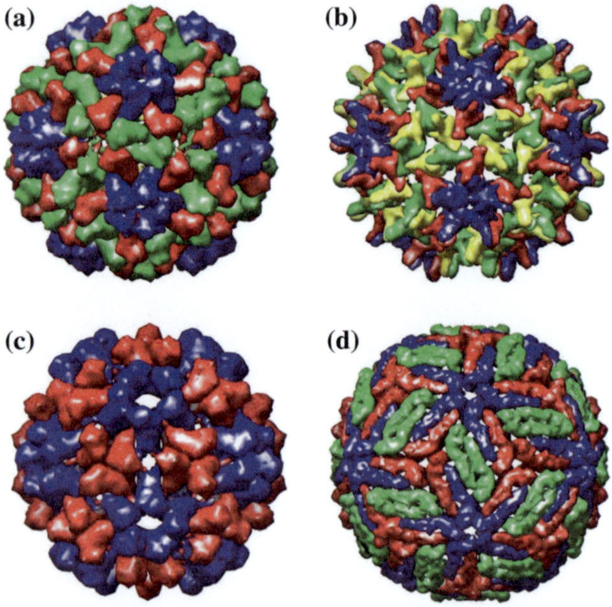

Fig. 4 Experimental cryoelectronic density for several viral capsids. Different positions of the same protein are given in different colors. Capsids (**a**) and (**b**) satisfy to the Caspar and Klug selection rules, while (**c**) and (**d**) are forbidden by them: **a** capsid of Cowpea Chlorotic Mottle Virus with $N = 3$ and "hexamer" organization of proteins; **b** capsid of Hepatitis B Virus with $N = 4$ and "hexamer" organization of proteins; **c** capsid of L-A virus with the forbidden $N = 2$; **d** capsid of Dengue Virus with $N = 3$ but without any "hexamer"

resolved structures which do not satisfy the CK selection rules nor their predictions about local proteins arrangement [6–8]. Figure 4 shows experimental cryoelectronic density distributions for several viruses. Among them Cowpea Chlorotic Mottle Virus (CCMV) is a typical example of virus with $N = 3$ verifying all predictions of the CK geometric model. The same remark concerns Hepatitis B Virus (HBV) with $N = 4$. On the other hand the capsid of L-A Virus (usually infecting yeast) is constituted by 120 proteins (i.e. $N = 2$), this number being forbidden by the CK selection rules. Capsid of Dengue Virus (DENV) shows local arrangement of proteins without any hexamer, also violating the CK geometric model. Thus, it becomes relevant to look for a theory which describes in a uniform way all experimentally observed small spherical viruses including those which can not be obtained using the CK geometric model.

3 Landau Density Wave Approach

In the present work we develop an approach which replaces the geometric notions by the notions of statistical physics. The positions of identical proteins in the shell are generated in the frame of the theory by one statistical protein density distribution function. To ensure "quasi-equivalence" of the positions the function is chosen irreducible under the action of the symmetry group I. The self-assembly thermodynamics is then described by the invariant free energy functional dependent on the distribution function. Similar approach was applied with success to describe crystallization process of atomic crystals [9–11] and quasicrystals [12–15]. Here we associate viral capsid formation with the unconventional crystallization process and propose to describe the capsid self-assembly using a generalization of the Landau theory of crystallization [9–11]. It is based on the successive application of methods of (a) representation theory of continuous and discrete groups; (b) invariant theory for the groups not-generated by reflections; (c) bifurcation theory of invariant functionals. To compare the predictions of the theory with the available cryoelectronic microscopy and AFM data we use the calculated irreducible density distribution functions which generate the protein positions on a spherical capsid.

3.1 Critical System of Density Waves

In the frame of the proposed theory the probability density ρ of protein distribution in the capsid in the vicinity of crystallization point is presented as:

$$\rho = \rho_0 + \Delta\rho, \tag{1}$$

where ρ_0 is an isotropic density in the solution of isolated proteins and $\Delta\rho$ corresponds to the density deviation induced by the ordering. The symmetry breaking during the crystallization is associated with one critical order parameter which spans an irreducible representation of the symmetry group of the disordered state. For the crystallization process the order parameter represents a critical system of density waves (CSDW) [9–11]. For the assembly process on a sphere considered as an unconventional crystallization the critical part $\Delta\rho_l$ of the density is determined by a CSDW with the *fixed* wave number l. The spherical harmonics Y_{lm} constituting CSDW on a sphere span one irreducible representation (IR) of the SO(3) symmetry group of the disordered state, thus $\Delta\rho_l$ is given by:

$$\Delta\rho_l(\theta,\phi) = \sum_{m=-l}^{m=l} A_{lm} Y_{lm}(\theta,\phi), \tag{2}$$

where l is the IR number, A_{lm} are the amplitudes of the spherical harmonics Y_{lm} and θ and ϕ are the angular variables of the spherical coordinate system.

3.2 Selection Rules: Representation Theory and Invariant Theory

The analysis based on the representation theory and the invariant theory shows that for any critical order parameter which drives the icosahedral assembly of asymmetric proteins the wave number l satisfies the relation:

$$l = 15 + 6i + 10j, \tag{3}$$

where i and j are positive integers or zero. Equation (3) defines the list of l numbers for which the restriction of an IR of the SO(3) group on the icosahedral group I contains at least one totally symmetric representation. To verify selection rule (3) one can use the well-known subduction criterion [16]. Active IR of the symmetry group $G_0 = SO(3)$ of the disordered state must subduce the identity representation of the symmetry group $G = I$ of the ordered state. The critical density function $\Delta\rho_l(\theta,\phi)$ is given by the basis functions $f_l^i(\theta,\phi)$ $(i = 1, 2...n_l)$ of all n_l totally symmetric representations of the icosahedral group I in the restriction of the 'active' IR of the $SO(3)$. The CSDW is a linear combination of these functions invariant with respect to the I group:

$$\Delta\rho_l(\theta,\phi) = \sum_{i=1}^{n_l} B_i f_l^i(\theta,\phi), \tag{4}$$

where B_i are arbitrary coefficients.

The frequency of subduction is defined as

$$n_l = (1/|G|)\sum_G \xi(\hat{g}) \tag{5}$$

where the sum runs over the elements $\hat{g}$ of the I group, $|G| = 60$ is the I group order, and $\xi(\hat{g})$ is the character of the SO(3) group element which reads as [16]:

$$\xi(l,\alpha) = \frac{sin((l+1/2)\alpha)}{sin(\alpha/2)},$$

where l is the IR number and the angle α is determined by the rotation $\hat{g}$. Then the explicit form of (5) becomes:

$$n_l = \frac{1}{60}(2l+1+15\xi(l,\pi)+20\xi(l,2\pi/3)+12\xi(l,2\pi/5)+12\xi(l,4\pi/5)). \tag{6}$$

From successive application of (6) one can deduce that $n_l \neq 0$ only for wave numbers l satisfying (3).

Selection rule (3) is justified in the frame of the theory of invariants [16, 17]. Any scalar function $F(\mathbf{r})$ invariant with respect to the group I, $F(\hat{g}\mathbf{r}) = F(\mathbf{r})$ for all $\hat{g} \in I$, can be expanded in formal polynomial series of x, y and z components of $\mathbf{r}$. All the terms in the series are the monomials of four following basis invariants (full set of generators for the ring of invariant polynomials: they constitute *integrity basis* [16, 17] of the I symmetry group):

$$J_0 = x^2+y^2+z^2,\ J_1 = \prod_{i=1}^{6}\mathbf{r}\cdot\mathbf{n}_i,\ J_2 = \prod_{i=1}^{10}\mathbf{r}\cdot\mathbf{p}_i,\ J_3 = \prod_{i=1}^{15}\mathbf{r}\cdot\mathbf{q}_i, \tag{7}$$

expressed here in terms of scalar products of the radius-vector $\mathbf{r} = \langle x, y, z\rangle$ with the vectors $\mathbf{n}_i$, $\mathbf{p}_i$ and $\mathbf{q}_i$ parallel to the icosahedron rotational axes. Namely, to the six 5-fold axes, to the ten 3-fold axes, and to the fifteen 2-fold axes, respectively. To verify the fullness of integrity basis (5) we use Molien's theorem [16, 17]. It gives the generating function $H(t)$ for invariants of the vector representation of a finite discrete group G in the form:

$$H(t) = (1/|G|)\sum_M \frac{1}{det(I - tM)} \tag{8}$$

where $M(\hat{g})$ is the representation matrix for the group element $\hat{g}$ and I is the identity matrix, the sum running over all $\hat{g} \in G$. The resulting generating function $H(t)$ can be presented in the Molien-Weyl form [16, 17]. For the rotational icosahedral group I it reads:

$$H(t) = \frac{1 + t^{15}}{(1 - t^2)(1 - t^6)(1 - t^{10})} \tag{9}$$

This indicates that the integrity basis for the I group contains algebraically independent basis invariants of degree 2, 6 and 10, and an additional invariant of degree 15. In contrast with the case of groups generated by reflections, the number m of invariants in the basis of a group not generated by reflections: (i) is greater than the vector space dimension ($m > 3$ for the group I); (ii) the product of degrees of basis invariants is not equal to the number of group elements $|G|$ ($\prod_{k=0}^{3} deg(J_k) \neq 60$ for the group I); (iii) invariants of the basis form a *syzygy*, an algebraic relation. In the case of the group I syzygy is the relation of the form $J_3^2 = P(J_0, J_1, J_2)$, where P is the homogeneous polynomial of 30th degree.

On the other hand, CSDW (4) is a linear combination of spherical harmonics with a given odd l. Even values of l are excluded from the consideration because the CSDW invariant with respect to the rotational group I, is not symmetric under the spatial inversion operation, Thus, it is a homogeneous function of a given odd degree l in $\mathbf{r} = \langle x, y, z \rangle$. Any irreducible scalar function $B_l(x, y, z)$ of a given odd degree l and invariant with respect to the rotational group I can be presented as:

$$B_l(x, y, z) = \sum_{2k+6i+10j+15=l} J_3(A_{k,i,j} J_0^k J_1^i J_2^j). \tag{10}$$

k, i and j being the degrees of the J_0, J_1 and J_2 invariants, respectively. Syzygy implies that the terms containing ${J_3}^2$ in any degree can be expressed in function of J_0, J_1, and J_2. Thus, all J_3-containing terms in any series invariant with respect to the group I are linear in J_3. Due to homogeneity of function (10) its wave number l verifies the relation $l = 2k + 6i + 10j + 15$ where k, i and j are non-negative integers. On the unit sphere surface ($r = 1$) the invariant J_0 becomes constant: $J_0 = 1$. Consequently, the form of the density function changes. Let us introduce a radial unit vector $\mathbf{e}_r = \mathbf{r}/r = \langle sin\theta \cos\phi, \sin\theta \sin\phi, \cos\theta \rangle$ depending on angular coordinates θ and ϕ in a standard way. Then, on the spherical surface the irreducible scalar function $B_l(\theta, \phi) = B_l(\mathbf{e}_r)$ has the form:

$$B_l(\mathbf{e}_r) = J_3(C + A_{1,0}J_1 + A_{0,1}J_2 + A_{2,0}J_1^2 + A_{1,1}J_1J_2 + \cdots \sum_{6i+10j+15=l} A_{i,j} J_1^i J_2^j) \tag{11}$$

where $C = A_{0,0}$ is constant. This gives, finally, that the wave number l for irreducible density functions (11) verifies the relation $l = 15 + 6i + 10j$, where i and j are non-negative integers.

3.3 Icosahedral Density Distribution Functions and Protein Arrangement in Viral Capsids

Selection rule (3) gives the possibility to obtain the explicit form of critical density (2). For small icosahedral capsids the practical construction of the protein density distribution is simplified because the CSDW contains only one function $f_l(\theta, \phi)$. Indeed, according to subduction criterion $n_l = 1$ for all $l \leq 43$. In this simplest case $\Delta\rho_l(\theta, \phi) = Bf_l(\theta, \phi)$, where B is a single arbitrary coefficient. The positions of maxima of the density function do not depend on the value of B. They are generated by a single universal function $f_l(\theta, \phi)$ which has no fitting parameter. In the following consideration the functions $f_l(\theta, \phi)$ possessing this properties are called irreducible icosahedral density functions. The explicit form of the irreducible density function $f_l(\theta, \phi)$ for a given value of l is obtained by averaging of $Y_{lm}(\theta, \phi)$ harmonics over the I symmetry group.

$$f_l(\theta, \phi) = \frac{1}{60} \sum_G Y_{l,m}(\hat{g}(\theta, \phi)). \tag{12}$$

The sequence of the wave number values l permitted by selection rule (3) is given by: $l = (15, 21, 25, 27, 31, 33, 35 \ldots)$. This sequence determines possible capsid shell structures for small icosahedral viruses. Figure 5 presents irreducible icosahedral

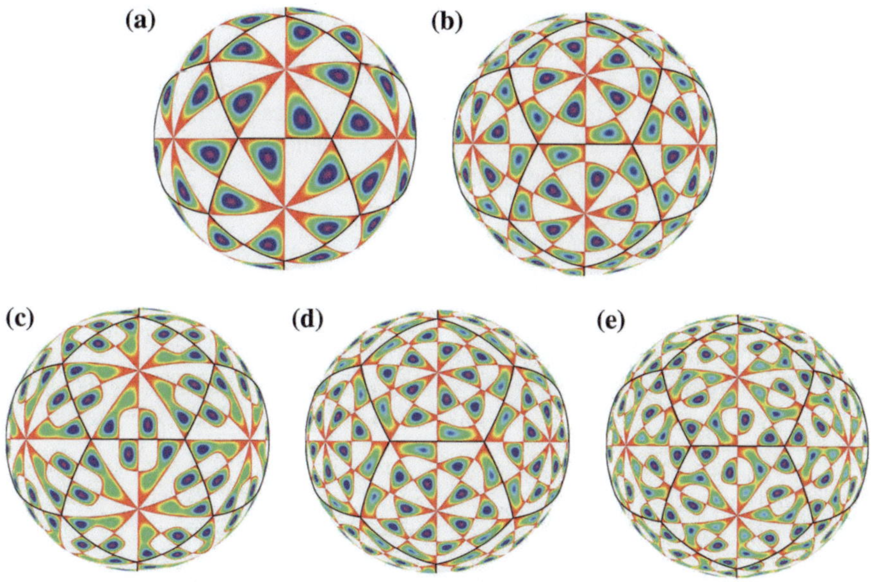

Fig. 5 Irreducible icosahedral density functions for the first five permitted values of the wave number l: **a** $l = 15$ resulting in $N = 1$ orbit of maxima; **b** $l = 21$ with $N = 2$ orbits; **c** $l = 25$ with $N = 3$; **d** $l = 27$ with $N = 3$; **e** $l = 31$ with $N = 4$

density functions for the first five permitted values of l. The value of $f_l(\theta, \phi)$ is represented using false colors: variation of colors from red to violet corresponds to the function growth. Note that all $f_l(\theta, \phi)$ functions are anti-symmetric: they change their sign under the inversion operation or under the action of mirror planes of a regular icosahedron. Thus, for the sake of clarity, we present the positive part $f_l(\theta, \phi) > 0$ only. The protein centers are associated with the positions of maxima of $\Delta\rho_l$ function (2). The number of maxima of the density functions is equal to $60N$, where N is the number of different regular 60-fold orbits of the I group. In the viral capsid N corresponds to the number of different positions occupied by the proteins.

The density wave approach replaces nonuniversal geometric CK model and describes in a uniform way capsid structures which can be obtained using CK geometric model and those which cannot. It is also worth noting that our approach shows that there exist viruses with capsids constituted by the same number of proteins but with qualitatively different protein distributions. For example, Fig. 5 shows two different distributions with $N = 3$. One distribution, generated by the irreducible icosahedral density function with $l = 27$, satisfies CK model and shows "pentamer" and "hexamer" local order, while the distribution with $l = 25$ does not satisfy the CK model and shows typical "rhombus" pattern without any "hexamer". The former distribution describes with a very good accuracy the experimental protein density distribution in CCMV virus (Fig. 4a), while the latter one fits the distribution in DENV virus (Fig. 4d), the capsids of both viruses being constituted of 180 proteins.

The proposed approach treats also the free energy of the transition from the isotropic aqueous solution of proteins to the assembled capsid shell (see the basic principles in [15, 18]). It gives an access to the thermodynamics of the virus self-assembly process and its relation with the symmetry of CSDW and irreducible icosahedral density functions. The invariant free energy functional Φ of the Landau crystallization theory adopted to describe the assembly process contains homogeneous part F dependent on local order parameter value and inhomogeneous part L expressing energy cost for deviation from spatial uniformity:

$$\Phi = \int_V [F((A_{l,m})) + L((\nabla A_{l,m}))] \mathrm{d}V \tag{13}$$

where CSDW amplitudes $(A_{l,m}) = \langle A_{l,-l}, A_{l,-l+1}, \ldots, A_{l,-l} \rangle$ are the order parameter components. The homogeneous free energy expansion can be taken in a standard for the crystallization theory form [15, 18] $F = F_0 + F_2 + F_3 + F_4 + \cdots$ of successive invariant terms

$$\begin{aligned}
F_2 &= A(T, c) \sum_{m=-l}^{m=l} A_{l,m} A_{l,-m}, \\
F_3 &= B(T, c) \sum_{m_l, m_2, m_3} a_{m_l, m_2, m_3} A_{l,m_1} A_{l,m_2} A_{l,m_3} \delta(m_l + m_2 + m_3) \equiv 0, \\
F_4 &= \sum_k C_k(T, c) \sum_{m_l, m_2, m_3, m_4} a^k_{m_l, m_2, m_3, m_4} A_{l,m_1} A_{l,m_2} A_{l,m_3} A_{l,m_4} \delta(m_l + m_2 + m_3 + m_4),
\end{aligned} \tag{14}$$

where a_i are weight coefficients of the SO(3) group (e.g. Clebsch-Gordan coefficients for the third order term F_3), $\delta(0) = 1$, $\delta(i \neq 0) = 0$, $A(T, c)$, $B(T, c)$, and $C_k(T, c)$ are temperature- and composition-dependent coefficients of the Landau theory. Because of the protein asymmetry and according to selection rule (3) free energy (14) can be written for an odd wave number l only. But for any odd wave number l the third-order term F_3 is identically zero. Then, the analysis of bifurcations of the invariant free energy functional shows [18] that in contrast with classical crystallization process the assembly of asymmetric proteins in an icosahedral shell can take place as a second-order phase transition without a free energy barrier between the isotropic and the icosahedral states.

3.4 Icosahedral Density Distribution Functions and Virus Infectivity

In the final part, we illustrate the relation between the parameters of the proposed theory and the problem of virus infectivity, extremely important for biology. We evidence the relation between the protein density distribution function and the probability of the host cell receptor molecule attachment in different sites of the viral capsid surface. This relation connects the proposed approach to the problem of virus entry into the infected cell. We take here an example of DENV virus interaction with dentritic host cells. Note that the DENV protein distribution obtained in the frame of our approach is the first DENV model based on physical principles and not on a simple empirical fit of the experimental data. Figure 6 shows the experimental cryoelectronic density of DENV virus (Fig. 6b) [19] with bound carbohydrate recognition domains (CRD) of a dendritic cell receptor. One can see a typical "rhombus" motif of the DENV protein arrangement identical to the pattern of maxima of the irreducible density function with $l = 25$ and $N = 3$ (Fig. 6a). Figure 6c [19] represents a schematic zoom of the "rhombus" motif with the fundamental domain of the I group ("asymmetric protein unit") shown by a triangle. Three different capsid protein environments are given in different colors. It also shows the positions of glycosylation sites Asn 67 and Asn 153. In in vitro experiments with CRD molecules and individual DENV coat proteins in solution at correct temperature, CRD binds to Asn 67 sites of all DENV proteins with extremely high probability. However, in vivo CRD (given in cyan in Fig. 6) binds to two Asn 67 sites leaving the third Asn 67 residue vacant, though all three DENV proteins occupying three different positions in the distribution are identical. This result shows that infectivity promoted by the interaction of cell receptors with virus surface depends not only on bio-specific binding properties of individual molecules but also on the capsid proteins distribution. The interaction of CRD with the proteins in the capsid represents a collective phenomenon.

Comparison of the calculated distribution with the experimental data allows to establish the relation between the minima of $f_l(\theta, \phi)$ and binding sites on the capsid

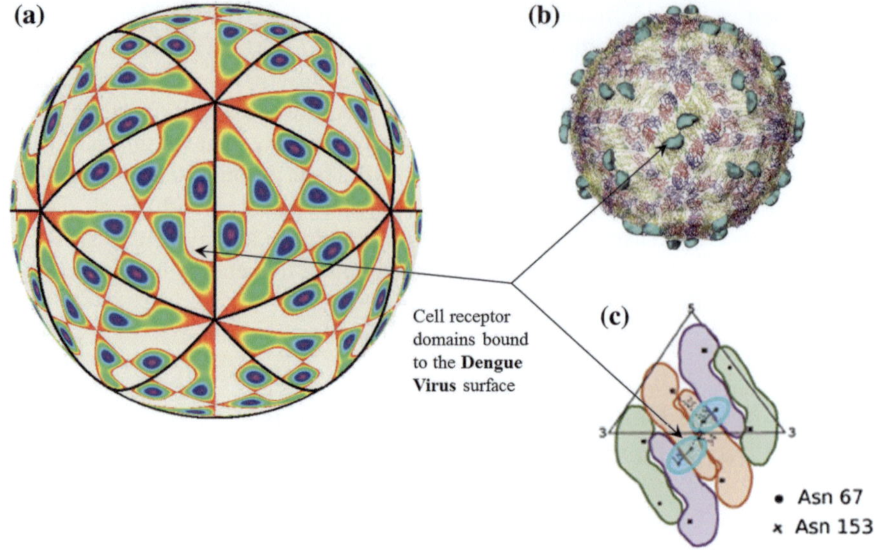

Fig. 6 Protein density distribution and Virus Infectivity. Interaction of the dendritic cell receptor molecules with the capsid of Dengue Virus: **a** irreducible density icosahedral density function with $l = 25$; **b** experimental cryoelectronique density of the Dengue Virus capsid with bound cell receptor domains; **c** schematic zoom of the "rhombus" motif in (**b**)

surface. In the present case it establishes a correspondence of the deepest minima of f_{25} and the binding sites for the carbohydrate recognition domains of the dendritic cell receptors on the DENV virus surface. Additional argument comes from the interpretation of the density function as a *probability* density distribution. From Fig. 6 it is evident that two of the three maxima of f_{25} are sharp picks with high amplitudes of occupation probability while the third one is a wide maximum with a low amplitude. Mean-square amplitude of protein position fluctuations is much high in the latter case than in two former ones. Thus, the attachment of CRD to the strongly fluctuating protein located in the position characterized by a wide maximum is much less probable than in two others.

Along the same line, the correspondence can be established between the positions of maxima for other irreducible density functions and peculiarities of protein distribution in other viruses. For example, sharp maxima of the function with $l = 33$ find themselves in positions situated very close one to another, and leaving between them a sequence of deep and narrow minima. This distribution is close to the experimental protein distribution in Human Rhino-Virus. Deep minima sequence at the surface of this virus is usual called a "canyon". This canyon is known to be too narrow for human antibodies to enter and to neutralize viral binding sites, but large enough for narrow human cell receptors (more precisely for intracellular adhesion molecules-1, or ICAM-1, of the cell receptors) to enter into the canyon and to bind to the virus protein domains.

4 Perspectives

The approach developed in this work is only the first step of the group theory methods application in virology. Further development of these methods can clarify mechanical properties of viral capsids which have in some cases elastic constants comparable with the hardest materials. One of the most interesting extensions is the mathematical (and physical) analysis of the viral genome (DNA or RNA) dense packing in capsids. Another field where these methods can be applied is the shape transformation usually undergone by viruses during their maturation process which makes them infectious.

Acknowledgments The authors would like to acknowledge financial support by the Laboratory of Excellence NUMEV and RFFR grant N° 13-02-12085. All graphic presentations of the experimental cryoelectronic density images were made with the Multiscale extension [20] to the Chimera interactive molecular graphics package [21, 22].

References

1. Flint, S.J., Enquist, L.W., Racaniello, V.R., Skalka, A.M.: Principles of Virology: Molecular Biology Pathogenesis and Control. ASM, Washington (2000)
2. Baker, T.S., Olson, N.H., Fuller, S.D.: Microbiol. Mol. Biol. Rev. **63**, 862 (1999)
3. Crick, F.H.C., Watson, J.D.: Nature (London) **177**, 473 (1956)
4. Caspar, D.L.D., Klug, A.: Cold Spring Harbor Symp. Quant. Biol. **27**, 1 (1962)
5. Wikipedia: http://en.wikipedia.org/wiki/Icosahedral_symmetry. Cited 7 Jan 2015
6. Naitow, H., et al.: Nat. Struct. Biol. **9**, 725 (2002)
7. Kuhn, R.J., et al.: Cell **108**, 717 (2002)
8. Mukhopadhyay, S., et al.: Science **302**, 248 (2003)
9. Landau, L.D.: Phys. Zs. Sowjet. **11**, 26 (1937)
10. Landau, L.D.: Phys. Zs. Sowjet. **11**, 545 (1937)
11. Alexander, S., McTague, J.: Phys. Rev. Lett. **41**, 702 (1978)
12. Janot, C.: Quasicrystals. A Primer, 2nd edn. Oxford University Press, New York (1994)
13. Bak, P.: Phys. Rev. Lett. **54**, 1517 (1985)
14. Rochal, S.B., Kozinkina, Y.A.: Phys. Rev. B **72**, 024210 (2005)
15. Konevtsova, O.V., Rochal, S.B., Lorman, V.L.: Phys. Lett. A **377**, 1215 (2013)
16. Elliot, J.P., Dawber, P.G.: Symmetry in Physics. Macmillan Press, London (1979)
17. Springer, T.A.: Invarient Theory. Lecture Notes in Mathematics, vol. 585. Springer, Berlin (1997)
18. Lorman, V.L., Rochal, S.B.: Phys. Rev. Lett. **98**, 185502 (2007)
19. Pokidysheva, E., et al.: Cell **124**, 485 (2006)
20. Goddard, T.D., Huang, C.C., Ferrin, T.E.: Structure **13**, 473 (2005)
21. UCSF Chimera: http://www.cgl.ucsf.edu/chimera. Cited 7 Jan 2015
22. Pettersen, E.F., et al.: J. Comput. Chem. **25**, 1605 (2004)

Applications of Nonlinear Programming to Packing Problems

Ernesto G. Birgin

Abstract The problem of packing items within bounded regions in the Euclidean space has multiple applications in a variety of areas, such as, Physics, Chemistry, and Engineering. Problems of this type exhibit various levels of complexity. Nonlinear programming formulations and methods had been successfully applied to a wide range of packing problems. In this review paper, a brief description of the state-of-the-art and an illustrated overview of packing nonlinear programming techniques and applications will be presented.

Keywords Cutting · Packing · Nonlinear programming

1 Introduction

The problem of packing items within bounded areas has been widely studied over the last decades. Due to the amount of publications dedicated to packing and cutting problems, making a comprehensive review of the recent literature may be impossible. A simple search by the words "circle packing" in the Google Images search engine displays a myriad of amazing pictures describing theoretical results and applications. In this work we focus on models and techniques based on nonlinear programming.

Among the packing problems that can be addressed by nonlinear models and methods, the problem of packing identical or non-identical circular and spherical items presents several applications, such as origami design [1], analysis of concrete properties [2], study of the properties of forest soils and its influence in the development of roots [3], Gamma Knife radiosurgery [4], molecular dynamics simulations [5, 6], and industrial problems like facility location [7] and container loading [8], among others. More recently, the packing of three-dimensional polygons, such as tetrahedra, has also gained attention due to its applications to model low-temperature states of

E.G. Birgin (✉)
Institute of Mathematics and Statistics, University of São Paulo, Rua do Matão, 1010, Cidade Universitária, São Paulo, SP 05508-090, Brazil
e-mail: egbirgin@ime.usp.br

R.S. Anderssen et al. (eds.), *Applications + Practical Conceptualization + Mathematics = fruitful Innovation*, Mathematics for Industry 11,
DOI 10.1007/978-4-431-55342-7_3

matter, including liquids, crystals, glasses, and powders (see, [9] and the references therein). In [10, 11] the concept of *sentinels* has been introduced as a tool to model the overlapping between polygons through the satisfaction of a finite set of continuous and differentiable constraints, that is one of the key ingredients to model packing problems as tractable nonlinear programming problems.

Nonlinear programming models include, in the objective function or the constraints, functions that "measure" the overlapping between every pair of items. As described in [12], these functions were named Φ-functions in [13]. Given a pair of items i and j, a function $\Phi_{ij}(\cdot)$ is said a Φ-function if its value is negative whenever items i and j overlap, null if they are tangent, and positive if they do not overlap. Every mathematical model related to a packing problem needs to deal with the items' overlapping and, in some way or another, makes use of and objective function or a set of constraints that can be seen as Φ-functions.

2 Packing of Circles and Spheres

The problem of packing a given set of items within an object with fixed dimensions may be modeled as a nonlinear (continuous and differentiable) feasibility problem. As an example, the problem of packing N circular items with radii $r_1, r_2, \ldots, r_N$ within a circular object with radius R (with $R \geq r_i$ for all i) can be modeled as finding $(x_i, y_i)^T \in \mathbb{R}^2$ (for $i = 1, \ldots, N$) such that

$$\begin{aligned} (x_i - x_j)^2 + (y_i - y_j)^2 &\geq (r_i + r_j)^2, \text{ for all } j > i, \\ x_i^2 + y_i^2 &\leq (R - r_i)^2, \text{ for all } i. \end{aligned} \tag{1}$$

In the problem described above, without loss of generality, it is assumed that the circular object is centered at the origin of the Cartesian coordinate system. The first set of equations says that the circular items must not overlap; while the second set of equations says that the items must be placed within the object. Both sets of equations are based on computing distances; and distances are squared to avoid the nondifferentiability of the square root at $x = 0$. Analogous models can be considered for different kinds of items and objects.

When the items to be packed are all identical, the considered goal may be maximizing the number of packed items within an object with fixed dimensions. In this case, the feasibility problem (1) may be used to solve the problem if an increasing number of items N is considered. More specifically, we may try to solve the feasibility problem (1) with $N = 1$. If we manage to find a solution, then we try with $N + 1$ until finding that for, let say, $N = \widehat{N}$, problem (1) is infeasible. In this case, the maximum number of identical items that can be packed is $\widehat{N} - 1$. In practice, guaranteeing that a nonlinear feasibility problem is infeasible is a very hard task. Therefore, we may fix some maximum effort we are able to do and, if within this limited amount of effort we are unable to find a solution, we may heuristically say that the problem is infeasible. At this point, many readers may think that it would be better to try some kind of bisection scheme instead of increasing N one by one. Assume that the maximum number of items that can be packed is N^*. Feasibility problems with $N < N^*$ are feasible and use to be simple (can be solved with a small computational effort). On the other hand, feasibility problems with $N > N^*$ are infeasible and detecting infeasibility may be very hard. For that reason, it is easier (cheaper) to approach N^* from below increasing N slowly. Of course, there is no need to start with $N = 1$ if it is known that the feasibility problem is feasible for some value $N = N_{\rm LB}$. The described strategy was considered in [14] to pack identical circles within circles, in [15] to pack identical circles within ellipses, and in [10, 16, 17] to pack identical rectangles (with different types of constraints related to the rectangular items' angle of rotation) within arbitrary convex regions.

If the items to be packed are different and each item has an associated value (that may be proportional to the item's area), the goal may be maximizing the value of the packed items [18]. In this case, packing an increasing number of items as described in the paragraph above does not provide an optimal strategy. Selecting the subsets of items for which a feasibility nonlinear (sub)problem may be modeled and solved is a combinatorial problem for which heuristic strategies may be considered in practice. On the other hand, independently of the items being identical or not, if the object dimensions are not fixed, the goal may be finding the smallest object of a certain type (circle, square, equilateral triangle, rectangle with smallest perimeter or area, etc.) within which a given set of items can be packed. In this case the problem can be modeled as an optimization problem. As an example, consider again the problem of packing a fixed number N circular items with radii $r_1, r_2, \ldots, r_N$ within a circular object with variable radius R and assume that the problem is to minimize R. This problem can be easily formulated as

$$\begin{array}{lll} \text{Minimize } R & & \\ \text{subject to } (x_i - x_j)^2 + (y_i - y_j)^2 \geq (r_i + r_j)^2, & \text{for all } j > i, & \\ \phantom{\text{subject to }} x_i^2 + y_i^2 \leq (R - r_i)^2, & \text{for all } i, & \\ \phantom{\text{subject to }} R \geq \bar{r}, & & \end{array} \qquad (2)$$

where $\bar{r} = \max_{1 \leq i \leq N}\{r_i\}$. Figure 1a shows the solution to the problem of finding the smallest circle within which $N = 6$ identical unitary-radius circular items can be packed. The solution corresponds to $R^* = 3$ [19, 20]. It is interesting to note that, within the circular object with $R = 3$, $N = 7$ identical unitary-radius circular items can also be packed (see Fig. 1b). This means that there is no equivalence between the problem of minimizing the object dimensions and the problem of packing as many items as possible. Variations of the model (2) were considered in [21, 22] for packing identical circular items within circles, squares, equilateral triangles, and strips, among others. Models and methodology presented in [22] also deal with 3D problems. See Fig. 2.

In [23, p. 171], sphere packing problems with up to 1,000,000 spheres are considered as an illustration of the capabilities of a nonlinear programming solver named Algencan [24, 25]. Those problems are an oversimplification of the problem of packing molecules considered in [5, 6]. One of the characteristics of this problem that

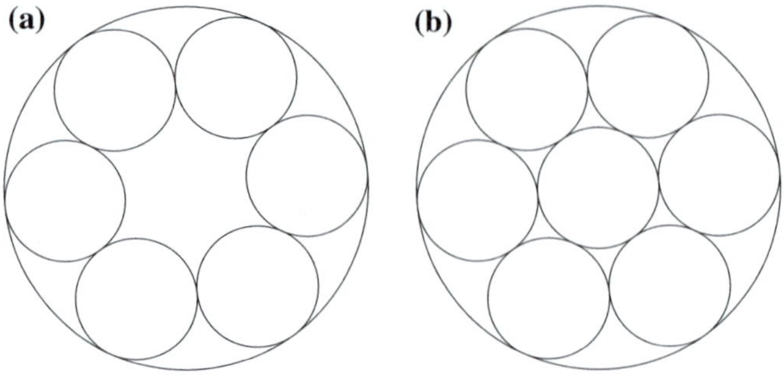

Fig. 1 Graphical representation of **a** the smallest circle (that has $R^* = 3$) within which 6 unitary-radius circles can be packed and **b** the maximum number of unitary-radius circles that can be packed within a circle with radius $R = 3$

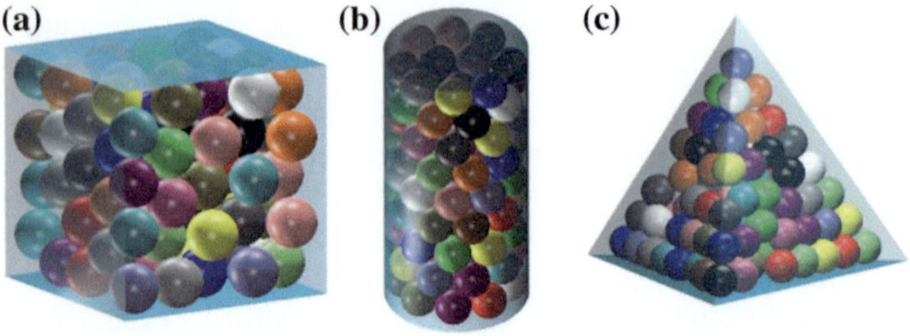

Fig. 2 Graphical representation of **a** smallest cube, **b** cylinder of minimal surface, and **c** smallest regular tetrahedron within which 100 unitary-radius spheres can be packed. See [22] for details

makes it tractable is that the density (proportion of occupied volume of the object) is relatively low. The first step in a molecular dynamics simulation consists of obtaining initial coordinates for all the atoms of the system. Since molecular dynamics force-fields contain repulsive terms that increase abruptly for short atom-to-atom distances, the distances between atoms from different molecules must be large enough so that repulsive potentials do not disrupt the simulations. Finding adequate initial configurations was modeled as a packing problem in [5, 6], giving rise to the software Packmol.[1]

Let us call $nmol$ the total number of molecules that we want to place in a region $\mathscr{R}$ of the three-dimensional space. For each $i = 1, \ldots, nmol$, let $natom(i)$ be the number of atoms of the ith molecule. Each molecule is represented by the orthogonal coordinates of its atoms. To facilitate the visualization, assume that the origin is the barycenter of all the molecules. For all $i = 1, \ldots, nmol$, $j = 1, \ldots, natom(i)$, let $A(i, j) = (a_1^{ij}, a_2^{ij}, a_3^{ij})$ be the coordinates of the jth atom in the ith molecule. Suppose that one rotates the ith molecule sequentially around the axes x_1, x_2, and x_3, being $\gamma^i = (\gamma_1^i, \gamma_2^i, \gamma_3^i)$ the angles that define such rotations. Moreover, suppose

[1]http://www.ime.unicamp.br/~martinez/packmol/.

that, after these rotations, the whole molecule is displaced so that its barycenter, instead of the origin, becomes $t^i = (t_1^i, t_2^i, t_3^i)$. These movements transform the atom of coordinates $A(i, j)$ in a displaced atom of coordinates $P(i, j) = (p_1^{ij}, p_2^{ij}, p_3^{ij})$. Observe that $P(i, j)$, $j = 1, \ldots, natom(i)$, is a function of (t^i, γ^i), the relation being $P(i, j) = t^i + R(\gamma^i)A(i, j)$, $j = 1, \ldots, natom(i)$, where

$$R(\gamma^i) = \begin{pmatrix} c_1^i c_2^i c_3^i - s_1^i s_3^i & s_1^i c_2^i c_3^i + c_1^i s_3^i & -s_2^i c_3^i \\ -c_1^i c_2^i s_3^i - s_1^i c_3^i & -s_1^i c_2^i s_3^i + c_1^i c_3^i & -s_2^i s_3^i \\ c_1^i s_2^i & s_1^i s_2^i & c_2^i \end{pmatrix}, \tag{3}$$

in which $s_k^i \equiv \sin \gamma_k^i$ and $c_k^i \equiv \cos \gamma_k^i$, for $k = 1, 2, 3$.

In [5, 6], the objective is to find angles γ_i and displacements t_i, $i = 1, \ldots, nmol$, in such a way that, whenever $i \neq i'$,

$$\|P(i, j) - P(i', j')\|_2^2 \geq d^2, \tag{4}$$

for all $j = 1, \ldots, natom(i)$, $j' = 1, \ldots, natom(i')$, where $d > 0$ is the required minimum distance, and

$$P(i, j) \in \mathscr{R}, \tag{5}$$

for all $i = 1, \ldots, nmol$, $j = 1, \ldots, natom(i)$. In other words, the rotated and displaced molecules must remain in the specified region and the distance between any pair of atoms must not be less than d. Problem (4), (5) is a nonlinear feasibility problem similar to (1). With some reformulations related to specific characteristics of the problem, using heuristics to construct initial guesses, and using a nonlinear programming solver [26], Packmol is able to solve practical problems like the ones illustrated in Fig. 3.

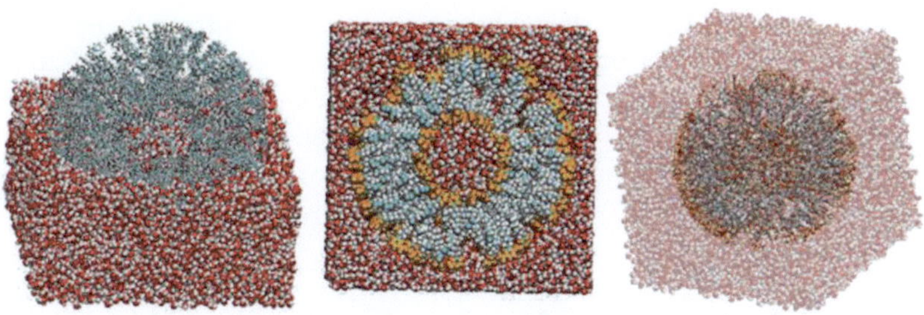

Fig. 3 Graphical representation of a double layered spherical vesicle with water inside and outside. The depicted configuration, obtained by solving a packing problem, can be used as initial point for molecular dynamics simulations. (This figure was extracted from the Packmol web site, where additional information can be found.) In practice, the region $\mathscr{R}$ in (5) is replaced for potentially different regions $\mathscr{R}_{ij}$ for each atom i of each molecule j

3 Packing of Polygons

The packing of rectangular items within rectangular objects is a particular case of the packing of polygons within polygons that has many practical applications in Logistics and Engineering. A large variety of problems exists, depending on the imposition of "cutting patterns", the possibility of allowing rotations of the items or not, the fact of the items being identical or not, etc. In many cases, the problem can be modeled as a mixed-integer linear programing problem and solved by dedicated exact or heuristic methods. On the other hand, when the rectangular items can be freely rotated or the items are other kind of polygons and the object within which the items must be placed is an arbitrary convex region, the problem may be modeled as a nonlinear programming problem.

In [10, 11] the concept of *sentinels* was introduced. Let I_1 and I_2 be nonempty, open, bounded, and convex sets of $\mathbb{R}^n$. Define $J_1 = \bar{I}_1$ the closure of I_1 and $J_2 = \bar{I}_2$ the closure of I_2 (J_1 and J_2 place the role of the items to be packed). Let $D_1, D_2 : \mathbb{R}^n \to \mathbb{R}^n$ be two displacement operators. So, D_1 and D_2 transform items in items preserving distances, angles, and orientation. If $D_1(I_1) \cap D_2(I_2) \neq \emptyset$ then we say that $D_1(J_1)$ and $D_2(J_2)$ (or $D_1(I_1)$ and $D_2(I_2)$) are superposed. Let S_1 and S_2 be finite subsets of J_1 and J_2, respectively. We say that S_1 and S_2 are *sentinels sets* relatively to J_1 and J_2 if the following property holds: *for all displacements D_1 and D_2, if $D_1(J_1)$ and $D_2(J_2)$ are superposed then $D_1(S_1) \cap D_2(I_2) \neq \emptyset$ or $D_2(S_2) \cap D_1(I_1) \neq \emptyset$*. Roughly speaking, if, after the displacements, the items J_1 and J_2 are superposed then at least one sentinel of J_1 becomes interior to J_2 or one sentinel of J_2 becomes interior to J_1.

The concept of sentinels can be easily extended to any family of (potentially non-identical) $m \geq 2$ "items", the key point being to define the (finite) sentinels sets for a given family of polygons (or to determine that they do not exist). In [11], minimal sets of sentinels for rectangles are exhibited (see Fig. 4), as well as sentinels sets for other types of polygons; while it is also shown that no finite sets of sentinels do not exist for triangles. In [10] it is shown how continuous and differentiable nonlinear programming models can be constructed that are based on the sentinels sets to define the constraints that avoid the overlapping between the items. A few examples of the obtained solutions can be seen in Fig. 5. In general, the resulting nonlinear programming problems are nonconvex with many spurious stationary points that are not

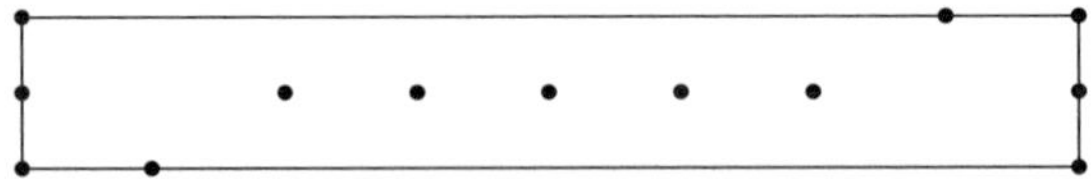

Fig. 4 Minimal set of sentinels for a family of m identical rectangular items with smallest side s and largest side $7s$. The distance between the sentinels in the smallest side is $s/2$ while the "horizontal distance" between the sentinels in the largest side and in the central line is smaller than s. (Enjoy yourself: make two copies of the rectangle in the figure and try to overlap them without having a sentinel of one rectangle in the interior of the other. Free rotations are allowed)

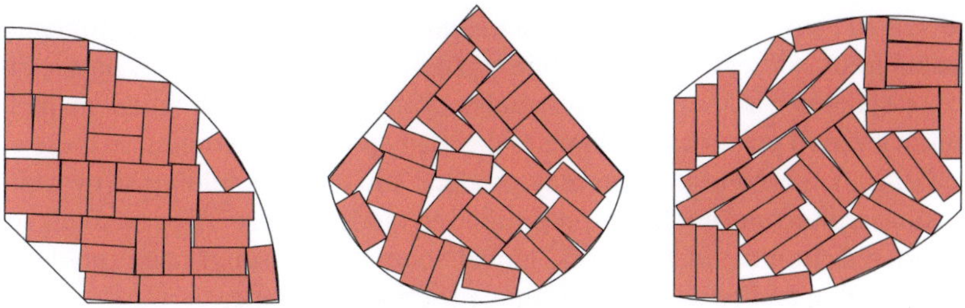

Fig. 5 Examples of solutions found in [10] to the problem of packing freely-rotated rectangular items within arbitrary convex regions

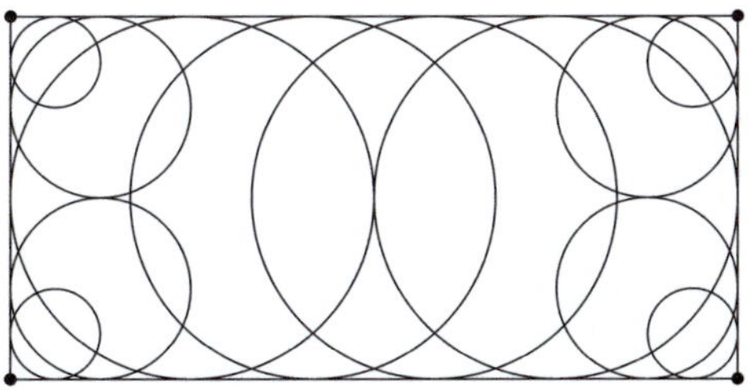

Fig. 6 Arbitrary partial covering of a rectangular item by *circles*. In a first phase, a packing problem with the circular items is solved (avoiding overlapping between the circles that cover different rectangles). The position of the rectangular items inherited by the solution of the problem of the first phase is used as initial guess to the original packing problem with rectangular items

solutions to the packing problem they represent (only global solutions are of interest). Because of that, clever heuristics need to be developed to determine promising starting points for the iterative optimization process. One possibility explored in [10] for packing rectangular items was to consider, for each rectangle, a partial covering with varied-sized circles (see Fig. 6). In a first phase, a circle packing problem is solved (with models similar to the one described in the previous section) to place the circles within the object and avoiding the overlapping of the circles that cover different items. In a second phase the circles are dismissed and the placement of the rectangular items is used as an initial guess to the rectangular items packing problem.

4 Concluding Remarks

There are packing problems that, by its nature, are combinatorial problems. Examples of those problems are one-dimensional bin packing problems, two-dimensional bin packing or stock cutting problems involving rectangular items (that can not be freely rotated) and rectangular objects, and some two- or three-dimensional puzzle problems, among many others. There are other packing problems, like for example finding the densest packing of circles or spheres in the infinite Euclidean space, that can be solved using lattices. On the other hand, there are also packing problems that can be naturally addressed with nonlinear programming techniques. The packing of circles or spheres within restricted domains or the packing of irregular shape items or arbitrary varied polygons are some of the many possible examples. In these cases, discretizing the problem domain may produce sub-optimal solutions (by reducing the feasible region) and nonlinear programming models and solution methods should be considered. In most of the cases, global solutions are sought and problems are very challenging and provide a nice source for benchmarking.

References

1. Lang, R.J.: A computational algorithm for origami design. In: 12th ACM Symposium on Computational Geometry, pp. 98–105 (1996)
2. Wong, H.C., Kwan, A.K.H.: Packing density: a key concept for mix design of high performance concrete. In: Proceedings of the Materials Science and Technology in Engineering Conference, HKIE Materials Division, Hong Kong, pp. 1–15 (2005)
3. Osman, K.T.: Foerst Soils—Properties and Management. Springer, Cham (2013)
4. Wu, Q.J., Bourland, J.D.: Morphology-guided radiosurgery treatment planning and optimization for multiple isocenters. Med. Phys. **26**, 2151–2160 (1999)
5. Martínez, L., Andrade, R., Birgin, E.G., Martínez, J.M.: Packmol: a package for building initial configurations for molecular dynamics simulations. J. Comput. Chem. **30**, 2157–2164 (2009)
6. Martínez, J.M., Martínez, L.: Packing optimization for automated generation of complex system's initial configurations for molecular dynamics and docking. J. Comput. Chem. **24**, 819–825 (2003)
7. Drezner, Z., Erkut, E.: Solving the continuous p-dispersion problem using non-linear programming. J. Oper. Res. Soc. **46**, 516–520 (1995)
8. Fraser, H.J., George, J.A.: Integrated container loading software for pulp and paper industry. Eur. J. Oper. Res. **77**, 466–474 (1994)
9. Chen, D., Jiao, Y., Torquato, S.: Equilibrium phase behavior and maximally random jammed state of truncated tetrahedra. J. Phys. Chem. B **118**, 7981–7992 (2014)
10. Birgin, E.G., Martínez, J.M., Mascarenhas, W.F., Ronconi, D.P.: Method of sentinels for packing items within arbitrary convex regions. J. Oper. Res. Soc. **57**, 735–746 (2006)
11. Mascarenhas, W.F., Birgin, E.G.: Using sentinels to detect intersections of convex and nonconvex polygons. Comput. Appl. Math. **29**, 247–267 (2010)
12. Stoyan, Y.G., Novozhilova, M.V., Kartashov, A.V.: Mathematical model and method of searching for a local extremum for the non-convex oriented polygons allocation problem. Eur. J. Oper. Res. **92**, 193–210 (1996)
13. Stoyan, Y.G.: On the generalization of the dense allocation function. Rep. Ukr. SSR Acad. Sci. Ser. A **8**, 70–74 (1980) (in Russian)
14. Birgin, E.G., Martínez, J.M., Ronconi, D.P.: Optimizing the packing of cylinders into a rectangular container: a nonlinear approach. Eur. J. Oper. Res. **160**, 19–33 (2005)
15. Birgin, E.G., Bustamante, L.H., Callisaya, H.F., Martínez, J.M.: Packing circles within ellipses. Int. Trans. Oper. Res. **20**, 365–389 (2013)
16. Birgin, E.G., Lobato, R.D.: Orthogonal packing of identical rectangles within isotropic convex regions. Comput. Ind. Eng. **59**, 595–602 (2010)
17. Birgin, E.G., Martinez, J.M., Nishihara, F.H., Ronconi, D.P.: Orthogonal packing of rectangular items within arbitary convex region by nonlinear optimization. Comput. Oper. Res. **33**, 3535–3548 (2006)
18. Martello, S., Toth, P.: Knapsack Problems: Algorithms and Computer Interpretations. Wiley, Chichester (1990)

19. Graham, R.L.: Sets of points with given minimum separation (solution to problem El921). Am. Math. Mon. **75**, 192–193 (1968)
20. Kravitz, S.: Packing cylinders into cylindrical containers. Math. Mag. **40**, 65–71 (1967)
21. Birgin, E.G., Gentil, J.M.: New and improved results for packing identical unitary radius circles within triangles, rectangles and strips. Comput. Oper. Res. **37**, 1318–1327 (2010)
22. Birgin, E.G., Sobral, F.N.C.: Minimizing the object dimensions in circle and sphere packing problems. Comput. Oper. Res. **35**, 2357–2375 (2008)
23. Birgin, E.G., Martínez, J.M.: Practical Augmented Lagrangian Methods for Constrained Optimization. Society for Industrial and Applied Mathematics, Philadelphia (2014)
24. Andreani, R., Birgin, E.G., Martínez, J.M., Schuverdt, M.L.: On augmented Lagrangian methods with general lower-level constraints. SIAM J. Optim. **18**, 1286–1309 (2008)
25. Andreani, R., Birgin, E.G., Martínez, J.M., Schuverdt, M.L.: Augmented Lagrangian methods under the constant positive linear dependence constraint qualification. Math. Program. **111**, 5–32 (2008)
26. Birgin, E.G., Martínez, J.M.: Large-scale active-set box-constrained optimization method with spectral projected gradients. Comput. Optim. Appl. **23**, 101–125 (2002)

Introduction to Mathematical Origami and Origami Engineering

Sachiko Ishida and Ichiro Hagiwara

Abstract This paper introduces the basics of origami-based foldable structure designs and gives examples of practical applications using these structures for readers who are not familiar with this research field. First, the mathematics associated with origami-based foldable models, such as the conditions to be flat-foldable around a single vertex (Kawasaki's theorem) and to form a foldable tube like a cylinder, is outlined. Next, referencing these conditions—examples of flat-foldable models, such as foldable winding tubes designed with regular folding patterns and folded along the winding axis of the tube, and circular disks that are wrapped and form polygonal pillars with uneven sides and angles—are demonstrated. Finally, recent application of these foldable models to engineering, as exemplified by studies on vibration isolators and energy absorbers, is discussed.

Keywords Origami · Mathematical modeling · Numerical simulation · Structural design · Vibration control · Buckling · Energy absorption

1 Introduction

Origami, the traditional paper craft and art, is well-known all over the world today. In recent years, scientists, designers, and engineers from around the world have been inspired to produce three-dimensional shapes by folding flat sheets. Origami generated by mathematical interpretation and computation is called *mathematical origami* and has led to the modeling of numerous novel origami-based structures beyond the aspect of the arts. Furthermore, mathematical origami provides impor-

S. Ishida (✉)
Department of Mechanical Engineering, Meiji University, Kanagawa, Japan
e-mail: sishida@meiji.ac.jp

I. Hagiwara
Meiji Institute for Advanced Study of Mathematical Sciences,
Meiji University, Tokyo, Japan
e-mail: ihagi@meiji.ac.jp

R.S. Anderssen et al. (eds.), *Applications + Practical Conceptualization + Mathematics = fruitful Innovation*, Mathematics for Industry 11,
DOI 10.1007/978-4-431-55342-7_4

tant ideas for the design of new structures and mechanisms in engineering. In the decade since Nojima [20] proposed *origami engineering*—the scientific study of the origami techniques and their application to engineering—the scope of practical applications to which origami has been applied has expanded. Current applications include space-efficient structure designs, robot designs, medical devices, and automobile components.

This paper serves as an introduction to mathematical origami and origami engineering for readers who are not familiar with this research field. First, mathematics used in the design of origami-based foldable models is presented. Next, representative mathematical origamis and recent applications to engineering studied by us are demonstrated.

2 Mathematical Origami

Origami-based models are geometrically designed with the assumption that the material to be folded has ideally zero thickness, and is not stretchable but flexible enough to allow bending. Paper is often used as the material to satisfy these prerequisites. One field of origami-based structure designs focuses on designing foldable models that change shape dramatically when pulled and compressed in accordance with given direction(s). For example, Fig. 1 shows a fundamental foldable model inspired from the twist buckling pattern of a cylinder [3, 8, 18, 19]. The model is constructed by folding according to the pattern shown in Fig. 1d. The solid and broken lines correspond to mountain and valley folds, respectively. After folding, the boundaries on the right and the left are glued to make a closed cylinder. Thus, it can be folded compactly or *flat-foldable* to the height direction of the cylinder.

Kawasaki's theorem specifies how such flat-foldable models are designed [7, 13]. Let $\alpha_1, \alpha_2, \ldots, \alpha_{2n-1}, \alpha_{2n}$ be successive angles around a single vertex. To be flat-

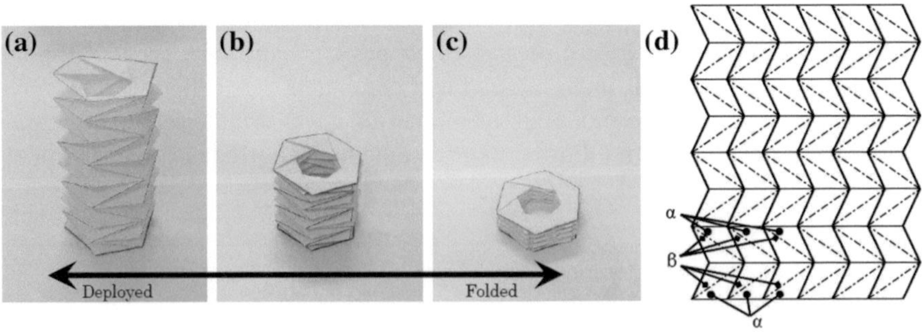

Fig. 1 Folding behaviors of the foldable cylindrical structure: **a** Initial spatial state; **b** folding/expanding state; **c** folded state; **d** folding pattern for the structure. This figure was extracted from Ishida et al. [10]

foldable around the single vertex, it is necessary and sufficient that the sum of the angles skipped one turn is equal to the other sum of the angles:

$$\alpha_1 + \alpha_3 + \cdots + \alpha_{2n-1} = \alpha_2 + \alpha_4 + \cdots + \alpha_{2n}.$$

For the case in which the sum of all the angles, $\alpha_1 + \alpha_2 + \alpha_3 + \cdots + \alpha_{2n}$, is equal to 2π or the vertex is positioned on a developable surface, Kawasaki's condition can be reduced to

$$\alpha_1 + \alpha_3 + \cdots + \alpha_{2n-1} = \alpha_2 + \alpha_4 + \cdots + \alpha_{2n} = \pi.$$

This condition can be obtained as follows. Consider four fold lines around a vertex, as shown in the example in Fig. 2. The sum of the angles around a vertex O is equal to 2π in the unfolded state. The angle goes forward and back in turn with the folding of lines OA, OB, OC, and OD. To be flat-foldable, it must return to the initial position or the sum of the angles must be zero in the folded state. Kawasaki's theorem for four fold lines around a vertex, $\alpha_1 + \alpha_3 = \alpha_2 + \alpha_4 = \pi$, is thus obtained. The structure shown in Fig. 1 satisfies Kawasaki's theorem for six fold lines around every vertex because of the symmetric design, and therefore, the whole structure is flat-foldable. For generalized designs, Tachi [22] provides a software application called *Freeform Origami* that facilitates control of the coordinates of vertices on the pattern independently for asymmetric designs, while still keeping the flat-foldability.

To form a closed tube such as a cylinder, another constraint must be applied in order that the position of the left boundary is identical to that of the right boundary when it is folded. For simple symmetric designs such as that shown in Fig. 1, the mountain and valley folds rotate the line segments by $2(\alpha + \beta)$ and -2α, respectively (Fig. 3). The sum of the rotation angles produced by N-time foldings, $2N\beta$, must be

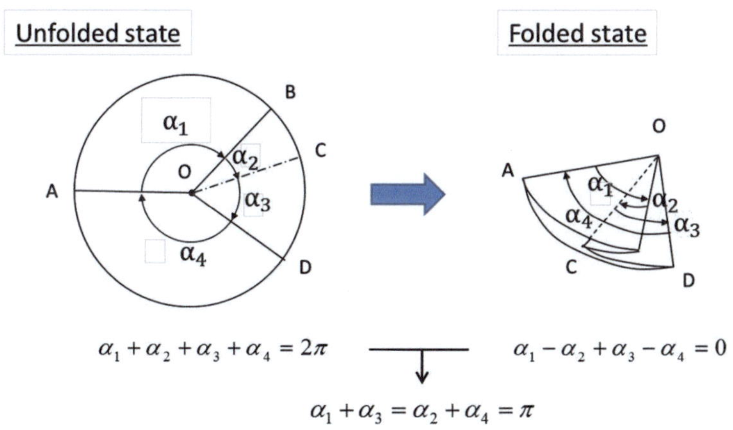

Fig. 2 Kawasaki's Theorem for four fold lines around a vertex

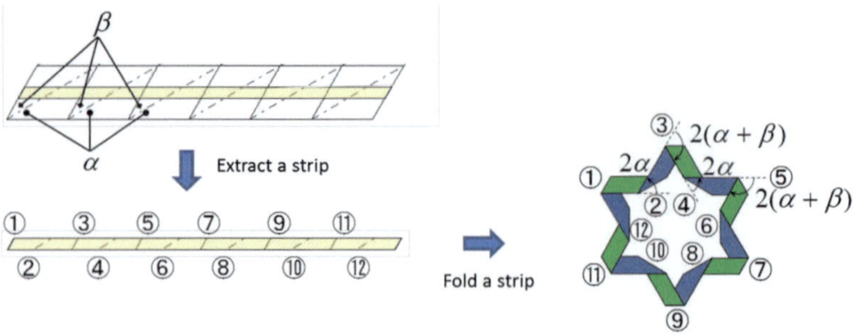

Fig. 3 Constraint of the angle to form the cylinder with thetorsional buckling pattern

equal to 2π [19]. Thus, the constraint of the angle, $2N\beta = 2\pi$, is obtained to form a cylinder. In this example, $N = 6$ so that β is determined as $\pi/6$.

Considering these conditions, a variety of origami-based structures have been studied around the world. Studies have been published, for example, in the book *Origami*[5], the post proceeding of the fifth meeting of origami science, mathematics, and education [23]. The next post proceeding, *Origami*[6], will be on the market soon.

Let us now look at some examples of flat-foldable structures. Ishida et al. [9] demonstrated the patterns for folding toroidal tubes with regular folding patterns and connected those patterns with different main curvatures while still maintaining the flat-foldability. Thus, suppose that arbitrary winding tubes are approximated to a combination of toroidal tubes with different main curvatures, it was clarified that not only straight cylinders but also even winding tubes, shown in Fig. 4, are flat-foldable.

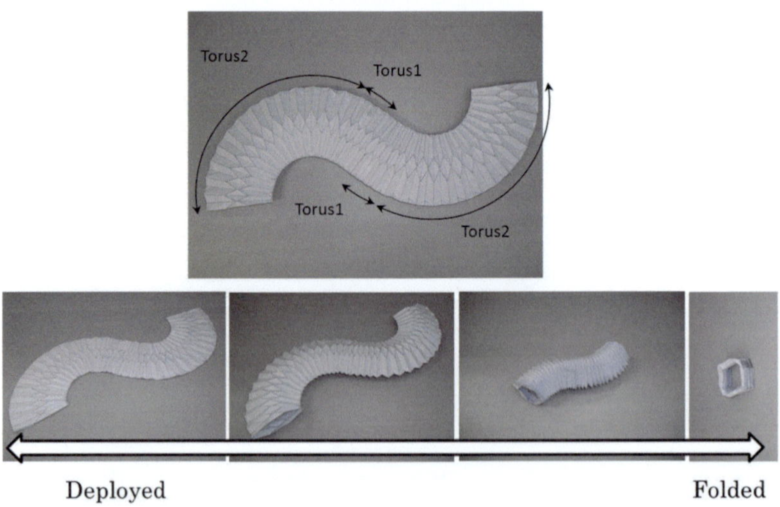

Fig. 4 Folding behaviors of the foldable S-shaped tube

Fig. 5 Flat-folding of the foldable circular disks: **a** Circular disk pattern to form a triangular prism; **b** folding process of the circular disk shown in (**a**); **c** variety of physical models. A part of the figure was extracted from Ishida et al. [11]

Further, the flat-folding of circular disks, shown in Fig. 5, is an interesting example. Those structures, designed by Nojima [18, 19], were folded as if they were wrapped around the center of the disk to form polygonal pillars. For example, the circular disk with three spirals from the circumference to the center of the disk was designed in accordance with Kawasaki's condition at every vertex (Fig. 5a). The disk was folded along the three spirals and formed a triangular prism, as shown in Fig. 5b. The cross-sectional shapes can be varied by changing the number of spirals, such as a square prism from four spirals and a hexagonal prism from six spirals. However, those were only regular polygonal prisms. Ishida et al. [11] widened the variation and produced prisms with unequal sides, such as a rectangular prism, and prisms with unequal angles, such as a rhombic prism, by arranging the positions and the angles of the spirals (Fig. 5c).

3 Origami Engineering

The application of origami-based structures to engineering devices, such as medical devices [14], self-assembling robots [2] and folding structures [1, 16], has been extensively researched globally. In this section, some recent examples studied by us are introduced.

3.1 Origami-Based Vibration Isolators

The foldable cylindrical model shown in Fig. 1 is known as a bi-stable structure; it is stable at the initial spatial state and the folded state, but unstable between both states

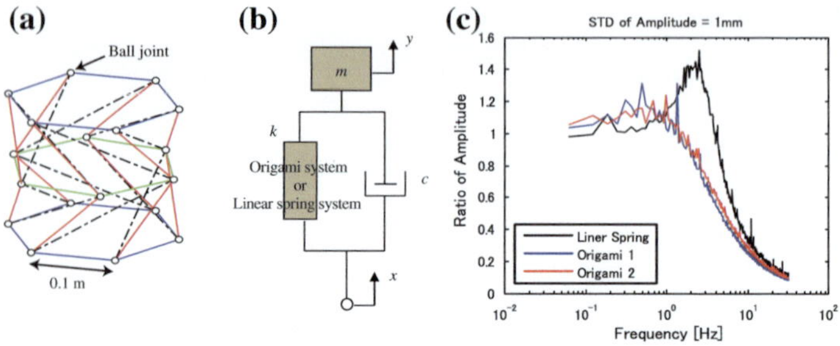

Fig. 6 Computational model for the origami-based vibration isolator and its vibration response: **a** Computational truss model consisting of horizontal bar elements, longitudinal bar elements, and diagonal bar elements, which are connected by ball joints; **b** vibration system consisting of the origami-based foldable model with linear springs, a damper, and a mass; **c** vibration response of the systems. This figure was extracted from Ishida et al. [10]

[17]. Hagiwara et al. [5] noticed that adding linear springs to the bi-stable structure results in a zero-stiffness spring. This is a key idea that led to the development of a novel origami-based vibration isolator that does not transmit any excitation to the other side of the structure around the region where the stiffness is ideally zero [10, 12]. Figure 6a shows the computational model of the isolator, which comprises metallic truss elements connecting each other by ball joints. If a linear spring is implemented in the spring-mass-damper system shown in Fig. 6b, the system will have a strong resonant frequency resulting from the eigen frequency of the spring that corresponds to 2 Hz in Fig. 6c. However, if the proposed isolator is used instead of the linear spring, the system will not have a strong peak because it is a nonlinear spring designed not to transmit excitation, and therefore, the amplification of the vibration can be reduced. To verify these computational results, experimental evaluation of the prototype models is currently in progress.

3.2 Origami-Based Energy Absorbers

Tubular shell structures are widely used as energy absorbers in passenger cars to save the lives of passengers when crushes occur. On the conventional absorbers with rectangular cross-sectional shells (upper left image in Fig. 7), small grooves called beads can be arranged in such a manner so as to prevent the absorber from bending to the outside and to efficiently absorb energy under axial loading [6]. However, these conventional absorbers with beads can deform only up to 70 % of their initial height and the other 30 % of the height remains because of the volume of the buckled absorber. Furthermore, they cannot mitigate peak shocks at the instant of the crush and so passengers can be severely injured. To solve these problems, tubular shell structures with foldable cylinders have been studied [15, 21, 24]. Numerical investigations have shown that optimization of the design parameters, such as the

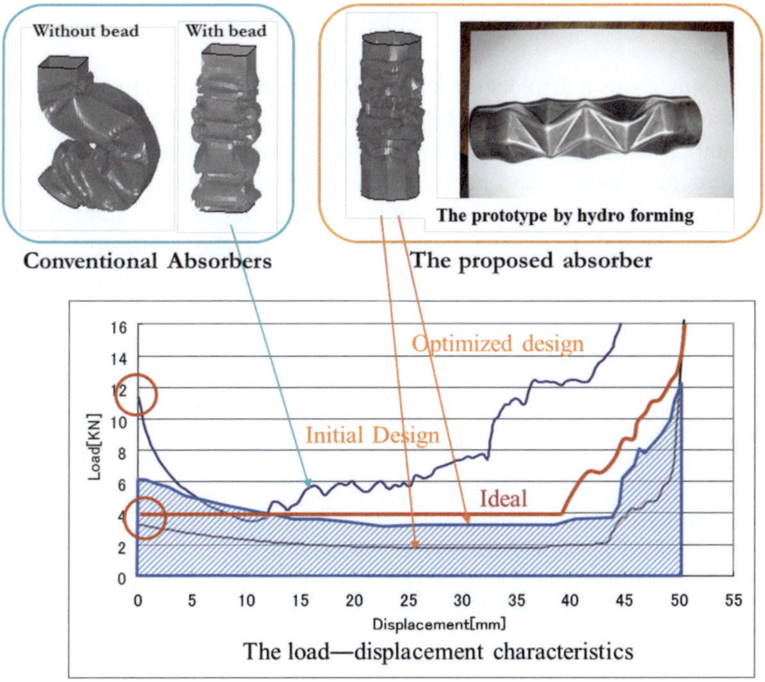

Fig. 7 Comparison of conventional energy absorbers and the proposed energy absorber. The figure was extracted from Hagiwara [4]

angles of fold lines, results in the proposed absorbers (upper right image in Fig. 7) being able to reduce the initial shock and also to buckle by up to 90 % of their initial height [4]. The optimized absorbers can also absorb 1.5 times more energy than the conventional ones. In addition, the hydro forming fabrication method that is currently used to mass produce various industrial parts can be applied to develop and produce the proposed absorbers.

Thus, the advantages of using foldable cylindrical models in absorber designs are clear. However, the proposed absorbers have not yet been adopted for actual use in cars because of the high forming cost incurred by the hydro forming method. Consequently, cost reduction is a challenge for the future.

4 Conclusion

To be flat-foldable around a vertex, it is necessary to satisfy the condition that the sum of angles skipped one turn is equal to the other sum of the angles (Kawasaki's theorem). To form a foldable tube, another constraint also has to be considered; the position of the left boundary should be identical to that of the right boundary when it is folded.

To date, numerous flat-foldable structures have been designed in accordance with the above-mentioned conditions. They are not limited to straight structures such as cylinders but also winding tubes and wrappable disks, for example. The varieties of foldable structure designs have expanded the applications to engineering. Although the remarkable characteristics of the foldable structures are the dramatic changes in the shapes through folding and deploying motions, the application is not only linked with the characteristics, but also with the mechanical characteristics that the folding patterns possess, such as bi-stability of vibration isolators. Further, foldable designs yield new solutions to practical problems, as exemplified by the case of energy absorbers. There is virtually no limit to the ideas for new applications using foldable structures. We hope that this paper stimulates the creativity of readers and helps them to produce new ideas in their research fields.

Acknowledgments We would like to acknowledge the kind advice and comments received from Dr. Taketoshi Nojima (Art Excel Co., Ltd.) regarding the designing of foldable structures. We also thank Prof. Hiroshi Uchida (Fukuyama University) for his substantial support in the numerical simulation on the origami-based vibration isolators and Prof. Xilu Zhao (Saitama Institute of Technology) for his insightful advice regarding numerical simulation on the origami-based energy absorbers.

References

1. Deng, D., Chen, Y.: Origami-based self-folding structure design and fabrication using projection based stereolithography. J. Mech. Des. **137**(2), 021701 (2015)
2. Firouzeh, A., Paik, J.: Robogami: a fully integrated low-profile robotic origami. J. Mech. Robot. **7**(2), 021009 (2015)
3. Guest, S.D., Pellegrino, S.: The folding of triangulated cylinders, part I: geometric considerations. ASME J. Appl. Mech. **61**, 773–777 (1994)
4. Hagiwara, I.: Application of Origami Techniques to Engineering. Adventure of Mathematical Modeling and Application, pp. 183–195. Meiji University Press (2015). ISBN: 4906811108 (in Japanese)
5. Hagiwara, I., Ishida, S., Uchida, H.: Nonlinear springs and anti-vibration device. Japanese patent pending, No. 2013-220548 (2013)
6. Hagiwara, I., Tsuda, M., Kitagawa, Y., Futamata, T.: Method of determining positions of beads. United States Patent, Patent Number 5048345 (1991)
7. Hull, T.: The combinatorics of flat folds: a survey. In: Origami^3, pp. 29–38. AK Peters (2002)
8. Hunt, G.W., Ario, I.: Twist bucking and the foldable cylinder: an exercise in origami. Int. J. Non-linear Mech. **40**, 833–843 (2005)
9. Ishida, S., Nojima, T., Hagiwara, I.: Origami-based foldable design technique for meandering tubes by using conformal transformation. Trans. Jpn. Soc. Ind. Appl. Math. **24**(1), 43–58 (2014) (in Japanese)
10. Ishida, S., Uchida, H., Hagiwara, I.: Vibration isolators using nonlinear spring characteristics of origami-based foldable structures. Trans. Jpn. Soc. Mech. Eng. **80**(820), DR0384 (2014) (in Japanese)
11. Ishida, S., Nojima, T., Hagiwara, I.: Design of deployable membranes using conformal mapping. J. Mech. Des. **137**(6), 061404 (2015)

12. Ishida, S., Uchida, H., Shimosaka, H., Hagiwara, I.: Design concepts and prototypes of vibration isolators using bi-stable foldable structures. In: Proceedings of ASME 2015 International Design Engineering Technical Conferences and Computers and Information in Engineering Conference. DETC2015- 46409 (2015)
13. Kawasaki, T.: On the relation between mountain-creases and valley-creases of a flat origami. In: Proceedings of the First International Meeting of Origami Science and Technology, pp. 229–237 (1989)
14. Kuribayashi, K., Tsuchiya, K., You, Z., Tomus, D., Umemoto, M., Ito, T., Sasaki, M.: Self-deployable origami stent grafts as a biomedical application of Ni-rich TiNi shape memory alloy foil. Mater. Sci. Eng. A **419**(1–2), 131–137 (2006)
15. Ma, J., You, Z.: A novel origami crash box with varying profiles. In: Proceedings of the ASME 2013 International Design Engineering Technical Conference and Computers and Information in Engineering Conference. DETC2013-13495 (2013)
16. Miyashita, S., DiDio, I., Ananthabhotla, I., An, B., Sung, C., Arabagi, S., Rus, D.: Folding angle regulation by curved crease design for self-assembling origami propellers. J. Mech. Robot. **7**(2), 021013 (2015)
17. Nagashima, G., Nojima, T.: Development of foldable triangulated cylinder. In: Proceedings of the 7th JSME Materials and Processing Conference (M&P), 153–154 (1999) (in Japanese)
18. Nojima, T.: Structure with folding lines, folding line forming mold, and folding line forming method. Patent No. WO 2001081821, A9 (2001)
19. Nojima, T.: Modelling of folding patterns in flat membranes and cylinders by origami. Int. J. Jpn. Soc. Mech. Eng. **45**(1), 364–370 (2002)
20. Nojima, T.: Structural modeling based on mathematical origami techniques. Kyoto University IIC fair, Kyoto Newspaper, (2002.11.26) (2002) (in Japanese)
21. Song, J., Chen, Y., Lu, G.: The thin-walled tubes with origami pattern under axial loading. In: Proceedings of International Conference on Simulation Technology (2013)
22. Tachi, T.: Freeform Origami (2011). http://www.tsg.ne.jp/TT/software/#ffo
23. Wang-Iverson, P., Lang, R.J., Yim, M. (eds.): *Origami*^{5} Fifth International Meeting of Origami Science, Mathematics, and Education. AK Peters/CRC Press (2011). ISBN 1568817142
24. Zhao, X., Hu, Y., Hagiwara, I.: Shape optimization to improve energy absorption ability of cylindrical thin-walled origami structure. J. Comput. Sci. Technol. **5**(3), 148–162 (2011)

An Interaction with Biologists: Insights into Development and Disease

Kerry A. Landman

Abstract The enteric nervous system (ENS) in our gastrointestinal tract is responsible for normal gut function and peristaltic contraction. Embryonic development of the ENS involves the colonisation of the gut wall from one end to the other by a growing population of motile neural crest cells. The colonisation wave is strictly timetabled and predictable, but individual neural crest cell movement is unpredictable in speed and direction. Failure of these cells to invade the whole gut results in the relatively common, potentially fatal birth defect. Continuum models of the population-level behaviour, based on the Fisher equation, are highly predictable. Discrete agent-based models, governed by agent probabilities, reproduce the population-level behaviour of the Fisher equation. However, individual agent contributions to the total population, measured by agent lineage, are highly variable. Both behaviours have been verified in a developmental invasion system. This work is the result of a rewarding long-standing and on-going collaboration between applied mathematicians and developmental biologists.

Keywords Mathematical biology · Cell invasion · Diffusion · Proliferation · Lineage

1 Introduction

The enteric nervous system in our gastrointestinal tract is responsible for normal gut function and peristaltic contraction. Embryonic development of the enteric nervous system involves the colonisation of the gut wall from one end to the other by a growing population of motile neural crest (NC) cells. Failure of these cells to invade the whole gut results in the relatively common, potentially fatal condition known as Hirschsprung disease.

K.A. Landman (✉)
Department of Mathematics and Statistics,
University of Melbourne, Victoria 3010, Australia
e-mail: kerryl@unimelb.edu.au

R.S. Anderssen et al. (eds.), *Applications + Practical Conceptualization + Mathematics = fruitful Innovation*, Mathematics for Industry 11,
DOI 10.1007/978-4-431-55342-7_5

During this colonisation process, there is a huge increase in NC cell numbers through cell division (proliferation) up to a preferred density, called the carrying capacity. At the same time, the cells making up the gut tissue also undergo cell division, resulting in massive elongation of the intestine. However, two styles of experiments on gut explants can be performed—one where the gut undergoes no elongation and one where the gut undergoes significant elongation.

Observations using time-lapse imaging in bird and mouse guts have emphasised that the speed and direction of movement of individual NC cells is unpredictable. In contrast, the wave-like spread of the entire population is predictable [11].

Here we discuss modelling and simulation of diffusive processes, both continuum and discrete, which have been used to provide insight into the key biological processes required for complete colonisation. Moreover, the modelling work has also generated experimentally testable predictions.

This modelling work was initiated by an approach by Dr. Donald F. Newgreen, a developmental biologist who heads an Embryology Laboratory at Murdoch Childrens Research Institute in Melbourne, Australia. He thought that mathematical modelling might help with understanding the colonisation process and the cause of Hirschsprung disease. Our research groups have collaborated for over ten years, addressing various aspects of ENS development and disease.

2 Colonisation Models

We have developed both macroscopic population-level models (partial differential equation models) based on diffusive processes and microscopic cellular-level stochastic models (cellular automata (CA)/agent-based models) based on random walks. These approaches are complementary, but the microscopic models include the variability inherent in biological phenomena. In the conclusions, we make some comments about interesting theoretical questions that arise in connecting the two approaches.

It is worth noting that there is no overriding directional signal in the gut. Placement of the NC cells at the anal end of the gut results in reverse-direction migration [2]. In the models discussed here, no directional signal is imposed. Colonisation proceeds from left to right only because the initial cell population is placed at the left-hand end.

In the population-level models, the cell density changes in time due to a flux term (e.g. linear diffusion due to NC cell random motion) and a logistic growth term (due to cell proliferation up to a preferred density). The Fisher equation [4, 7] is the typical continuum description of an invading population. Here we modify that equation by including an extra advective term to describe domain growth. The invading NC cell population is restricted to a cylindrical surface, so a one-dimensional spatial description in terms of the cylindrical axis is appropriate. Let $u(x, t)$ be the density as a function of longitudinal distance x and time t. Then the continuum model is

$$\frac{\partial u}{\partial t} = \mathscr{D}\frac{\partial^2 u}{\partial x^2} + \lambda u\,(1-u) - \frac{\partial}{\partial x}\left(v(x,t)u\right), \tag{1}$$

where $\mathscr{D}$ is the diffusivity, λ is the proliferation rate, unity is the normalised carrying capacity, and $v(x,t)$ is the advective velocity due to the underlying domain growth induced by the elongating tissue. For uniform growth

$$v(x,t) = \frac{dL}{dt}\frac{x}{L}. \tag{2}$$

If there are host $H(x,t)$ and donor $D(x,t)$ NC cells, with the same properties, then the host and donor populations are governed by

$$\frac{\partial H}{\partial t} = \mathscr{D}\frac{\partial^2 H}{\partial x^2} + \lambda H\Big(1-(H+D)\Big) - \frac{\partial}{\partial x}\left(v(x,t)H\right), \tag{3}$$

$$\frac{\partial D}{\partial t} = \mathscr{D}\frac{\partial^2 H}{\partial x^2} + \lambda H\Big(1-(H+D)\Big) - \frac{\partial}{\partial x}\left(v(x,t)D\right), \tag{4}$$

respectfully.

In the CA model, a two-dimensional square lattice is used. At each time step, randomly chosen agents are offered opportunities to move and proliferate, governed by local rules and probabilities. These are illustrated in Fig. 1b, c. Each lattice site can be occupied by at most one agent, representing a NC cell, due to volume exclusion. Movement or proliferation events which attempt to place a NC agent on an already occupied site are aborted.

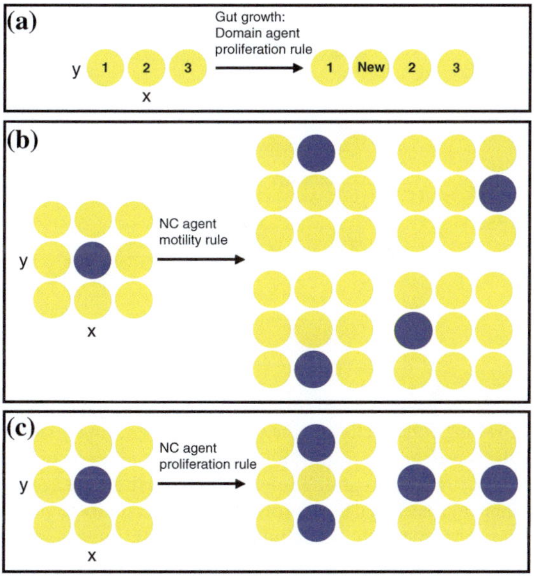

Fig. 1 CA mechanisms for domain agent (*yellow*) proliferation and NC agent (*blue*) motility and proliferation. **a** Domain agent proliferation rule produces a growing domain. **b** A NC agent can move to one of the four configurations shown with equal probability. **c** A NC agent can divide into two daughter agents, placed at two possible configurations (to replicate experimental observations) with equal probability

We developed a discrete growing domain model [1], where the underlying lattice sites represent the gut tissue cells. To model gut elongation, new lattice sites are randomly inserted into each row of the lattice, thus growing the lattice, as illustrated in Fig. 1a.

The domain has periodic boundary conditions on the two horizontal boundaries, and there will be no-flux conditions on the vertical boundaries.

2.1 No Gut Growth

In the absence of gut growth, we can determine experimentally how NC cells behave in different parts of the colonisation wave using quail and chick NC cells, since they are functionally equivalent but label differently using antibodies. We suggested experiments to test against our models.

In tissue culture where there is no gut growth, a segment of chick host gut tissue is removed and replaced with quail donor tissue containing NC cells. Three days later the resulting host and donor cells are identified. Donor cells can be placed at or behind the wavefront (Fig. 2a, b).

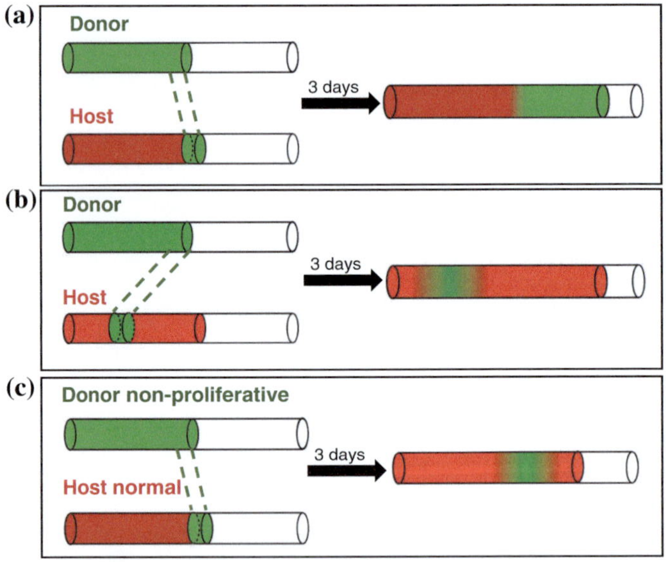

Fig. 2 Schematic profiles of chick-quail graft experiments. Host cells (*red*), donor cells (*green*). **a** Grafting donor cells into the wavefront, **b** grafting donor cells behind the wavefront. In (**a**, **b**), cells at the wavefront are responsible for driving further colonisation, called frontal expansion. **c** Grafting donor cells with no proliferative ability into the wavefront. Host cells overtake the non-proliferative donor cells and then have proliferative opportunity to drive further colonisation [9]. (*Note* Variants were experimentally tested where the donor tissue is removed from behind the wavefront instead of at the wavefront as shown. The outcomes were the same)

Both models (Eqs. (3) and (4) with appropriate initial conditions [9] and discrete CA) and experiments give the same results. Cells at the wavefront are responsible for an increase in cell numbers into unoccupied regions, thereby driving the colonisation wave. We call this frontal expansion. We conclude that proliferation at the wavefront is the driver of colonisation [9, 10].

What happens if NC cell proliferation is eliminated at the wavefront (Fig. 2c)? Donor cells can be treated to eliminate cell division but preserving cell motility. The experimentalists thought there were two possible outcomes. Either nothing happens (stall) or the host cells proliferate and push the donor cells forward (shunting). In contrast, our models predict that the donor cells diffuse, leave space for the host cells to move into and eventually overtake the donor cells, proliferate at the wavefront and drive further colonisation. The experimentalists were surprised at their results—they matched the model predictions [10].

2.2 *With Gut Growth*

The models predict that colonisation again proceeds by frontal expansion; however, now there is also proliferation behind the wavefront due to additional space produced by domain growth. Successful colonisation only occurs if the NC cell proliferation rate is sufficiently large, compared with the domain growth rate (Fig. 3). The predictions were confirmed experimentally [12]. Incomplete colonisation corresponds to Hirschprung disease.

For the deterministic PDE models, there is a critical value of the NC cell proliferation rate that determines whether colonisation is complete or not. Therefore, if the

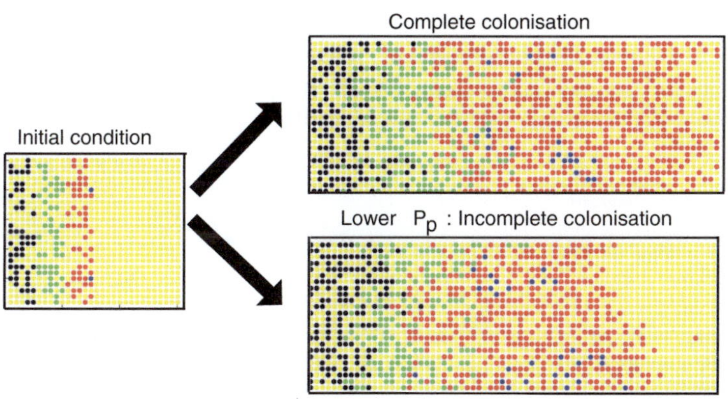

Fig. 3 With gut growth. Two realisations of the CA algorithm for two proliferation rates P_p. The subpopulations mix more than they do in the non-growing case. If the proliferation rate is sufficiently high, then there is complete colonisation. If the proliferation rate is sufficiently small, then there is incomplete colonisation, which emulates Hirschprung disease

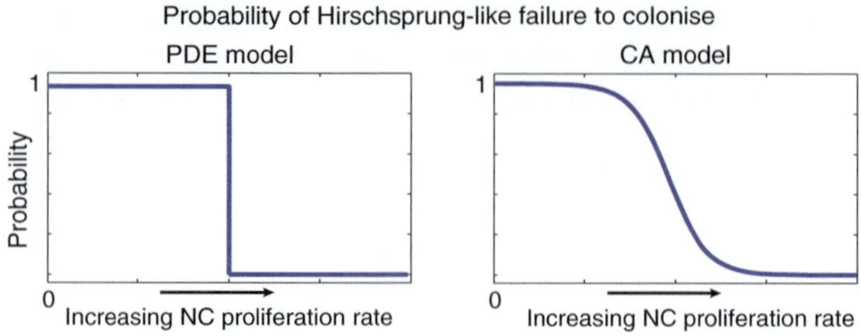

Fig. 4 The probability of Hirschsprung-like failure to complete colonisation for the PDE (deterministic) and CA (stochastic) models

proliferation rate is below the critical value, then a Hirschsprung-like failure occurs (Fig. 4). However, the CA approach is stochastic. This means that there is a range of NC cell proliferation rates where the probability of incomplete colonisation is nonzero, but not unity (Fig. 4). In particular, there exists a value of the proliferation rate where there is a 50 % chance of a Hirschsprung-like failure.

These results have important implications for Hirschprung disease. We now understand Hirschsprung disease to be in most cases a NC cell proliferation defect and not, as thought, a NC cell motility defect. Furthermore, we have identified proliferation as the common link between the known Hirschsprung disease genes. In particular, stochastic effects can determine success or failure of the colonisation process for certain range of NC cell proliferation rates. This may explain the existence of identical twins where only one of the pair has Hirschsprung disease [6].

3 Cell Lineage

We know investigate the contribution of individual agents/cells to the colonisation wave. A lineage tree can be constructed for each agent. Agents either divide into two at a successive generation, or do not divide, as illustrated in Fig. 5.

Fig. 5 Schematic diagram representing the lineage of a single agent at some time t. The total number of filled discs represents the agent lineage tracing, while an empty disc represents a division event

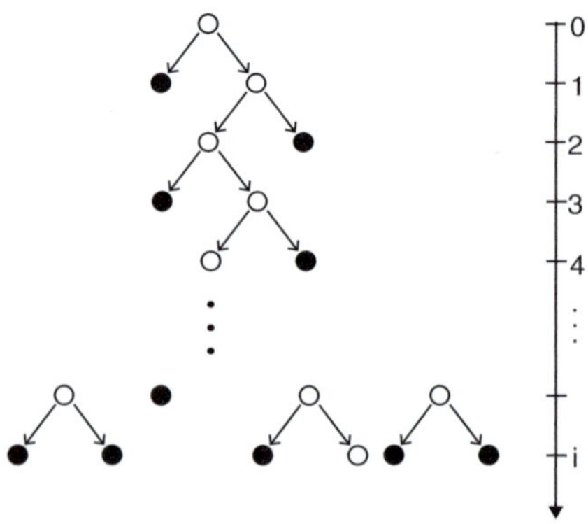

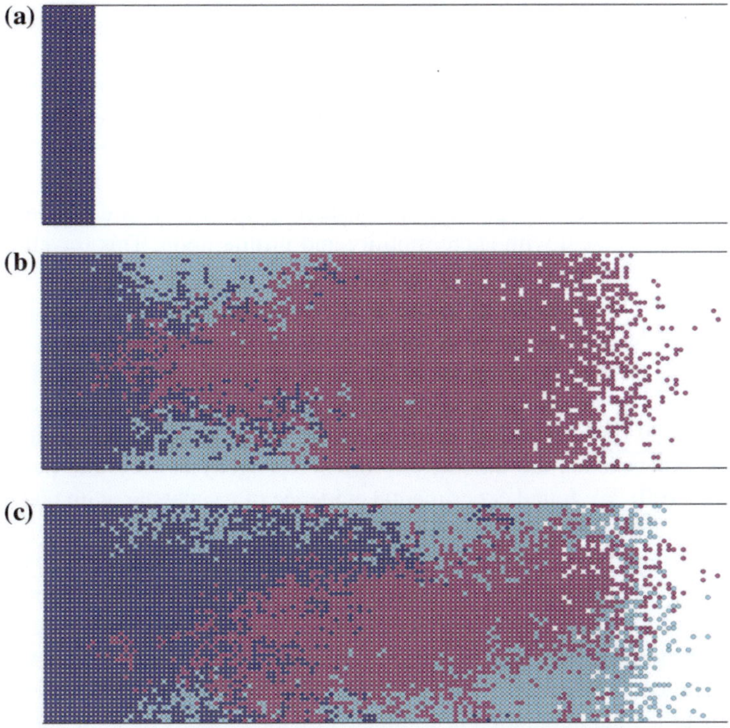

Fig. 6 Invasion wave and spatial distribution of agent tracings in two realisations for no gut growth case. **a** Initial condition for all simulations. **b, c** Two realisations of the travelling wave that moves progressively to the *right*. The largest and second largest single agent lineage tracing (*pink* and *turquoise* respectively) and the 498 other agent lineage tracings (all collected together in *blue*) are illustrated. In **(b)** there are significant differences in the agent numbers between the two largest tracings, while in **(c)** the two largest tracings have a similar number of agent numbers

Starting with a localised number of agents at maximum density at the left-hand end of the lattice (Fig. 6a), the agent density evolves and moves progressively to the right as illustrated in the two realisations presented in Fig. 6b, c. Stochastic competition for resources (space, growth factor, nutrients) between agents can be visualised through the spatial distribution of the single progeny lineage within the invasion wave. Realisations with the largest and second largest lineage tracing (pink and turquoise respectively), together with all progeny from the remaining initial agents (blue) are illustrated in Fig. 6b, c. These two realisations illustrate the large individual variability of the spatial distribution of single agent lineage. Furthermore, in every realisation a small proportion of identical initial agents accounts for a substantial proportion of the total final agent population. We term these individuals superstars. A full analysis of the complete set of lineage tracings can be found in Cheeseman et al. [3]. (Note, that the proliferation rule is slightly modified here. Instead of the two daughters being placed in opposing nearest-neighbour sites as in Fig. 1c, the two daughters are placed in nearest neighbour sites. The results are qualitatively the same for both cases.)

In the two realisations, the total population has advanced approximately the same extent and is therefore highly predictable at this level. When lattice-site occupancy of the agents is column-averaged for a single realisation and averaged over many identically prepared realisations, a smooth density wave evolves into a travelling wave moving from left to right with a constant speed, which is dependent on the probabilities associated with agent motility and proliferation. This travelling wave emulates the Fisher wave [7].

Experiments were performed where a single starting NC cell is labelled green (green fluorescent protein GFP labelled cell) is placed within a starting population of red labelled cells. GFP+ve progeny inherit the green label, while all other NC cells label red.

After several days the whole gut is full of NC cells (thousands). However in numerous experiments, the number of GFP+ve progeny is very low (many <10, almost all <100). We found experimental evidence of a superstar, with one GFP+ve cell lineage tracing being of disproportionately huge number.

Unfortunately, there are insufficient cell-labelling colours to uniquely identify every starting NC cell and therefore uniquely determine the progeny of every starting cell (if indeed all the hues are inheritable), as can be performed with the CA model. However, the modelling results explain why most of the determined lineages had few progeny, and contribute very little to the total final population. The models support the fact that the final population is dominated by the progeny of a few superstars.

This suggests that stochastic competition for resources, e.g. space, is the source of the large variability and therefore inequality to the individual lineages. All agents/cells are identical but only a very few contribute significantly, due to most progeny being unable to contribute further progeny because neighbouring sites are already occupied. Note that for a growing domain, the contribution of superstars is not as large as for a non-growing domain, as all agents experience new empty lattice sites/tissue, as it is uniformly growing [3].

4 Conclusions

The model and experimental results imply that Hirschsprung disease is chiefly a NC cell proliferation defect and not, as previously thought, a NC cell motility defect [9, 12]. Furthermore, we have identified proliferation as the common link between the known Hirschsprung disease genes.

Cell lineage tracing is a powerful tool for understanding how proliferation of individual cells contribute to population behaviour. Single neural crest cells labelled to follow progeny in the intestine reveal extraordinary and unpredictable variation in number and position of descendant cells, even though ENS development is highly predictable at the population level. Lattice agent-based model explains these puzzling results [3].

The experiments and model suggest that stochastic competition for resources is an important concept when understanding biological processes which feature high levels of cell proliferation. The results have implications for cell fate processes in the ENS and other developmental systems.

The ENS is a fascinating developmental system with many interesting modelling challenges, combining diffusive processes and simulation. This work has provided much insight into a real biological process.

Finally, this biological collaboration involves two mathematical approaches. An interesting mathematical problem is to investigate the connection between the two. Our agent-based CA models form an interacting particle system. A mean field approach has been successfully implemented to transform the discrete model to a continuum model. We have developed a robust framework for taking general cellular probabilistic rules and converting them into a partial differential equation (non-linear diffusion equation for single species, nonlinear advection-diffusion equations for multi-species) in a systematic way [8]. Furthermore, growing domains can also be accounted for in a proper way [5]. This is a growing era of research which combines probability theory, statistical physics and mathematical biology.

Acknowledgments This work is the result of collaborations with many people, including research fellows, students and other colleagues. We express our thanks to Ben Binder, Bevan Cheeseman, Barry Hughes, Anthony Fernando, Emily Hackett-Jones, Jack Hywood, Michael Mariani, Donald Newgreen, Catherine Penington, Mat Simpson and Dong Zhang. This work was supported by Australian Research Council and National Health and Medical Research Council grants.

References

1. Binder, B.J., Landman, K.A., Simpson, M.J., Mariani, M., Newgreen, D.F.: Modeling proliferative tissue growth: a general approach and an avian case study. Phys. Rev. E **78**, 031912 (2008)
2. Burns, A.J., Delalande, J.M., Le Douarin, N.M.: In ovo transplantation of enteric nervous system precursors from vagal to sacral neural crest results in extensive hindgut colonisation. Development **129**, 2785–2796 (2002)
3. Cheeseman, B.L., Zhang, D., Binder, B.J., Newgreen, D.F., Landman, K.A.: Cell lineage tracing in the developing enteric nervous system: superstars revealed by experiment and simulation. J. R. Soc. Int. **11**, 20130815 (2014)
4. Fisher, R.A.: The wave of advance of advantageous genes. Ann. Eugen. **7**, 355–369 (1937)
5. Hywood, J.D., Hackett-Jones, E.J., Landman, K.A.: Modeling biological tissue growth: discrete to continuum representations. Phys. Rev. E **88**, 032704 (2013)
6. Jung, P.M.: Hirschsprung's disease: one surgeon's experience in one institution. J. Pediatr. Surg. **30**, 646–651 (1995)
7. Murray, J.D.: Mathematical Biology, 3rd edn. Springer, New York (2002)
8. Penington, C.J., Hughes, B.D., Landman, K.A.: Building macroscale models from microscale probabilistic models: a general probabilistic approach for nonlinear diffusion and multi-species phenomena. Phys. Rev. E **84**, 041120 (2011)
9. Simpson, M.J., Landman, K.A., Hughes, B.D., Newgreen, D.F.: Looking inside an invasion wave of cells using continuum models: proliferation is the key. J. Theor. Biol. **243**, 343–360 (2006)

10. Simpson, M.J., Mariani, M., Zhang, D., Landman, K.A., Newgreen, D.F.: Cell proliferation drives neural crest cell invasion of the intestine. Dev. Biol. **302**, 553–568 (2007)
11. Young, H.M., Bergner, A.J., Anderson, R.B., Enomoto, H., Milbrandt, J., Newgreen, D.F., Whitington, P.M.: Dynamics of neural crest-derived cell migration in the embryonic mouse gut. Dev. Biol. **270**, 455–473 (2004)
12. Zhang, D., Brinas, I.M., Binder, B.J., Landman, K.A., Newgreen, D.F.: Neural crest regionalisation for enteric nervous system formation: implications for Hirschsprung's disease and stem cell therapy. Dev. Biol. **339**, 280–294 (2010)

Modeling Biochemical Reaction Systems with Markov Chains

Arnab Ganguly

Abstract A biochemical reaction system is a complex network of several species interacting through different reaction channels. Modern experimental techniques have conclusively testified to the stochastic nature of these systems, particularly those with small molecular counts of reactant species. The stochastic nature of such systems can be successfully modeled by an appropriate continuous time Markov chain. The aim of this note is to describe the framework of this Markov modeling approach and then focus on different simulation techniques that are particularly beneficial to the practitioners. Our discussion includes certain approximate simulation schemes, which are computationally efficient for simulating large multiscale systems compared to the statistically exact ones.

Keywords Stochastic simulation · Biochemical reactions · Markov chains · Tau leaping · Diffusion approximation

1 Introduction

A biochemical reaction system involves multiple reactions and several molecular species. Clear evidences of randomness present in these systems are demonstrate due to recent advances in single cell imaging, microfluidic techniques etc. [7, 11]. As a result, stochastic modeling of biomolecular reaction networks has increased dramatically over the last decade. The randomness, often called biochemical noise in the biology literature, has been put forward as the explanation for lysis/lysogeny decision of the bacteria λ-phage and the loss of synchrony of Circadian clocks. Multiple features of experimentally observed variability including transcription bursts, different steady state distributions of RNA and protein counts and random transitions

A. Ganguly (✉)
Department of Mathematics, Louisiana State University,
222 Lockett, Fieldhouse Dr, Baton Rouge, LA 70803, USA
e-mail: aganguly@lsu.edu

R.S. Anderssen et al. (eds.), *Applications + Practical Conceptualization + Mathematics = fruitful Innovation*, Mathematics for Industry 11,
DOI 10.1007/978-4-431-55342-7_6

between two metastable states have been successfully captured by modeling reaction network dynamics by appropriate Markov processes (e.g., see [24, 25]).

One of the important areas in mathematical modeling of reaction networks involves designing efficient simulation schemes. Oftentimes conducting some particular experiments is expensive and it is desirable for a biologist to run a computer simulation instead. This is particularly the case when, for example, a biologist is trying to discover if certain reactions are part of a cellular process. In this case it is best to run a series of computer simulations with the hypothesized reaction set describing the proposed cellular process to test the biologist's belief rather than directly starting with expensive laboratory experiments.

A Markov process representation of the state of the system is particularly beneficial in this regard. Popular algorithms for statistically exact simulations of such reaction systems include Gillespie's stochastic simulation algorithms, next reaction method and their more efficient variants [14]. These algorithms track every molecular reaction event and thus become computationally expensive when reaction system becomes larger or more complex. In many of these examples, where the reaction systems include fast reactions and species with high copy numbers, substantial gain in simulation speed with good accuracy is often obtained by resorting to approximate algorithms like tau-leaping (which is essentially an Euler-type approximation technique [15]) or a suitable diffusion approximation. The review article [16] by Gillespie is a nice exposition to some of these algorithms. While standard tau leaping algorithms and usual diffusion approximations work well under classical scaling (which requires all the species and reaction rates to be of the same order), they work poorly for reaction systems with varying orders of abundance of species and varying orders of magnitudes of the reaction rates. In these scenarios suitable hybrid methods need to be implemented to gain speed and efficiency while maintaining a certain level of accuracy. One typical technique in such a multi-scale algorithm involves partitioning the reaction set into 'fast' and 'slow' reactions. The idea is to simulate the members of the fast set by a continuous process (diffusion approximation or ODE approximation) and keep the discrete counting process formulation for the slow ones.

The goal of this article is to discuss some of these popular simulation techniques in the context of biochemical reaction networks. We will also discuss some error analysis which provides deeper insight into the accuracy of some of the approximate algorithms. The rest of the article is organized as follows. Section 2 provides background on Markov modeling of reaction systems. Section 3 presents statistically exact simulation algorithms, while approximate algorithms for faster speed are discussed in Sect. 4. Multiscale reaction systems are treated in Sect. 5.

2 Markov Processes and Biological Modeling

Consider a reaction system consisting of M chemical species, $S_1, \ldots, S_M$ and R reactions, $R_1, \ldots, R_R$

$$R_k : \sum_{i=1}^{M} \nu_{ik}^{-} S_i \rightarrow \sum_{i=1}^{M} \nu_{ik}^{+} S_i, \quad k = 1, \ldots, R. \tag{1}$$

Here, ν_{ik}^{-} and ν_{ik}^{+} respectively denote the number of molecules of the species S_i consumed and created due to one occurrence of reaction R_k. Let $X(t) \in \mathbb{N}^M$ denote the state of the reaction system at time t. If ν_k denotes the vector with ith component $\nu_{ik}^{+} - \nu_{ik}^{-}$, then an occurrence of R_k at time t updates the state by the following equation

$$X(t) = X(t-) + \nu_k.$$

The process X is a continuous time Markov chain (CTMC) with transition probabilities governed by

$$\mathsf{P}[X(t+h) = x + \nu_k | X(t) = x] = a_k(x)h + o(h),$$

where a_k is the intensity (called propensity in biochemical literature) function associated with reaction R_k and is calculated by the law of mass action kinetics in the present article. In other words,

$$a_k(x) = \theta_k \prod_{i=1}^{M} \binom{x_i}{\nu_{ik}^{-}},$$

where θ_k is the stochastic reaction rate constant for reaction R_k. A path-wise representation of the process X is given by the stochastic equation

$$X(t) = X(0) + \sum_{k=1}^{R} \xi_k\left(\int_0^t a_k(X(s))ds\right)\nu_k, \tag{2}$$

where the ξ_k are independent unit Poisson processes. It should be noted that the quantities $\xi_k(\int_0^t a_k(X(s))ds)$ count the number of occurrences of the reaction R_k by time t. Equation (2) is an example of a random time change representation in which stochastic equations involve random time changes of other Markov processes (for details, see [9, Chap. 6]). The generator of the Markov process X is given by

$$\mathscr{A} f(x) = \sum_{k=1}^{R} a_k(x)(f(x+\nu_k) - f(x)),$$

that is, for a bounded-measurable function f, the quantity

$$f(X(t)) - f(X(0)) - \int_0^t \mathscr{A} f(X(s))\, ds$$

is a martingale. Consequently, taking $f(y) = 1_{\{x\}}(y)$, it follows that the probability mass function of $X(t)$ satisfies the following Kolmogorov forward equation (also known as the master equation in chemical literature)

$$\frac{\partial p(x,t)}{\partial t} = \sum_{k=1}^{R} [a_k(x - \nu_k)p(x - \nu_k, t) - a_k(x)p(x,t)], \tag{3}$$

where $p(x,t) = \mathsf{P}(X(t) = x)$. Anderson and Kurtz [3] is a good source for more details on Markov chain modeling of reaction networks, and it includes some necessary background on general theory of stochastic processes.

3 Exact Stochastic Simulation

Except for some simple systems, (3) cannot be solved in closed form and is often hard to solve numerically. Instead, Monte-Carlo type algorithms like Gillespie's stochastic simulation algorithm (SSA) or Gibson Bruck's next reaction method and their variants have proved to be quite popular for exact statistical simulation of smaller reaction networks. Given the current state, these algorithms calculate the waiting time for the occurrence of next reaction and identify the reaction which happens at that time. Using the fact that X is a non-homogeneous Markov chain, one concludes that the waiting time, Δ, for the next event after time t is an exponential random variable with parameter $a_0(X(t)) \equiv \sum_{k=1}^{R} a_k(X(t))$. Moreover, the probability of the next reaction being R_k is given by $a_k(X(t))/a_0(X(t))$. These observations form the backbone of Gillespie's stochastic simulation algorithm whose details are outlined below.

Algorithm 1: Gillespie's algorithm.

Input: The initial state vector X, stoichiometric matrix $S = \{\nu_{ij}\}$,
$i = 1, 2, \ldots, M,\ j = 1, 2, \ldots, R$, end of the simulation time $T > 0$.
Output: The number of molecules of each species in time interval $t \in [0, T]$.

```
Set t = 0, n = 0.
while t < T and Σ_{i=1}^{R} a_i > 0 do
    Calculate the propensity function, a_k, for each k.
    Set a_0 = Σ_{k=1}^{R} a_k.
    Generate u, v independently from Uniform(0, 1).
    Set Δ = −log u/a_0 (Δ is distributed as Exponential(a_0)).
    Find k such that Σ_{j=1}^{k−1} a_j < v a_0 ≤ Σ_{j=1}^{k} a_j. This implies that the reaction
    R_k is chosen with probability a_k/a_0.
    Set t = t + Δ and X = X + ν_k.
end
```

Alternative approaches to statistically exact simulation include Gibson and Bruck's next reaction method which calculates the next firing time based on the minimum of the firing times of individual reactions. After the first step, the next reaction method is known to use one random number per iteration. In contrast the Gillespie's algorithm uses two random numbers per iteration, and thus the next reaction method can be more efficient for certain biochemical reaction systems. Some improvements over these classical methods of exact simulation have since been obtained (for example see [1]).

4 Approximate Algorithms

Tracking occurrence of every reaction events will be computationally expensive for large complex systems. For instance, simulation speed is crucial for getting an accurate picture of equilibrium distribution in the cellular processes. Thus, it is imperative that proper approximate techniques are designed which generate faster sample paths without significant loss in accuracy. Limit theorems of Markov processes provide crucial insights into validity of these approximate algorithms.

4.1 τ-leaping Scheme

Here instead of tracking every reaction, the modeler wants to directly find the number of reactions happening in intervals of length τ. So, he divides the interval $[0, t]$ into subintervals $[t_{i-1}, t_i)$ of length τ and assumes that the intensity functions a_k essentially stays the same over $[t_{i-1}, t_i)$. This immediately implies that the number of occurrence of reaction k in $[t_{i-1}, t_i)$ is distributed as a Poisson random variable with parameter $a_k(X(t_{i-1})\tau$. Similar to (2), a pathwise representation of the approximating 'τ-leap' process is given by

$$Z(t) = X(0) + \sum_k \xi_k \left(\int_0^t a_k(Z \circ \eta(s))ds \right) (\nu_k^+ - \nu_k^-),$$

where $\eta(s) = t_n$ if $t_n \leq s < t_{n+1}$ and $t_{n+1} - t_n = \tau$. The main steps of a τ-leaping algorithm are outlined below.

Algorithm 2: τ-leaping algorithm.

Input: The initial state vector X, stoichiometric matrix $S = \{\nu_{ij}\}$, $i = 1, 2, \ldots, M,\ j = 1, 2, \ldots, R$, leap size τ, end of the simulation time $T > 0$.

Output: The number of molecules of each species in time interval $t \in [0, T]$.

Set $t = 0, n = 0$.

while $t < T$ ***and*** $\sum_{i=1}^{R} a_i > 0$ **do**

    Calculate the propensity function, a_k, for each k.

    For each k, generate N_k from Poisson$(a_k\tau)$. N_k is the approximate number of occurrences of the reaction R_k.

    Set $t = t + \tau$ and update $X = X + \sum_{k=1}^{R} N_k \nu_k$.

end

It is immediately clear that the strength of the approximation depends on τ. But it should be noted that unlike the Euler approximation techniques in ODE, here τ cannot be chosen small independently of the system. This is because the modeler wants to make sure that he is gaining substantial speed by using a tau leaping scheme. This essentially implies that he wants the expected number of occurrence of reaction k in every subinterval $[t_{i-1}, t_i)$ which is simply $a_k(X(t_{i-1})\tau)$ to be large. This necessitates that τ is chosen such that (i) τ is small but, (ii) $a_k(X(t_{i-1})\tau)$ is large. Algorithm 3 is the most basic version of the τ-leaping numerical scheme. The subtlety of such a numerical scheme lies in the choice of τ, and in practice τ is chosen in an adaptive way such that the leap conditions (i) and (ii) are respected in some sense. Some of these adaptive τ-leaping methods can be found, for example, in [6].

As mentioned before, the leap size τ cannot be chosen arbitrarily small, and hence a rigorous error analysis of a τ-leaping method needs to consider the right scaling regime where τ leaping is appropriate and efficient. Such an analysis under the so-called classical scaling has been performed in [2], where the authors proved that the standard τ-leap is an order $1/2$ method. Anderson et al. [2] also derived the limiting form of the error process proving the optimality of the error bound.

4.2 Diffusion Approximation

Let N be some measure of volume of the system. Then notice that the species concentration vector $X^N \equiv X/N$ can be represented by

$$
\begin{aligned}
X^N(t) &= X^N(0) + \frac{1}{N}\sum_{k=1}^{R} \xi_k\Big(N\int_0^t A_k(X^{N(s)})ds\Big)\nu_k,\\
&= X^N(0) + \sum_{k=1}^{R}\int_0^t A_k(X^N(s))ds\ \nu_k + \frac{1}{N}\sum_{k=1}^{R}\tilde{\xi}_k\Big(N\int_0^t A_k(X^N(s))ds\Big)\nu_k.
\end{aligned}
\tag{4}
$$

Here $\tilde{\xi}_k(u) = \xi_k(u) - u$ denotes the centered Poisson process and $a_k(X(s)) \approx N A_k(X^N(s))$ with $A_k(u) = O(1)$. The A_k are essentially the scaled versions of a_k. A diffusion approximation of X^N involves replacing the rescaled centered Poisson processes $\frac{1}{\sqrt{N}}\tilde{\xi}_k(N\cdot)$ by independent Brownian motions W_k. This leads to an approximation of the process X^N by the process Z^N defined by

$$
Z^N(t) = X^N(0) + \sum_{k=1}^{R}\int_0^t A_k(Z^N(s))ds\ \nu_k + \frac{1}{\sqrt{N}}\sum_{k=1}^{R} W_k\Big(\int_0^t A_k(Z^N(s))ds\Big)\nu_k.
$$

As could be seen that under some mild conditions Z^N is equivalent (in the sense of distribution) to the solution of the Itô stochastic differential equation (SDE):

$$
Z^N(t) = X^N(0) + \sum_{k=1}^{R}\int_0^t A_k(Z^N(s))ds\ \nu_k + \frac{1}{\sqrt{N}}\sum_{k=1}^{R}\int_0^t \sqrt{A_k(Z^N(s))}dW_k(s)\nu_k.
$$

The latter is also known as the Langevin approximation in the chemical physics literature (see [17]), and in practice the above SDE is simulated by some standard discretization technique. For example, a basic Euler-Maruyama scheme of simulating the SDE leads to the following algorithm.

Algorithm 3: Diffusion approximation algorithm.

Input: The initial state vector X, stoichiometric matrix $S = \{\nu_{ij}\}$, $i = 1, 2, \ldots, M,\ j = 1, 2, \ldots, R$, discretization step Δ, end of the simulation time $T > 0$.

Output: The number of molecules of each species in time interval $t \in [0, T]$.

Set $t = 0, n = 0$.

while $t < T$ ***and*** $\sum_{i=1}^{R} a_i > 0$ **do**

  Calculate the propensity function, a_k, for each k.

  For each k, set $N_k = a_k + \sqrt{a_k}Z_k$, where Z_k is a standard normal random variable.

  Set $t = t + \Delta$ and update $X = X + \sum_{k=1}^{R} N_k\nu_k$.

end

4.3 Approximation by Ordinary Differential Equations (ODE)

Randomness in biochemical reaction system is most evident when the molecular count of the species is low. If the abundance of the species is sufficiently high, then one expects that the randomness vanishes and it is enough to model the system by coupled (deterministic) differential equations. In fact Kurtz [22] proved that if $X^N(0) = X(0)/N \to x_0$, then under some mild conditions on the intensity functions, X^N converges to the solution of

$$x(t) = x(0) + \sum_{k=1}^{R} \int_0^t A_k(x(s))ds \, \nu_k.$$

This essentially follows from (4) after an application of law of large numbers for Poisson processes. The ODE approximation is usually the fastest of the three algorithms mentioned, but as always one has to balance the error of approximation with the speed. Kurtz's result on convergence implies that the ODE approximation gets better as N, which captures the abundance of species in the system, gets larger. The ODE approximation can be thought of as the zeroth order expansion of the process X^N, while the diffusion approximation Z^N represents the first order approximation with the randomness captured in the stochastic integral part; the latter is thus expected to be more accurate.

5 Multiscale Systems

However, most cellular reactions are multi-scale in nature. In other words, such a system involves species with varying orders of abundance and reactions with varying orders of magnitudes of reaction rates. The approximate methods of simulating the system state, that are mentioned before, will fare poorly in these situations. These realistic complex cellular processes need new approximation techniques for formulation of faster algorithms which will also maintain a certain level of accuracy. Two main ideas, that are often put forward for simulation of multiscale systems, include approximation by a hybrid diffusion or a hybrid ODE process and approximation by averaging techniques.

The central idea in building an approximate model of a multiscale reaction system is to partition the reaction set into 'fast' and 'slow' reactions. The averaging techniques (also known as quasi steady state or partial equilibrium analysis in chemistry literature) investigate conditions under which the fast set will have a quasi-stationary distribution. Since the reactions in the fast set happen rapidly, it is reasonable to assume that this stationary distribution will be reached quickly (compared to the firing times from the slow reaction set). The averaging principle suggests that in such a case the trajectory of the slow system can be approximated by averaging the parameters in the fast variable with respect to its stationary distribution. For general

Markov processes, there exists a vast amount of literature on rigorous analysis of appropriate averaging principles (e.g. see [10, 19, 23, 28]). Crudu et al. [8] used some basic averaging techniques to get simplifications for multiscale gene networks. A broad mathematical framework for treating complex multiscale reaction systems has been developed by Kang and Kurtz in [21] (also see [4, 20]), where the author introduces several scaling exponents to capture variations in different parameters of the system. The general framework adopted in [21] can handle multiple time scales and uses sophisticated limit theorems including averaging principle to derive the limiting model. The advantage is that the limiting models are usually much simpler, thereby saving enormous computational time for simulation purposes. Although this approach does yield a reduced limiting model, it should be noted that the derivation of the final limiting model is not algorithmic. The mathematical techniques involved are often quite advanced and also vary from system to system.

In contrast, a so-called hybrid diffusion or a hybrid ODE approximation is quite mechanical. The resulting approximate model can be obtained by suitable algorithms, and thus they are extremely valuable to the practitioners for simulation of complex systems. Below we briefly discuss their central principles.

5.1 Hybrid ODE and Diffusion Approximation

As before, these types of approximations hinge on partitioning the reaction set into subsets of 'fast' and 'slow' reactions. The fast set is then modeled by a continuous process (diffusion approximation or ODE approximation), while the discrete counting process formulation is maintained for the slow one. Several hybrid models have been proposed in [5, 8, 18, 26, 27]. However, in much of the existing literature this partitioning is mostly done in a subjective way. The difficulty probably lies in quantifying the actual speed of the reaction. As a result, a computerized encoding of the method, which could be readily implemented for a wide class of multi-scale reaction systems, is extremely difficult to formulate.

A solution to this problem was obtained in [12], where the authors used appropriate scaling parameters (introduced by Kang and Kurtz in [20, 21]) to capture the varying orders of magnitude of the species and rate constants. A rigorous derivation of the error bound for a general hybrid diffusion approximation is then carried out in the article, and the result for the corresponding hybrid ODE approximation has been obtained in [13]. The crucial significance of the error bounds in [12, 13] lied in its utilization for engineering an efficient algorithm which can do the above-mentioned partitioning by itself and also check the validity of the partitioning dynamically over the course of time.

We now briefly discuss the scaling technique used in [12]. In a typical multi-scale model, the abundance of various species in the reaction system can vary over different orders of magnitude. This can be captured by introducing scaling exponents α_i such that $\bar{X}_i^N \equiv X_i/N^{\alpha_i} = O(1)$; in other words, α_i measures the order magnitude in abundance for species S_i. Similarly, defining $d_k = c_k/N^{\beta_k}$ such that $d_k = O(1)$; one captures the variations in the rate constants for individual reactions. Under the above scaling of the state vector and the rate constants, the propensities a_k scale as

$$a_k(X) = N^{\beta_k + \nu_k^- \cdot \alpha} \lambda_k(\bar{X}^N),$$

where $\alpha = (\alpha_1, \ldots, \alpha_M)$. Note that with these choices of exponents, the functions $\lambda_k(\cdot)$ are $O(1)$. Oftentimes, it is beneficial to scale time as well by $t \to tN^\gamma$. It readily follows from (2) that X^N, defined by $X^N(t) = \bar{X}^N(tN^\gamma)$, satisfies

$$X^N(t) = X^N(0) + \sum_{k=1}^{R} \xi_k(N^{\rho_k} \int_0^t \lambda_k(X^N(s))ds)\, \nu_k^N, \tag{5}$$

where $\rho_k = \gamma + \beta_k + \nu_k^- \cdot \alpha$ and $\nu_{ki}^N = \nu_{ki}/N^{\alpha_i}$. If the reaction R_1 is fast and its occurrences is simulated by a diffusion approximation, while the others are counted by the usual discrete processes, then the resulting approximating state of the system is given by the equation

$$Z^N(t) = X^N(0) + N^{\rho_1} \int_0^t \lambda_1(Z^N(s))ds\, \nu_1^N + W_1(N^{\rho_1} \int_0^t \lambda_1(Z^N(s))ds)\nu_1^N \tag{6}$$

$$+ \sum_{k>1} \xi_k(N^{\rho_k} \int_0^t \lambda_k(Z^N(s))ds)\, \nu_k^N. \tag{7}$$

It has been proved in [12], that the strong error between X^N in (5) and Z^N in (6) is bounded by $O(\log N^{\rho_1}/N^{m_1} + 1/N^{2\rho_1 + m_1})$, where the constant involved in $O(\cdot)$ is independent of the reaction chosen to be simulated by a diffusion approximation. It has been argued in [12] that the above error bound implies that a suitable partitioning criteria of the reaction set can be obtained by comparing the quantity $\hat{\Upsilon}_k(\Delta) \equiv \sum_i \frac{\log(a_k(X)\Delta)}{X_i} + \frac{1}{(a_{k(X)}\Delta)^2 X_i}$ to a user defined threshold. Here, the summation is over all the species i whose molecular count is altered by the reaction R_k. Specifically, a reaction R_k is classified as a continuous reaction and simulated by diffusion approximation, if $\hat{\Upsilon}_k \le \varepsilon$ for a pre-chosen ε. For a hybrid-ODE approximation, it has been proved in [13] that a corresponding appropriate partitioning could be obtained according as $\hat{\eta}_k(\Delta) \equiv \sum_i \frac{\sqrt{a_k(X)\Delta}}{X_i}$ to a given threshold. This leads to the following efficient hybrid diffusion algorithm.

Algorithm 4: Dynamic partitioning algorithm.

Input: The state vector X, the error bound ε, a discretization time step Δ, stoichiometric matrix $S = \{\nu_{ij}\}, i = 1, 2, \ldots, M, \; j = 1, 2, \ldots, R$, a positive number P to check repartitioning of reactions, end of the simulation time $T > 0$.

Output: The number of molecules of each species in time interval $t \in [0, T]$.

1 Set $t = 0, n = 0$.
2 Calculate $\hat{\Upsilon}_k$ (respectively, $\hat{\eta}_k$) for all reactions.
3 Partition the reaction set into $\mathscr{C}$ (continuous) and $\mathscr{D}$ (discrete) sets such that for each $R_k \in \mathscr{C}$, $\hat{\Upsilon}_k \leq \varepsilon$ (respectively, $\hat{\eta}_k \leq \varepsilon$) and for each $R_m \in \mathscr{D}$, $\hat{\Upsilon}_m > \varepsilon$ (respectively, $\hat{\eta}_m > \varepsilon$).
4 For each $R_m \in \mathscr{D}$, set $T_m = 0$.
5 For each $R_m \in \mathscr{D}$, draw $J_m \sim -\log(z), \quad z \sim \mathscr{U}(0, 1)$.
6 **while** $t < T$ ***and*** $\sum_{i=1}^{R} a_i > 0$ **do**
7     $n = n + 1$.
8   For each $R_m \in \mathscr{D}$, calculate $h_m = \frac{(J_m - T_m)}{a_m}$.
9   Choose α such that $h \equiv h_\alpha \equiv \min_{R_m \in \mathscr{D}} h_m$.
10   **if** $\Delta > h$ **then**
11   Update X by a suitable numerical scheme for simulating the diffusion part (respectively, ODE part) until the h.
12   Carry out reaction R_α and update $X = X + \nu_\alpha$.
13   Update $J_\alpha = J_\alpha - \log(u)$, $u \sim \mathscr{U}(0, 1)$.
14   For each $R_m \in \mathscr{D}$, put $T_m = T_m + a_m h$.
15         $t = t + h$.
16   **else**
17   Update X by a suitable numerical scheme for simulating the diffusion part (respectively, ODE part) until the Δ.
18   For each $R_m \in \mathscr{D}$, $T_m = T_m + a_m \Delta$.
19         $t = t + \Delta$.
20   **end**
21   Recalculate the propensities of all reactions.
22   **if** $n \equiv 0 \pmod{P}$ **then**
23   Recalculate errors of all reactions $\hat{\Upsilon}_k$.
24   Repartition reactions as in Step 3.
25   **end**
26 **end**

5.2 Example

The well known Michaelis-Menten model for enzymatic substrate conversion consists of four species, the enzyme (E), the substrate (S), the enzyme-substrate complex (ES) and the product (P).

$$E + S \underset{d_1}{\overset{c_1}{\rightleftharpoons}} ES, \quad ES \xrightarrow{c_2} E + P. \tag{8}$$

The state of the system is given by the vector of copy numbers $X = (E, S, ES, P)^T$. This is a classical multi-scale problem because the reversible reaction $E + S \underset{d_1}{\overset{c_1}{\rightleftharpoons}} ES$ is much faster than the reaction $ES \xrightarrow{c_2} E + P$ by orders of magnitude. In our example, the initial number of molecules is taken as $X(0) = (48, 298, 2, 0)^T$ and the rate constants of reactions R_1, R_2, R_3 are given by $c_1 = 0.02$ molec^{-1}s^{-1}, $d_1 = 0.5$ s^{-1}, $c_2 = 0.1$ s^{-1}. The picture below was obtained in [12], and it indicates the accuracy of the above hybrid algorithm by comparing the probability distributions and the Q-Q plots obtained from it with that of the exact Gillespie's algorithm. More examples can be found in [12, 13] (Fig. 1).

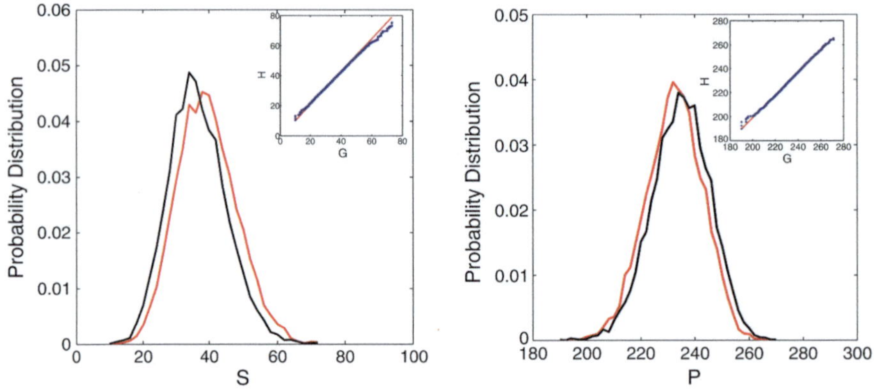

Fig. 1 Probability distributions of S (*left*), P (*right*) at time $t = 60$ s from 30000 samples constructed with (i) the Gillespie's algorithm (*black line*) and by (ii) the hybrid diffusion algorithm (*red line*). *Insets* show Q-Q plot of 30000 samples comparing the Gillespie's algorithm (G) and the hybrid diffusion algorithm (H)

References

1. Anderson, D.F.: A modified next reaction method for simulating chemical systems with time dependent propensities and delays. J. Chem. Phys. **127**(21), 214107 (2007)
2. Anderson, D.F., Ganguly, A., Kurtz, T.G.: Error analysis of tau-leap simulation methods. Ann. Appl. Probab. **21**(6), 2226–2262 (2011)
3. Anderson, D.F., Kurtz, T.G.: Continuous time Markov chain models for chemical reaction networks. In: Koeppl, H., Setti, G., di Bernardo, M., Densmore, D. (eds.) Design and Analysis of Biomolecular Circuits. Springer (2011)
4. Ball, K., Kurtz, T.G., Popovic, L., Rempala, G.: Asymptotic analysis of multiscale approximations to reaction networks. Ann. Appl. Probab. **16**(4), 1925–1961 (2006)
5. Cao, Y., Gillespie, D.T., Petzold, L.R.: The slow-scale stochastic simulation algorithm. J. Chem. Phys. **122**, 014116 (2005)
6. Cao, Y., Gillespie, D.T., Petzold, L.R.: Efficient step size selection for the tau-leaping simulation method. J. Chem. Phys. **124**, 044109 (2006)
7. Coulon, A., Chow, C.C., Singer, R.H., Larson, D.R.: Eukaryotic transcriptional dynamics: from single molecules to cell populations molecules to cell populations. Nat. Rev. Genet. **14**, 572–584 (2013)
8. Crudu, A., Debussche, A., Radulescu, O.: Hybrid stochastic simplifications for multiscale gene networks. BMC Syst. Biol. **3**(89) (2009)
9. Ethier, S.N., Kurtz, T.G.: Markov Processes: Characterization and Convergence. Wiley Series in Probability and Mathematical Statistics: Probability and Mathematical Statistics. Wiley, New York (1986)
10. Freidlin, M.I., Wentzell, A.D.: Random Perturbations of Dynamical Systems. Grundlehren der Mathematicischen Wissenschaften [Fundamental Principles of Mathematical Sciences], vol. 260, 2nd edn. Springer, New York (1998). Translated from the 1979 Russian original by Joseph Szücs
11. Friedman, N., Cai, L., Xie, X.S.: Stochasticity in gene expression as observed by single-molecule experiments in live cells. Isr. J. Chem. **49**, 333–342 (2010)
12. Ganguly, A., Altintan, D., Koeppl, H.: Jump-diffusion approximation of stochastic reaction dynamics: error bounds and algorithms. Submitted to SIAM: Multiscale Modeling and Simulation, Sept 2014
13. Ganguly, A., Altintan, D., Koeppl, H.: Error bound and simulation algorithm for piecewise deterministic approximations of stochastic reaction systems. In: Accepted to Proceedings of American Control Conference, Sept 2015 (2015)
14. Gillespie, D.T.: The chemical Langevin equation. J. Chem. Phys. **113**(1), 297–306 (2000)
15. Gillespie, D.T.: Approximate accelerated simulation of chemically reaction systems. J. Chem. Phys. **115**(4), 1716–1733 (2001)
16. Gillespie, D.T.: Stochastic simulation of chemical kinetics. Annu. Rev. Phys. Chem. **58**(1), 35–55 (2007)
17. Gillespie, D.T.: The chemical langevin equation. J. Chem. Phys. **113**(1), 297–306 (2000)
18. Haseltine, E.L., Rawlings, J.B.: Approximate simulation of coupled fast and slow reactions for stochastic chemical kinetics. J. Chem. Phys. **117**(15), 6959–6969 (2002)
19. Has'minskiĭ, R..Z.: On the principle of averaging the Itô's stochastic differential equations. Kybernetika (Prague) **4**, 260–279 (1968)
20. Kang, H.W., Kurtz, T.G., Popovic, L.: Central limit theorems and diffusion approximations for multiscale Markov chain models. Ann. Appl. Probab. **24**(2), 721–759 (2014)
21. Kang, H.W., Kurtz, T.G.: Separation of time-scales and model reduction for stochastic reaction networks. Ann. Appl. Probab. **23**(2), 529–583 (2013)
22. Kurtz, T.G.: The relationship between stochastic and deterministic models for chemical reactions. J. Chem. Phys. **57**(7), 2976–2978 (1972)
23. Kurtz, T.G.: Averaging for martingale problems and stochastic approximation. Applied Stochastic Analysis (New Brunswick, NJ, 1991) **177**, 186–209 (1992)

24. Ozdudak, E.M., Thattai, M., Kurtser, I., Grossman, A.D., van Oudenaarden, A.: Regulation of noise in the expression of a single gene. Nat. Genet. **31**(1), 69–73 (2002)
25. Paulsson, J.: Summing up the noise in gene networks. Nature **427**(6973), 415–418 (2004)
26. Rao, C.V., Arkin, A.P.: Stochastic chemical kinetics and the quasi-steady-state assumption: application to the gillespie algorithm. J. Chem. Phys. **118**(11), 4999–5010 (2003)
27. Salis, H., Kaznessis, Y.: Accurate hybrid stochastic simulation of a system of coupled chemical or biochemical reactions. J. Chem. Phys. **122**, 054103 (2005)
28. Yin, G.G., Zhang, Q.: Discrete-Time Markov Chains: Two-Time-Scale Methods and Applications, Stochastic Modelling and Applied Probability. Applications of Mathematics (New York), vol. 55. Springer, New York (2005)

Analytical Solutions of Basic Models in Quantum Optics

Daniel Braak

Abstract The recent progress in the analytical solution of models invented to describe theoretically the interaction of matter with light on an atomic scale is reviewed. The methods employ the classical theory of linear differential equations in the complex domain (Fuchsian equations). The linking concept is provided by the Bargmann Hilbert space of analytic functions, which is isomorphic to $L^2(\mathbb{R})$, the standard Hilbert space for a single continuous degree of freedom in quantum mechanics. I give the solution of the quantum Rabi model in some detail and sketch the solution of its generalization, the asymmetric Dicke model. Characteristic properties of the respective spectra are derived directly from the singularity structure of the corresponding system of differential equations.

Keywords Quantum optics · Bargmann space · Differential equations · Singularity theory · Integrable systems

1 Introduction

The interaction of matter with light forms a major subject of theoretical and applied physics [1]. It is essentially characterized by the quantum nature of both constituents, studied within Quantum Optics [2]. The quantum features of the processes occurring in recently realized nano-sized devices can be used to control the generation of entangled states [3], thereby allowing to construct the basic elements of quantum information technology [4].

The technological advances in nanofabrication made it possible to reach very large coupling strengths between the light (usually confined to a single or few modes in a cavity) and an (artificial) atom described by a discrete set of energy levels in the deep quantum limit [5–7]. The prototypical system consists of a "matter" part

D. Braak (✉)
EP VI and Center for Electronic Correlations and Magnetism,
Institute of Physics, University of Augsburg, Augsburg, Germany
e-mail: daniel.braak@physik.uni-augsburg.de

R.S. Anderssen et al. (eds.), *Applications + Practical Conceptualization + Mathematics = fruitful Innovation*, Mathematics for Industry 11,
DOI 10.1007/978-4-431-55342-7_7

with two possible states coupled to the dipole component of a single radiation mode. The Hamiltonian of the atom can thus be expressed via Pauli spin matrices σ_j and the radiation through a harmonic oscillator with frequency ω. The Hilbert space of the total system reads then $\mathscr{H} = \mathbb{C}^2 \otimes L^2(\mathbb{R})$ and the Hamiltonian

$$H_R = \omega a^\dagger a + g\sigma_x(a + a^\dagger) + \frac{\omega_0}{2}\sigma_z. \tag{1}$$

Here $a^\dagger$ and a are the creation and annihilation operators of the bosonic mode and energy is measured in units of frequency ($\hbar = 1$). ω_0 denotes the energy splitting of the two-level system (the "qubit") which is coupled linearly to the electric field ($\sim (a + a^\dagger)$) with interaction strength g. This model was studied semiclassically already in 1936 by Rabi [8] and the fully quantized version (1) has been introduced in 1963 by Jaynes and Cummings [9]. It is therefore called the quantum Rabi model (QRM). Despite its apparent simplicity, the QRM is difficult to solve analytically because it does not exhibit invariant subspaces of finite dimension like the following model,

$$H_{JC} = \omega a^\dagger a + g(\sigma^+ a + \sigma^- a^\dagger) + \frac{\omega_0}{2}\sigma_z, \tag{2}$$

with $\sigma^\pm = (\sigma_x \pm i\sigma_y)/2$, which corresponds to the "rotating-wave" approximation of (1) [9]. The Jaynes-Cummings model (JCM) (2) can be justified close to resonance, $\omega \sim \omega_0$ and small coupling $g/\omega \ll 1$ [10] and has been the standard model for typical quantum optical applications with $g/\omega \leq 10^{-8}$ for many years. The major simplification arising in (2) as compared to (1) consists in the fact that the operator $\hat{C} = a^\dagger a + \sigma^+\sigma^-$ commutes with H_{JC}, which means that $\mathscr{H}$ decays into infinitely many H_{JC}-invariant subspaces in which $\hat{C}$ takes constant values $\in \mathbb{N}_0$, the set of non-negative integers. Each of these spaces is two-dimensional entailing trivial diagonalization of H_{JC}. The fact that the polynomial algebra of $\hat{C}$ has infinite dimension means that $\hat{C}$ generates a continuous $U(1)$-symmetry of H_{JC} [10, 11]: defining $U(\phi) = \exp(i\phi\hat{C})$, we have $U^\dagger(\phi)aU(\phi) = e^{i\phi}a$, $U^\dagger(\phi)a^\dagger U(\phi) = e^{-i\phi}a^\dagger$ and $U^\dagger(\phi)\sigma^\pm U(\phi) = e^{\mp i\phi}\sigma^\pm$. This abelian symmetry associated with the integral of motion $\hat{C}$ renders the JCM integrable, because it has only two degrees of freedom, the continuous one of the radiation mode with Hilbert space $L^2(\mathbb{R})$ and the discrete one of the qubit with Hilbert space $\mathbb{C}^2$.

The concept of integrability underlying this argument amounts to a direct transfer of Liouville's definition from classical mechanics to quantum mechanics: A system with N degrees of freedom is integrable if it exhibits N independent phase space functions which are in involution with respect to the Poisson bracket [12]. In the JCM these are the Hamiltonian H_{JC} and $\hat{C}$. But because independence of operators cannot be defined in analogy to functions on phase space, this definition is not feasible as *any* Hamiltonian system would be integrable according to it [13].

The continuous $U(1)$-symmetry of the JCM is broken down to a discrete symmetry by the counter-rotating term $a^\dagger\sigma^+ + a\sigma^-$ in the QRM. $U^\dagger(\pi)H_R U(\pi) = H_R$ and H_R commutes with $\hat{P} = (-1)^{a^\dagger a}\sigma_z = -U(\pi)$. Because $\hat{P}^2 = \mathbb{1}$, its polynomial algebra is two-dimensional and $\hat{P}$ generates a $\mathbb{Z}_2$-symmetry of H_R, usually called parity. The eigenvalues ± 1 of $\hat{P}$ characterize two H_R-invariant subspaces (parity-chains), each of them infinite-dimensional [14]. Therefore, the problem appears to be only marginally simplified by using the parity symmetry and it was widely held that the QRM is not integrable [11]. However, it could be demonstrated that the weak parity symmetry is indeed sufficient for integrability of the QRM because it possesses only one *continuous* degree of freedom, whereas the Hilbert space dimension of the discrete degree of freedom matches the dimension of the polynomial algebra generated by $\hat{P}$, rendering the QRM integrable according to the level-labeling criterion for quantum integrability [15]. The detailed understanding of the spectrum (and dynamics) of the QRM beyond the rotating-wave approximation has been necessitated by the recent experimental access to the ultra-strong and deep-strong coupling regime in circuit QED [6, 7] and through quantum simulations.

This survey is organized as follows: In Sect. 2, the analytical solution of the QRM is presented based on a formulation of the problem in Bargmann's space of analytical functions, Sect. 3 deals with multi-qubit models and the last section contains some remarks on possible future research directions.

2 The Quantum Rabi Model

The $\mathbb{Z}_2$-symmetry of the QRM can be used to eliminate the discrete degree of freedom from the problem, just as the $U(1)$-symmetry of the JCM allows elimination of the continuous degree of freedom. Each parity-chain $\mathscr{H}_\pm$ is isomorphic to $L^2(\mathbb{R})$ and H_R reads in $\mathscr{H}_\pm$

$$H_\pm = \omega a^\dagger a + g(a + a^\dagger) \pm \Delta(-1)^{a^\dagger a}, \tag{3}$$

with $\Delta = \omega_0/2$. The complication of this reduced Hamiltonian comes from the last term $(-1)^{a^\dagger a}$. On the other hand, this term is instrumental for the analytical solution of the model. To elucidate its meaning, it is convenient to represent (3) in Bargmann's space of analytical functions which is isometrically isomorphic to $L^2(\mathbb{R})$ [16]. The space $\mathscr{B}$ is spanned by functions $f(z)$ of a complex variable z which have finite norm $\langle\phi|\phi\rangle$ with respect to the scalar product

$$\langle\psi|\phi\rangle = \frac{1}{\pi}\int \mathrm{d}z\mathrm{d}\bar{z} e^{-z\bar{z}}\overline{\psi(z)}\phi(z) \tag{4}$$

and are analytic in all $\mathbb{C}$ ($\mathrm{d}z\mathrm{d}\bar{z} = \mathrm{d}\Re(z)\mathrm{d}\Im(z)$). The criterion for being an element of $\mathscr{B}$ is therefore two-fold: $\phi(z) \in \mathscr{B}$ if both of the following conditions are satisfied:

(B-I): $\langle\phi|\phi\rangle < \infty$
(B-II): $\phi(z)$ is holomorphic everywhere in the open domain $\mathbb{C}$.

The isometry $\mathscr{I}$ maps $f(q) \in L^2(\mathbb{R})$ to an analytic function $\phi(z) \in \mathscr{B}$,

$$\phi(z) = \mathscr{I}[f](z) = \frac{1}{\pi^{1/4}} \int_{-\infty}^{\infty} \mathrm{d}q\, e^{-\frac{1}{2}(q^2+z^2)+\sqrt{2}qz} f(q). \tag{5}$$

The operators $a^\dagger$, a are mapped to z and $\mathrm{d}/\mathrm{d}z$, respectively,

$$\mathscr{I} a \mathscr{I}^{-1} = \frac{\mathrm{d}}{\mathrm{d}z}, \qquad \mathscr{I} a^\dagger \mathscr{I}^{-1} = z. \tag{6}$$

The normalized vacuum $|0\rangle$ with $a|0\rangle = 0$ is mapped to the constant function $\phi_0(z) = 1$. On infers from (4) that all polynomials in z are elements of $\mathscr{B}$. Especially the nth eigenstate of the harmonic oscillator $|n\rangle \sim e^{-q^2/2} H_n(q)$ is mapped onto the monomial $z^n/\sqrt{n!}$. Moreover, all functions which have the asymptotic expansion

$$\phi(z) = e^{\alpha_1 z} z^{-\alpha_0}(c_0 + c_1 z^{-1} + c_2 z^{-2} + \cdots) \quad \text{for} \quad z \to \infty, \tag{7}$$

with arbitrary $\alpha_1 \in \mathbb{C}$ satisfy (B-I) [17]. Equation (7) is the asymptotic form of the normal solutions of a differential equation having an unramified irregular singular point of s-rank two at infinity [18, 19]. The coherent state $|\alpha\rangle = e^{\alpha a^\dagger}|0\rangle$ in $L^2(\mathbb{R})$ is mapped obviously to the exponential function $e^{\alpha z}$. Functions behaving as

$$\phi(z) = \exp\left(\frac{\alpha_2}{2} z^2 + \alpha_1 z\right) z^{-\alpha_0}(c_0 + c_1 z^{-1} + c_2 z^{-2} + \cdots) \tag{8}$$

asymptotically, satisfy (B-I) only if $|\alpha_2| < 1$ and correspond to an irregular singularity of s-rank three. The limiting value $|\alpha_2| = 1$ belongs to the (not normalizable) plane wave states $f_p(q) = \exp(ipq)/\sqrt{2\pi}$,

$$\mathscr{I}[f_p](z) = \frac{e^{-p^2/2}}{\pi^{1/4}} \exp\left(\frac{1}{2} z^2 + i\sqrt{2}pz\right). \tag{9}$$

The Hamiltonian H_+ reads in $\mathscr{B}$ (with $\omega = 1$),

$$H_+ = z\frac{\mathrm{d}}{\mathrm{d}z} + g\left(z + \frac{\mathrm{d}}{\mathrm{d}z}\right) + \Delta\hat{T}, \tag{10}$$

where $\hat{T}$ denotes the reflection operator $\hat{T}[\psi](z) = \psi(-z)$. The Schrödinger equation $(H_+ - E)\psi(z) = 0$ corresponds to a linear but non-local differential equation in the complex domain,

$$z\frac{\mathrm{d}}{\mathrm{d}z}\psi(z) + g\left(\frac{\mathrm{d}}{\mathrm{d}z} + z\right)\psi(z) = E\psi(z) - \Delta\psi(-z). \tag{11}$$

The theory of these equations initiated by Riemann and Fuchs [18] can now be applied to (11). First, one obtains with the definition $\psi(z) = \phi_1(z)$ and $\psi(-z) = \phi_2(z)$ the coupled local system,

$$(z+g)\frac{\mathrm{d}}{\mathrm{d}z}\phi_1(z) + (gz - E)\phi_1(z) + \Delta\phi_2(z) = 0, \tag{12a}$$

$$(z-g)\frac{\mathrm{d}}{\mathrm{d}z}\phi_2(z) - (gz + E)\phi_2(z) + \Delta\phi_1(z) = 0. \tag{12b}$$

This system has two regular singular points at $z = \pm g$ and an (unramified) irregular singular point of s-rank two at $z = \infty$ [19]. The normal solutions of (12) behave asymptotically as (7) with $\alpha_1 = g$ or $\alpha_1 = -g$. The two Stokes rays are the positive and negative real axis. We infer that *all* solutions of (12) satisfy (B-I). It follows that the discrete spectrum $\{E_n\}$, $n \in \mathbb{N}_0$ will be determined by (B-II), because not all solutions of (12) are analytic in $\mathbb{C}$. Define $x = E + g^2$. Then the exponents of $\phi_1(z)$ at the regular singular point g $(-g)$ are $\{0, 1+x\}$ $(\{0, x\})$, while for $\phi_2(z)$ the exponents at g $(-g)$ are $\{0, x\}$ $(\{0, 1+x\})$.

If E belongs to the spectrum of H_+, ϕ_1 and ϕ_2 must be analytic in $\mathbb{C}$, especially at both points $\pm g$. This leads naturally to a division of the spectrum into two parts:
(1) The *regular* spectrum σ_{reg} consisting of those values E_n for which $x_n = E_n + g^2$ is not a non-negative integer.
(2) The *exceptional* spectrum σ_{exc} for which $x_n \in \mathbb{N}_0$.

2.1 The Regular Spectrum

If $x \notin \mathbb{N}_0$, the only allowed exponent at both points $\pm g$ is 0. We consider the case $-g$ and define $y = z + g$, $\phi_{1,2} = e^{-gy+g^2}\bar{\phi}_{1,2}$. Then,

$$y\frac{\mathrm{d}}{\mathrm{d}y}\bar{\phi}_1 = x\bar{\phi}_1 - \Delta\bar{\phi}_2, \tag{13}$$

$$(y-2g)\frac{\mathrm{d}}{\mathrm{d}y}\bar{\phi}_2 = (x - 4g^2 + 2gy)\bar{\phi}_2 - \Delta\bar{\phi}_1. \tag{14}$$

A local Frobenius solution for $\bar{\phi}_2(y)$, analytic at $y = 0$, reads $\bar{\phi}_2(y) = \sum_{n=0}^{\infty} K_n(x)y^n$ with coefficients $K_n(x)$ to be determined. Integration of (13) yields

$$\bar{\phi}_1(y) = cy^x - \Delta\sum_{n=0}^{\infty} K_n(x)\frac{y^n}{n-x}. \tag{15}$$

Because $x \notin \mathbb{N}_0$, c must be zero. This determines $\bar{\phi}_1(z)$ uniquely in terms of $\bar{\phi}_2(z)$. Setting $K_0 = 1$, the following three-term recurrence relation for the $K_n(x)$ is obtained from (14),

$$nK_n = f_{n-1}(x)K_{n-1} - K_{n-2}, \tag{16}$$

with

$$f_n(x) = 2g + \frac{1}{2g}\left(n - x + \frac{\Delta^2}{x-n}\right), \tag{17}$$

and initial condition $K_0 = 1$, $K_1(x) = f_0(x)$. $\bar{\phi}_2(y)$ will usually develop a branch-cut singularity at $y = 2g$. The radius of convergence of the Frobenius solution around $y = 0$ is $R = 2g$, which can be deduced from the asymptotic value $1/(2g)$ of $f_{n-1}(x)/n$ for $n \to \infty$. Due to the relation $\phi_2(z) = \psi(-z)$, the formal solution of (13) and (14) yields two expansions for $\psi(z)$, one analytic at $z = g$ and the other analytic at $z = -g$,

$$\psi(z) = \phi_2(-z) = e^{gz}\sum_{n=0}^{\infty} K_n(x)(-z+g)^n, \tag{18}$$

$$\psi(z) = \phi_1(z) = e^{-gz}\sum_{n=0}^{\infty} K_n(x)\Delta\frac{(z+g)^n}{x-n}. \tag{19}$$

The two circles centered at $z = \pm g$ are shown in Fig. 1. Because the vectors $(\phi_1(z), \phi_2(z))^T$ and $(\phi_2(-z), \phi_1(-z))^T$ satisfy both the homogeneous first-order system (12), they coincide in a neighborhood of z_0 if

$$\phi_1(z_0) = \phi_2(-z_0), \quad \phi_1(-z_0) = \phi_2(z_0) \tag{20}$$

for any z_0 in the intersection of their domain of convergence. That means that $\phi_2(-z)$ is the analytic continuation of $\phi_1(z)$ and itself analytic at $z = g$, therefore $\psi(z)$ is analytic at both singular points. Both conditions in (20) are equivalent if $z_0 = 0$ [20]. This leads to the definition of the G-function for the regular spectrum of H_+ [15, 21],

$$G_+(x) = \phi_2(0) - \phi_1(0) = \sum_{n=0}^{\infty} K_n(x)\left[1 - \frac{\Delta}{x-n}\right]g^n. \tag{21}$$

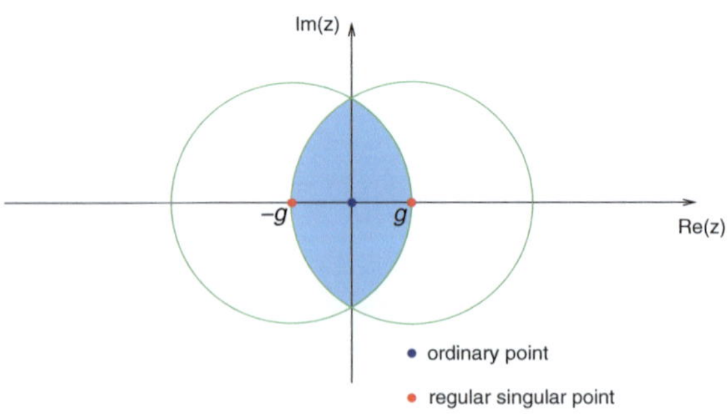

Fig. 1 The singularity structure of (11) and (12). Two local Frobenius solutions analytic at $z = g$ and $z = -g$ respectively are defined by (16) and (17). If they coincide in the intersection of their domains of convergence, they describe the same (analytic) function in $\mathbb{C}$

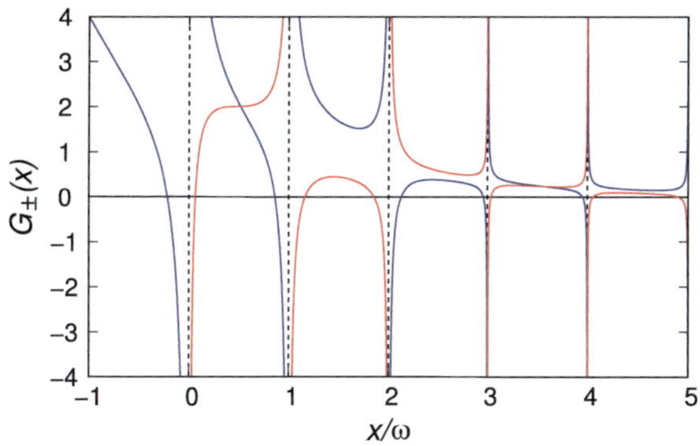

Fig. 2 The G-functions for odd (*blue*) and even (*red*) parity for $g = \omega = 1$ and $\Delta = 0.4$

If $G_+(E_n + g^2) = 0$, the corresponding formal solution $\psi(z)$ is analytic everywhere and an element of $\mathscr{B}$ because it satisfies (B-I) and (B-II), entailing that $E_n \in \sigma_{reg}(H_+)$. $G_-(x)$ for H_- is obtained from $G_+(x)$ by replacing Δ with $-\Delta$ in (21). It follows from (16), (17) and (21) that $G_\pm(x)$ has simple poles at $x \in \mathbb{N}_0$. The zeros of $G_\pm(x)$ are distributed between these poles. Figure 2 shows $G_\pm(x)$ and Fig. 3 the corresponding spectrum of H_R for both parities. It is easy to see that the regular spectrum is never degenerate, neither within each parity chain nor among states with different parity [15]. The G-function can be expressed in terms of known special functions as follows

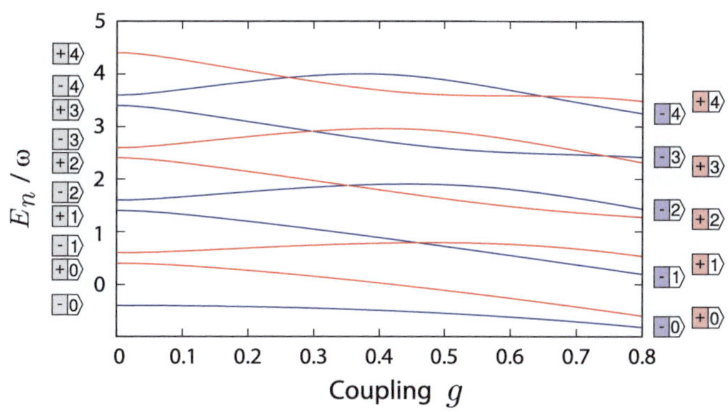

Fig. 3 Rabi spectrum for the same Δ, ω as in Fig. 2 and $0 \le g \le 0.8$. The intersections between the spectra of different parity indicate the degenerate part of the exceptional spectrum. The two-fold labeling of states on the *left* corresponds to the uncoupled system (the $\pm$ denotes the spin quantum number) and on the *right* to the coupled case ($\pm$ denotes the parity quantum number)

$$G_{\pm}(x) = \left(1 \mp \frac{\Delta}{x}\right) H_c(\alpha, \gamma, \delta, p, \sigma; 1/2) - \frac{1}{2x} H_c'(\alpha, \gamma, \delta, p, \sigma; 1/2). \quad (22)$$

$H_c(\alpha, \gamma, \delta, p, \sigma; z)$ denotes a confluent Heun-function [22] and $H_c'(\alpha, \gamma, \delta, p, \sigma; z)$ its derivative with respect to z. The parameters are given as

$$\alpha = -x, \quad \gamma = 1 - x, \quad \delta = -x,$$
$$p = -g^2, \quad \sigma = x(4g^2 - x) + \Delta^2.$$

The functional form (21) of $G_{\pm}(x)$ leads to the following conjecture about the distribution of its zeros along the positive real axis.

Conjecture: The number of zeros in each interval $[n, n+1]$, $n \in \mathbb{N}_0$ is restricted to be 0, 1, or 2. Moreover, an interval $[n, n+1]$ with two roots of $G_{\pm}(x) = 0$ can only be adjacent to an interval with one or zero roots; in the same way, an empty interval can never be adjacent to another empty interval.

2.2 The Exceptional Spectrum

We shall demonstrate in the following that the presence of the exceptional spectrum σ_{exc} poses certain constraints on the model parameters g and Δ such that for given g, Δ, at most two eigenvalues are exceptional. Furthermore, $\sigma_{exc} = \sigma_{exc}^d \cup \sigma_{exc}^{nd}$, where σ_{exc}^d comprises the values $E = m - g^2$ with $m \in \mathbb{N}$. Each eigenvalue in σ_{exc}^d is doubly degenerate among states with different parity. σ_{exc}^{nd} is not degenerate and may take values $E = m - g^2$ with $m \in \mathbb{N}_0$.

We begin with σ_{exc}^{nd}. The poles of $G_+(x)$ at $x \in \mathbb{N}_0$ indicate that an integer x can only signify an eigenvalue of H_+ if the corresponding pole in $G_+(x)$ is lifted for special values of the parameters g and Δ. If $x \in \mathbb{N}_0$, not only the exponent 0 but also the exponents x, respectively $x + 1$ guarantee analyticity of $\phi_1(z)$ and $\phi_2(z)$ at $z = -g$. However, as the difference of the two exponents at both singular points is a positive integer if $x > 0$ (for $x = 0$ this difference is positive at one singular point), the local analytic Frobenius solutions around $z = -g$ will develop a logarithmic branch-cut at $z = g$ in general. For $x = m \in \mathbb{N}_0$, there exist always a solution for $\bar{\phi}_2(y)$ analytic at $y = 0$ of the form

$$\bar{\phi}_2(y) = \sum_{n=m+1}^{\infty} K_n y^n, \quad (23)$$

because the largest exponent of $\bar{\phi}_2(y)$ at $y = 0$ is $x + 1$ [18]. Integration of (13) yields for $\bar{\phi}_1(y)$,

$$\bar{\phi}_1(y) = cy^m - \Delta \sum_{n=m+1}^{\infty} K_n \frac{y^n}{n-m}. \quad (24)$$

In this case, the constant c may be different from zero because $\bar{\phi}_1(y)$ is then analytic at $y = 0$. Solving now (14) with the ansatz (23), we obtain for $n \geq m + 2$ the recurrence (16), (17) and the initial conditions

$$K_{m+1} = \frac{c\Delta}{2(m+1)g}, \qquad K_m = 0. \tag{25}$$

c is fixed in terms of K_{m+1}. Setting $K_{m+1} = 1$ we obtain for $\psi(z)$ the two expressions

$$\psi(z) = \phi_2(-z) = e^{gz} \sum_{n=m+1}^{\infty} K_n(m; g, \Delta)(-z+g)^n, \tag{26}$$

$$\psi(z) = \phi_1(z) = e^{-gz} \left(\frac{2(m+1)g}{\Delta}(z+g)^m - \Delta \sum_{n=m+1}^{\infty} K_n(m; g, \Delta) \frac{(z+g)^n}{n-m} \right) \tag{27}$$

and the G-function follows as

$$G_+^{(m)}(g, \Delta) = -\frac{2(m+1)}{\Delta} + \sum_{n=m+1}^{\infty} K_n(m; g, \Delta) \left(1 + \frac{\Delta}{n-m}\right) g^{n-m-1}. \tag{28}$$

The zeros of the function $G_+^{(m)}(g, \Delta)$ determine those values of the parameters g and Δ for which H_+ has the exceptional eigenvalue $m - g^2$ with $m \in \mathbb{N}_0$. For odd parity, we have $G_-^{(m)}(g, \Delta) = G_+^{(m)}(g, -\Delta)$. It follows that $G_+^{(m)}(g, \Delta)$ and $G_-^{(m)}(g, \Delta)$ have no common zeros, so this part of the exceptional spectrum is not degenerate, just as the regular spectrum. It was computed by a related method in [23].

To obtain σ_{exc}^d, we consider now the smaller exponent, zero, of $\bar{\phi}_2(y)$ at $y = 0$, leading to the expansion

$$\bar{\phi}_2(y) = \sum_{n=0}^{\infty} K_n y^n. \tag{29}$$

After integration of (13), $\bar{\phi}_1(y)$ reads

$$\bar{\phi}_1(y) = cy^m - \Delta \sum_{n \neq m}^{\infty} K_n \frac{y^n}{n-m} - \Delta y^m K_m \ln(y). \tag{30}$$

The K_n are determined again with (16) and initial conditions $K_0 = 1$, $K_1 = f_0(m)$. Therefore $K_m(m; g, \Delta)$ is uniquely fixed. The logarithmic term in (30) vanishes if $K_m(m; g, \Delta) = 0$ [15]. In this case there exist two local solutions analytic at $y = 0$, (23) and (29). If $m \neq 0$, they are linearly independent and span the whole solution space for $\bar{\phi}_2(y)$. Because of the reflection symmetry mapping y to $2g - y$, these solutions describe the solution space in a neighborhood of $y = 2g$ as well and *all* solutions of (12) are analytic at both g and $-g$, thus in all of $\mathbb{C}$, if $K_m(m; g, \Delta) = 0$

and no further condition is necessary. Moreover, the K_m's are the same for odd parity, so the eigenvalue $E = m - g^2$ obtained via this condition is always doubly degenerate between states of different parity. The presence of the spectrum σ^d_{exc} and its "quasi-exact" nature [24] may be explained more generally in terms of the representation theory of $\mathfrak{sl}_2(\mathbb{R})$ [25]. A special situation arises for $x = 0$. The condition $K_0(0) = 0$ renders the ansatz (29) equivalent to (23) and both solutions are linearly dependent. Thus only one local solution may be analytic at $y = 0$ and $\bar{\phi}_1(y)$ is given by (24) for $m = 0$. If $E = -g^2$ is an eigenvalue of H_+, $G^{(0)}_+(g, \Delta)$ must vanish. It follows that for these parameter values $G^{(0)}_-(g, \Delta) \neq 0$, the eigenvalue $E = -g^2$ is never degenerate and an element of σ^{nd}_{exc}.

2.3 Methods Based on Continued Fractions

The Bargmann space formalism has been applied to the QRM as early as 1967 by Schweber [26]. He did not make use of the $\mathbb{Z}_2$-symmetry but derived the coupled system (12) directly from (1) with the ansatz $\boldsymbol{\psi} = (\varphi_1(z), \varphi_2(z))^T \in \mathscr{B} \otimes \mathbb{C}^2$ for the wave function with energy E. (12) is then satisfied by $\phi_1 = \varphi_1 + \varphi_2$ and $\phi_2 = \varphi_1 - \varphi_2$. He obtained the local Frobenius solution for $\bar{\phi}_2(y)$ given by (16) and (17). The convergence radius of the series (29) is $2g$ for arbitrary x and the discrete set of eigenvalues is selected by determining those x for which (29) has infinite convergence radius. The problem is equivalent to compute the *minimal solution* of the recurrence (16) [27]. The spectral condition obtains then by equating the minimal $K_1(x)/K_0$ with $f_0(x)$ from the initial conditions. The equation has the form $F(x) = 0$, where $F(x)$ is represented by a continued fraction [26]. This method, while formally correct, has several conceptual shortcomings:

- The function $F(x)$ has an unknown singularity structure and it is impossible to infer qualitative aspects on the distribution of its zeros from it.
- The actual computation of the continued fraction makes a truncation at some order necessary which is equivalent to define the model on a finite-dimensional Hilbert space, which is the starting point for other work employing continued fractions to compute the Rabi spectrum [28, 29].
- The zeros of $F(x)$ correspond to $\sigma_{reg} \cup \sigma^d_{exc}$, but there is no possibility to discern both types of spectra, especially the double degeneracy of σ^d_{exc} cannot be detected with this method.
- The spectrum σ^{nd}_{exc} is not accessible because the expansion (29) with $K_0 \neq 0$ is assumed in the derivation of $F(x)$.

Besides these conceptual problems, the method is numerically feasible only for the first low-lying eigenvalues. The continued fraction has a pole in close vicinity of each zero and their distance approaches zero exponentially for growing x, so that at most ten energy levels may be resolved within a double precision calculation. On the other hand, the equivalence of the continued fraction approach to exact diagonalization in finite-dimensional Hilbert spaces proves the validity of the latter for the QRM [17].

Schweber's technique is confined to problems reducible to three-term recurrence relations for the local Frobenius solutions and implements then (B-II) as the spectral condition. It fails for models with more than a single qubit because the ensuing recurrence relations have more than three terms. The next section is devoted to the application of the theory presented above to models with $N > 1$ qubits.

3 The Dicke Models

The natural generalization of the Hamiltonian (1) couples several two-level systems to the same mode of the radiation field,

$$H_{DN} = \omega a^\dagger a + \sum_{i=1}^{N} \frac{\omega_{0i}}{2}\sigma_{iz} + (a + a^\dagger)\frac{1}{\sqrt{N}}\sum_{i=1}^{N} g_i'\sigma_{ix}. \tag{31}$$

This model assumes different qubit frequencies ω_{0i} and couplings g_i to the field and is therefore called the asymmetric Dicke model (ADM$_N$) with N qubits. Dicke introduced the permutation-invariant version of (31) in 1954 and predicted the (later observed) phenomenon of "superradiance" for large N [30]. Its rotating-wave approximation is integrable for all N [31], while the full model is non-integrable for any $N > 1$ according to the level-labeling criterion [15]. Applications in quantum information technology mandate the study of (31) without approximations for small N, because it describes the implementation of quantum gates within circuit QED [32].

The following section treats the asymmetric model for $N = 2$ and Sect. 3.2 the symmetric model for $N = 3$.

3.1 ADM$_2$ and Exceptional States

The Hamiltonian of the ADM$_2$ reads in slightly different notation,

$$H_{D2} = \omega a^\dagger a + g_1\sigma_{1x}(a + a^\dagger) + g_2\sigma_{2x}(a + a^\dagger) + \Delta_1\sigma_{1z} + \Delta_2\sigma_{2z}. \tag{32}$$

This model has a $\mathbb{Z}_2$-symmetry similar to (1), generated by $\hat{P} = \exp(i\pi a^\dagger a)\sigma_{1z}\sigma_{2z}$. However, because it has only two irreducible one-dimensional representations, the discrete degrees of freedom cannot be labeled with a $\mathbb{Z}_2$-quantum number (their Hilbert space is $\mathbb{C}^4$). The symmetry of the model is not sufficient to make it integrable as the $N = 1$ case (1).

Nevertheless, the same methods as above can be used to solve (32) exactly [33], at the expense of a more complicated G-function, which is no longer a linear combination of formal solutions as (21). After application of the symmetry, the remaining Hilbert space is not $\mathscr{B}$, but $\mathscr{B} \otimes \mathbb{C}^2$. The Hamiltonian reads ($\omega = 1$),

$$H_{\pm} = z\frac{\mathrm{d}}{\mathrm{d}z} + (g_1 + g_2\sigma_z)\left(z + \frac{\mathrm{d}}{\mathrm{d}z}\right) + (\Delta_2 \pm \Delta_1 \hat{T})\sigma_x, \tag{33}$$

with $\hat{T} = (-1)^{z\frac{\mathrm{d}}{\mathrm{d}z}}$. An eigenfunction of (33) with eigenvalue E is the vector $\boldsymbol{\psi} = (\varphi_1(z), \varphi_2(z))^T$. Defining $\varphi_3(z) = \varphi_1(-z)$ and $\varphi_4(z) = \varphi_2(-z)$, we obtain a coupled system of four ordinary first order differential equations with four regular singular points located at $g = g_1 + g_2$, $g' = g_1 - g_2$, $-g$ and $-g'$. Moreover, $z = \infty$ is an unramified irregular singular point of s-rank two; we conclude that again all formal solutions fulfill (B-I) and E is determined by postulating analyticity of the solution at all regular singular points. The corresponding coupled recurrence relations for the Frobenius expansions around each of the points 0, g and g' cannot be reduced to a three-term recurrence, except in the case $g' = 0$, which allows a treatment similar to the QRM. It turns out that eight initial conditions determine functions $\phi_k(z), k = 1, \ldots, 32$, describing the $\varphi_j(z)$ around different expansion points. Overall analyticity is then equivalent to the vanishing of the determinant of a 8×8-matrix $M_{\pm}(E)$, whose entries are composed of the ϕ_k, evaluated at the ordinary points z_0 and z_0', whose location depends on the geometry of the analytic regions [33]. The G-function can then be defined as $G_{\pm}(E) = \det(M_{\pm}(E))$. This function has poles at integer values of $E + g'^2$ and $E + g^2$, defining the exceptional spectrum, besides the regular, given by the condition $G(E_n) = 0$.

The spectra of H_{D2} obtained in this way are depicted in Fig. 4, as function of g and various levels of asymmetry. For the completely asymmetric cases in Fig. 4a, b, we observe level crossings between states of different parity, whereas states with equal parity show avoided crossings, some of them quite narrow as the insets demonstrate. There is no relation between degeneracies and the exceptional spectrum as in the case of the QRM, because the G-functions $G_+(E)$ and $G_-(E)$ are not simply related. The set of lines with $E + g^2 = n, n \in \mathbb{N}_0$, give the asymptotic (parity degenerate) spectrum in the deep strong coupling limit. Figure 4c shows the completely symmetric case, $\Delta_1 = \Delta_2$, $g_1 = g_2$. The invariance under permutation symmetry of (32) leads to separation of the Hilbert space of the spin-1/2 qubits into singlet and triplet sector according to $\frac{1}{2} \otimes \frac{1}{2} = 0 \oplus 1$. The total Hilbert space becomes thus $\mathscr{B} \otimes \mathbb{C}^4 \rightarrow \mathscr{B} \oplus \mathscr{B} \otimes \mathbb{C}^3$. The singlet subspace is isomorphic to $\mathscr{B}$ and the Hamiltonian describes the decoupled radiation mode. Thus the eigenenergies are just integer multiples of ω, independent of the coupling. They are seen as horizontal lines in Fig. 4c. The triplet subspace is coupled to the radiation field and the spectrum shows a nontrivial dependence on g.

An interesting situation obtains for equal couplings $g_1 = g_2$ but different qubit energies, $\Delta_1 \neq \Delta_2$. The full permutation symmetry is broken, but there is a certain remnant of it. For $g_1 = g_2$, $z = 0$ is a regular singular point and there exist "quasi-exact" eigenstates, belonging to the exceptional spectrum with $E_n \in \mathbb{N}$ for certain parameter values of g, Δ_j. These states contain a finite number N of photons (contrary

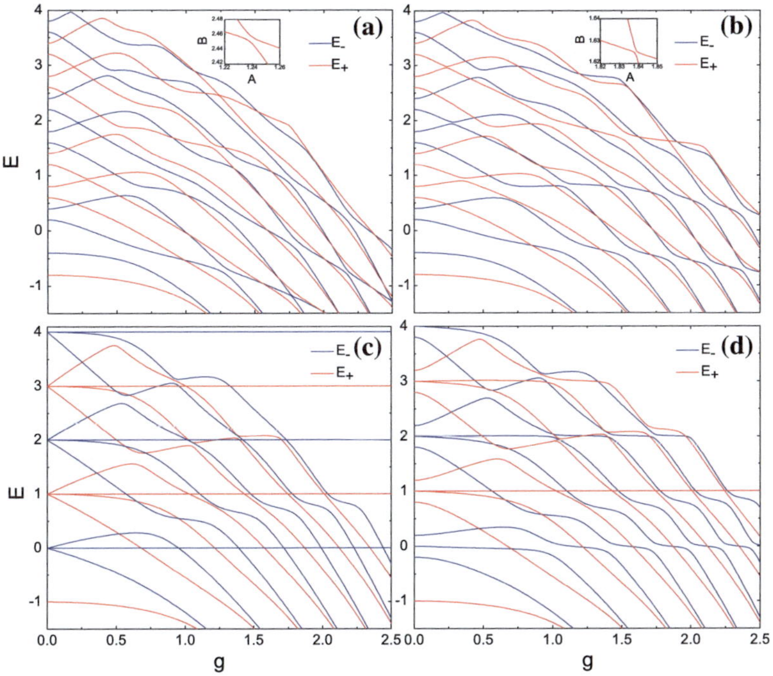

Fig. 4 The spectra of H_{D2} with **a** $\Delta_1 = 0.6$, $\Delta_2 = 0.2$, $\omega = 1$, $0 \le g = g_1 + g_2 \le 2.5$, $g_1 = 4g_2$. **b** $\Delta_1 = 0.6$, $\Delta_2 = 0.2$, $\omega = 1$, $0 \le g = g_1 + g_2 \le 2.5$, $g_1 = 2g_2$. **c** $\Delta_1 = \Delta_2 = 0.5$, $\omega = 1$, $0 < g = g_1 + g_2 < 2.5$, $g_1 = g_2$. **d** $\Delta_1 = 0.6$, $\Delta_2 = 0.4$, $\omega = 1$, $0 \le g \le 2.5$, $g_1 = g_2$. *Blue lines* are eigenvalues with odd parity, while *red lines* are eigenvalues with even parity

to the likewise quasi-exact elements of σ^d_{exc} in the QRM) and are determined by a polynomial equation for g, Δ_1, Δ_2 depending on the energy value $E = N$. It reads for $N = 2$,

$$\left[\left(2 - \frac{(\Delta_2 \pm \Delta_1)^2}{2}\right)\left(1 - (\Delta_2 \mp \Delta_1)^2\right) - g^2\right](\mp\Delta_1 - \Delta_2) = 0, \tag{34}$$

where the + ($-$) corresponds to even (odd) parity. The condition comprises the symmetric case $\Delta_1 = \Delta_2$ for odd parity, but otherwise determines Δ_1, Δ_2 in terms of g. This is true for all $N \ge 2$. However, for $N = 1$ we find,

$$(\pm\Delta_1 - \Delta_2)\left[1 - (\Delta_2 \pm \Delta_1)^2\right] = 0. \tag{35}$$

There exists an eigenstate with energy $E = 1$ and even (odd) parity, if $\Delta_1 + \Delta_2 = 1$, ($|\Delta_1 - \Delta_2| = 1$). This state contains at most one photon and the condition for its existence does *not* depend on g. It was first discovered by Chilingaryan and Rodríguez-Lara [34]. With the notation $|\psi\rangle = |n, s_1, s_2\rangle$ for a basis element in $\mathscr{H}$, where we have used the occupation number basis for the boson mode and $s_j \in \{g, e\}$ denotes the state of the jth qubit, the exceptional state with even parity ($\Delta_1 + \Delta_2 = 1$) reads,

$$|\psi_e\rangle = \frac{1}{\mathcal{N}} \left(\frac{2(\Delta_1 - \Delta_2)}{g} |0, e, e\rangle - |1, e, g\rangle + |1, g, e\rangle \right), \tag{36}$$

with a normalization factor $\mathcal{N}$. This state becomes the singlet state

$$\frac{1}{\sqrt{2}} |1\rangle \otimes (|g, e\rangle - |e, g\rangle) \tag{37}$$

in the symmetric case $\Delta_1 = \Delta_2$. The spectrum for $\Delta_1 + \Delta_2 = 1$ is shown in Fig. 4d, exhibiting the quasi-exact state with $E = 1$ as a g-independent line. In contrast to the fully decoupled singlet state (37), the state (36) is strongly coupled to the radiation field, as its components depend on g. It is quite remarkable that states with finite maximal photon number exist for arbitrary strong coupling without making the rotating wave approximation. This feature cannot be realized in the QRM, where each eigenstate contains always an infinite number of photons. Due to its very simple structure, the state (36) could be useful for quantum computing applications, especially as the condition for its existence depends only on the (easily controllable) qubit energies Δ_1 and Δ_2 and not on the coupling strength. Similar states are expected to exist in all models ADM$_N$ with $g_i \equiv g$ and even N.

3.2 ADM$_3$

The Hilbert space of the symmetric model ADM$_3$ with $g_i' \equiv g' = \sqrt{3}g$, $\omega_{0i} \equiv \omega_0 = 2\Delta$ may be splitted according to $\frac{1}{2} \otimes \frac{1}{2} \otimes \frac{1}{2} = \frac{1}{2} \oplus \frac{1}{2} \oplus \frac{3}{2}$. Each of the two spin-1/2 components are equivalent to the QRM, while the last component describes a single spin-3/2 coupled to the radiation mode with Hamiltonian,

$$H_{D3} = a^\dagger a + 2\Delta \hat{J}_z + 2g(a + a^\dagger)\hat{J}_x, \tag{38}$$

where $\hat{J}_z$ and $\hat{J}_x$ are generators of $SU(2)$ in the (four-dimensional) spin-3/2 representation. The $\mathbb{Z}_2$-generator has here the form $\hat{P} = \exp(i\pi a^\dagger a)\hat{R}$. $\hat{R}$ is an involution acting in spin-space as $\hat{R}\hat{J}_z\hat{R} = \hat{J}_z$, $\hat{R}\hat{J}_x\hat{R} = -\hat{J}_x$. Application of $\hat{P}$ gives the following differential operator in each parity subspace,

$$H_\pm = z\frac{\mathrm{d}}{\mathrm{d}z} + \Delta \begin{pmatrix} 0 & \sqrt{3} \\ \sqrt{3} & \pm 2\hat{T} \end{pmatrix} - g \begin{pmatrix} 3 & 0 \\ 0 & 1 \end{pmatrix} \left(\frac{\mathrm{d}}{\mathrm{d}z} + z \right). \tag{39}$$

Employing now the same machinery as in the previous section, we obtain again four coupled first order equations having regular singular points at $\pm g$, $\pm 3g$ and an irregular singular point (s-rank two) at infinity [35]. Because of the stronger

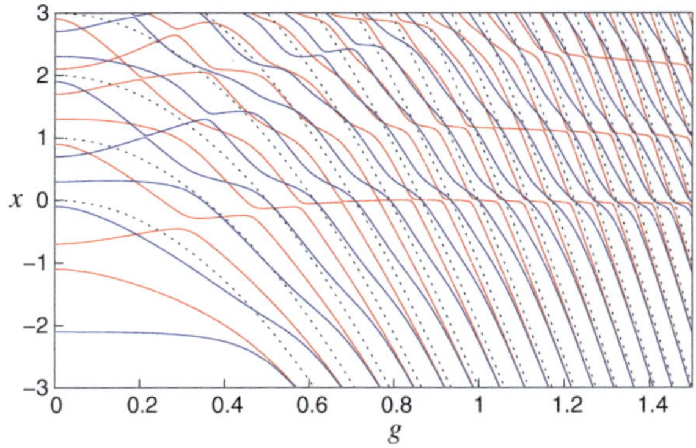

Fig. 5 The spectrum of the Dicke model for even (*red*) and odd (*blue*) parity at $\Delta = 0.7$ and for varying g. The y-axis shows $x = E + g^2$. The ground state has odd parity as in the QRM. The two ladders of eigenvalues with different parity intersect within the regular spectrum. There are no degeneracies (but narrow avoided crossings) for fixed parity in this parameter window. *Dashed lines* denote the set $E + 9g^2 \in \mathbb{N}_0$ and emerge as limiting values in the deep strong coupling regime $g \gg 1$

symmetry of (38) compared to (32), the matrix $M(E)_{\pm}$, whose determinant gives the G-function is only 6×6. It contains 24 functions evaluated at the points $z_0 = 2g$ and $z_0' = 0$. The poles of $G_{\pm}(E)$ (giving σ_{exc}^{nd}) are located at $E + g^2 \in \mathbb{N}$ and $E + 9g^2 \in \mathbb{N}_0$. The curves determined by the latter set are also the limiting values for the spectrum for very large coupling. Figure 5 shows the spectral graph of the model as function of the coupling g. As in the $N = 2$ case, the degeneracies occur within the regular spectrum between states of different parity. However, the fact that we have now a determinant as G-function means that in principle degeneracies within the *same* parity chain are not excluded as the corresponding matrix $M_{\pm}(E)$ could have a higher-dimensional kernel at a specific energy. Up to now there is no numerical evidence for this scenario and the existence of these novel degeneracies is an open question.

4 Conclusions

We have seen in the previous sections that the classical theory of linear differential equations in the complex domain may be used to solve exactly elementary but important problems in the field of theoretical and applied physics, contributing in this way to a better understanding of the basic models of circuit QED, which has been envisioned as a promising environment to implement devices capable of performing quantum computations.

The mathematical technique relies on the Bargmann Hilbert space $\mathcal{B}$, which allows to represent the Hamiltonians from Quantum Optics as differential operators acting on functions of a complex variable z. Of central importance is here Bargmann's two-fold spectral condition, which demands not only normalizability with respect to the scalar product but also analyticity in $\mathbb{C}$ for any function $\phi(z)$ being an element of $\mathscr{B}$. In this way it becomes possible to use the easily accessible singularity structure of the corresponding differential equations to implement the spectral condition *without* recourse to a polynomial ansatz for the wave functions, which works for elementary integrable systems like the harmonic oscillator, the hydrogen atom and the Jaynes-Cummings model, but fails already for the quantum Rabi model, which is nevertheless integrable in a well-defined sense [15] (for a recent comparison with Yang-Baxter integrability see [36]).

The method can be extended to models of central relevance for quantum technology, the Dicke models with a small number of qubits. These models are no longer integrable in view of the level labeling criterion [15] but exactly solvable with the presented technique, although many characteristic simplifications of the quantum Rabi model are absent. Further applications concern models with a single irregular singular point as the two-photon Rabi model [37], or the anharmonic oscillator. One may also try to extend the formalism to multi-mode models [38]. Notwithstanding these generalizations, we note that the already solved systems give rise to a set of open mathematical problems like the conjecture on the level distribution presented in Sect. 2.1, or the question whether a novel class of degeneracies exists in the Dicke models (Sect. 3.2). Thus, a future research direction will be the exploration of the recently observed connection [39] between the quantum Rabi model and the non-commutative harmonic oscillator [40]. It should be possible to transfer mathematical methods used for the study of the latter to the problems mentioned above. The techniques used in [41–44] could be applied f.e. to investigate the level crossing appearing in the ground state of the anisotropic quantum Rabi model [45].

Acknowledgments This work was supported by Deutsche Forschungsgemeinschaft through TRR 80.

References

1. Cohen-Tannoudji, C., Dupont-Roc, J., Grynberg, G.: Atom-Photon Interactions: Basic Processes and Applications. Wiley, Weinheim (2004)
2. Haroche, S., Raymond, J.M.: Exploring the Quantum. Oxford University Press, New York (2006)
3. Romero, G., Ballester, D., Wang, Y.M., Scarani, V., Solano, E.: Ultrafast quantum gates in circuit QED. Phys. Rev. Lett. **108**, 120501 (2012)
4. Nielsen, M.A., Chuang, I.L.: Quantum Computation and Quantum Information. Cambridge University Press, Cambridge (2000)
5. Wallraff, A., Schuster, D.I., Blais, A., Frunzio, L., Huang, R.S., Majer, J., Kumar, S., Girvin, S.M., Schoelkopf, R.J.: Strong coupling of a single photon to a superconducting qubit using circuit quantum electrodynamics. Nature **431**, 162 (2004)
6. Niemczyk, T., Deppe, F., Huebl, H., Menzel, E.P., Hocke, F.: Circuit quantum electrodynamics in the ultrastrong-coupling regime. Nat. Phys. **6**, 772 (2010)

7. Forn-Diaz, P., Lisenfeld, J., Marcos, D., García-Ripoll, J.J., Solano, E., Harmans, C.J.P.M., Mooij, J.E.: Observation of the Bloch-Siegert shift in a qubit-oscillator system in the ultrastrong coupling regime. Phys. Rev. Lett. **105**, 237001 (2010)
8. Rabi, I.I.: On the process of space quantization. Phys. Rev. **49**, 324 (1936)
9. Jaynes, E.T., Cummings, F.W.: Comparison of quantum and semiclassical radiation theories with application to the beam maser. Proc. IEEE **51**, 89 (1963)
10. Klimov, A.B., Sainz, I., Chumakov, S.M.: Resonance expansion versus rotating-wave approximation. Phys. Rev. A **68**, 063811 (2003)
11. Amico, L., Frahm, H., Osterloh, A., Ribeiro, G.A.P.: Integrable spin-boson models descending from rational six-vertex models. Nucl. Phys. B **787**, 283 (2007)
12. Arnold, V.I.: Mathematical Methods of Classical Mechanics. Springer, New York (1989)
13. Caux, J.S., Mossel, J.: Remarks on the notion of quantum integrability. J. Stat. Mech. P02023 (2011)
14. Casanova, J., Romero, G., Lizuain, I., García-Ripoll, J.J., Solano, E.: Deep strong coupling regime of the Jaynes-Cummings model. Phys. Rev. Lett. **105**, 263603 (2010)
15. Braak, D.: Integrability of the Rabi model. Phys. Rev. Lett. **107**, 100401 (2011)
16. Bargmann, V.: On a Hilbert space of analytic functions and an associated integral transform, part I. Commun. Pure Appl. Math. **14**, 187 (1961)
17. Braak, D.: Continued fractions and the Rabi model. J. Phys. A: Math. Theor. **46**, 175301 (2013)
18. Ince, E.L.: Ordinary Differential Equations. Dover, New York (1956)
19. Slavyanov, S.Y., Lay, W.: Special Functions. A Unified Theory Based on Singularities. Oxford University Press, New York (2000)
20. Braak, D.: A generalized G-function for the quantum Rabi model. Ann. Phys. (Berlin) **525**, L23 (2013)
21. Zhong, H., Xie, Q., Batchelor, M.T., Lee, C.: Analytical eigenstates for the quantum Rabi model. J. Phys. A: Math. Theor. **46**, 14 (2013)
22. Ronveaux, A. (ed.): Heun's Differential Equations. Oxford University Press, New York (1995)
23. Maciejewski, A.J., Przybylska, M., Stachowiak, T.: Full spectrum of the Rabi model. Phys. Lett. A **378**, 16 (2014)
24. Turbiner, A.V.: Quasi-exactly-solvable problems and $sl(2)$ algebra. Commun. Math. Phys. **118**, 467 (1988)
25. Wakayama, M., Yamasaki, T.: The quantum Rabi model and the Lie algebra representations of $\mathfrak{sl}_2$. J. Phys. A: Math. Theor. **47**, 335203 (2014)
26. Schweber, S.: On the application of Bargmann Hilbert spaces to dynamical problems. Ann. Phys., NY **41**, 205 (1967)
27. Gautschi, W.: Computational aspects of three-term recurrence relations. SIAM Rev. **9**, 24 (1967)
28. Swain, S.: A continued fraction solution to the problem of a single atom interacting with a single radiation mode in the electric dipole approximation. J. Phys. A: Math. Nucl. Gen. **6**, 192 (1973)
29. Tur, E.A.: Energy spectrum of the Hamiltonian of the Jaynes-Cummings model without rotating-wave approximation. Opt. Spectrosc. **91**, 899 (2001)
30. Dicke, R.H.: Coherence in spontaneous radiation processes. Phys. Rev. **93**, 99 (1954)
31. Tavis, M., Cummings, F.W.: Exact solution for an N-molecule-radiation-field Hamiltonian. Phys. Rev. **170**, 379 (1968)
32. Haack, G., Helmer, F., Mariantoni, M., Marquardt, F., Solano, E.: Resonant quantum gates in circuit quantum electrodynamics. Phys. Rev. B **82**, 024514 (2010)
33. Peng, J., Ren, Z.Z., Braak, D., Guo, G.J., Ju, G.X., Zhang, X., Guo, X.Y.: Solution of the two-qubit quantum Rabi model and its exceptional eigenstates. J. Phys. A: Math. Theor. **47**, 265303 (2014)
34. Chilingaryan, S.A., Rodríguez-Lara, B.M.: The quantum Rabi model for two qubits. J. Phys. A: Math. Theor. **46**, 335301 (2013)
35. Braak, D.: Solution of the Dicke model for $N = 3$. J. Phys. B: At. Mol. Opt. Phys. **46**, 224007 (2013)

36. Batchelor, M.T., Zhou, H.Q.: Integrability versus exact solvability in the quantum Rabi and Dicke models. Phys. Rev. A **91**, 053808 (2015)
37. Travěnec, I.: Solvability of the two-photon Rabi Hamiltonian. Phys. Rev. A **85**, 043805 (2012)
38. Fan, J., Yang, Z., Zhang, Y., Ma, J., Chen, G., Jia, S.: Hidden continuous symmetry and Nambu-Goldstone mode in a two-mode Dicke model. Phys. Rev. A **89**, 023812 (2014)
39. Wakayama, M.: Equivalence between the eigenvalue problem of non-commutative harmonic oscillators and existence of holomorphic solutions of Heun differential equations, eigenstates degeneration and the Rabi model. Kyushu University, MI-preprint series (2013)
40. Ichinose, T., Wakayama, M.: Zeta functions for the spectrum of the non-commutative harmonic oscillators. Commun. Math. Phys. **258**, 697 (2005)
41. Hirokawa, M.: The Dicke-type crossing among eigenvalues of differential operators in a class of non-commutative oscillators. Indiana Univ. Math. J. **58**, 1493 (2009)
42. Wakayama, M.: Simplicity of the lowest eigenvalue of non-commutative harmonic oscillators and the Riemann scheme of a certain Heun's differential equation. Proc. Jpn. Acad. Ser. A **89**, 69 (2013)
43. Hiroshima, F., Sasaki, I.: Spectral analysis of non-commutative harmonic oscillators: the lowest eigenvalue and no crossing. J. Math. Anal. Appl. **105**, 595 (2014)
44. Hirokawa, M., Hiroshima, F.: Absence of energy level crossing for the ground state energy of the Rabi model. Comm. Stoch. Anal. (to appear)
45. Xie, Q.T., Cui, S., Cao, J.P., Amico, A., Fan, H.: Anisotropic Rabi model. Phys. Rev. X **4**, 021046 (2014)

Quantum Simulation of Spin Chains Coupled to Bosonic Modes with Superconducting Circuits

U. Las Heras, L. García-Álvarez, A. Mezzacapo, E. Solano and L. Lamata

Abstract We propose the implementation of a digital quantum simulation of spin chains coupled to bosonic field modes in superconducting circuits. Gates with high fidelities allow one to simulate a variety of Ising magnetic pairing interactions with transverse field, Tavis-Cummings interaction between spins and a bosonic mode, and a spin model with three-body terms. We analyze the feasibility of the implementation in realistic circuit quantum electrodynamics setups, where the interactions are either realized via capacitive couplings or mediated by microwave resonators.

Keywords Quantum optics · Quantum information · Quantum simulations · Superconducting circuits · Bosons · Digital techniques

1 Introduction

A two-level system coupled with a single radiation mode is modeled by the ubiquitous and paradigmatic quantum Rabi model [29] that describes the most fundamen-

U. Las Heras (✉) · L. García-Álvarez · A. Mezzacapo · E. Solano · L. Lamata (✉)
Department of Physical Chemistry, University of the Basque Country UPV/EHU,
Apartado 644, E-48080 Bilbao, Spain
e-mail: ulasheras@gmail.com

L. Lamata
e-mail: lucas.lamata@gmail.com

L. García-Álvarez
e-mail: garcia.alvarez.la@gmail.com

A. Mezzacapo
e-mail: ant.mezzacapo@gmail.com

E. Solano
IKERBASQUE, Basque Foundation for Science, Maria Diaz de Haro 3,
48013 Bilbao, Spain
e-mail: enr.solano@gmail.com

Present Address: A. Mezzacapo
IBM T. J. Watson Research Center, Yorktown Heights, NY 10598, USA

R.S. Anderssen et al. (eds.), *Applications + Practical Conceptualization + Mathematics = fruitful Innovation*, Mathematics for Industry 11,
DOI 10.1007/978-4-431-55342-7_8

tal interaction between quantum light and quantum matter. There have been many efforts, both in theory and experiments, to capture the features of this model in different quantum technologies [13, 19]. These analysis have an impact on understanding about different quantum phenomena [7, 14, 26, 30, 35, 38].

The concept of a quantum simulator can be attributed to Feynman [10], and it refers to a controllable quantum platform that mimics the behaviour of another quantum system. Analog quantum simulators have been proposed and implemented in several quantum technologies, such as trapped ions [6, 22], ultracold atoms [5], or superconducting circuits [2, 25, 27, 36]. Digital methods based on discrete-time gate sequences [20] in order to simulate dynamics of quantum systems have been proposed and realized in trapped ions [15], photonic systems [16], spin-photon hybrid systems [8] and superconducting circuits [4, 12, 17, 18, 24, 31].

Here, we analyze the quantum simulation of arbitrary and generic models, where spin chains alone or coupled to bosonic modes are emulated in superconducting circuits [9]. We use digital techniques in order to imitate systems whose dynamics in principle may differ from the ones of the experimental setups. Finally, we study the feasibility and efficiency of the implementation of three generic models in a realistic circuit quantum electrodynamics setup.

2 Digital Quantum Simulations

The goal of simulating diverse and generic models involving spin interactions and bosonic modes leads us to consider digital techniques, due to their suitability and flexibility for mimicking different dynamical structures. Hamiltonian dynamics can be approximated by the digital decomposition of the exact unitary evolution into discrete stepwise unitary operations, implemented by using quantum gates in an efficient way [20, 33]. Digital methods are based on the Trotter formula, which allows us to expand the evolution operator of Hamiltonians that are written as a sum of terms, $H = \sum_{j=1}^{N} H_j$, into a product of evolution operators for the interactions given by the summands of the Hamiltonian, H_j. The Trotter expansion can be written as

$$e^{-iHt} = \left(e^{-iH_1 t/s} \cdots e^{-iH_N t/s}\right)^s + \sum_{i<j} \frac{[H_i, H_j]\,t^2}{2s} + \sum_{k=3}^{\infty} E(k), \qquad (1)$$

where the total time of the simulated dynamics is divided into s intervals in which the evolution associated to each summand of the complete Hamiltonian are applied. The error scales with t^2/s for short times, as can be observed in the second term, and the upper bound for higher order error contributions is $s\|Ht/s\|_{\mathrm{sup}}^k/k! \geq \|E(k)\|_{\mathrm{sup}}$.

Our goal is to propose a systematic procedure using digital methods for simulating efficiently different models, namely spin-spin interaction and spins coupled to bosonic modes. First, we employ gates that commute with each other and do not produce digital error. For those that do not commute, we apply several Trotter steps because the more Trotter steps one applies, the smaller the digital error produced is. In realistic experiments, one has to take into account decoherence times and gate errors. Therefore, we have to regulate the number of steps in order to be able to perform the simulation before decoherence effects take place, and in order to reduce the accumulated gate error. Consequently, once the digital error is small enough applying a certain number of Trotter steps, the error coming from the experimental setup always must be smaller than the digital one.

3 Quantum Simulation of Spin Chains Coupled to Bosonic Modes with Superconducting Circuits

In this section, we present a method to implement the dynamics of several spin models, coupled with bosonic modes, in circuit quantum electrodynamics setups. For this purpose, we take under consideration two different architectures of superconducting circuits. We show how to use linear arrays of superconducting qubits with capacitive coupling between nearest neighbors [3] to simulate the Ising model with transverse field. Then we simulate the behavior of a spin-chain coupled to a bosonic mode via a Tavis-Cummings interaction [34]. Moreover, we show how to implement many-body spin dynamics with highly nonlinear terms in superconducting qubits coupled to transmission line resonators acting as a quantum bus [37].

In the following, we propose digital quantum simulations based on quantum gates implemented in superconducting architectures. Capacitive coupling setups allow one to implement ZZ gates, $\exp(-i\theta\sigma_j^z\sigma_k^z)$, for nearest-neighbor superconducting qubits by the sequence of two single qubit rotations along the z axis, $Z(\phi)$, and a c-phase gate, $CZ(\phi)$, as shown in Fig. 1, where

$$Z(\phi) = \begin{pmatrix} 1 & 0 \\ 0 & e^{i\phi} \end{pmatrix}, \quad CZ(\phi) = \begin{pmatrix} 1 & 0 & 0 & 0 \\ 0 & 1 & 0 & 0 \\ 0 & 0 & 1 & 0 \\ 0 & 0 & 0 & e^{-i2\phi} \end{pmatrix}. \tag{2}$$

The current achievable fidelities in superconducting circuits are of [3] 99.9 and 99.4 % for the single and two-qubit (CZ) gates, respectively. They enable circuit QED setups to be great candidates for digital quantum simulators where the stroboscopic application of gates is necessary. Notice that $ZZ_{12}(\theta) = (Z_1(\phi) \otimes Z_2(\phi))CZ_{12}(\phi)$ for $\theta = \phi/2$.

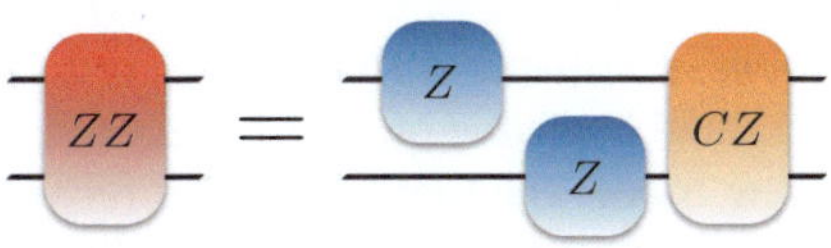

Fig. 1 Protocol for decomposing a ZZ interaction between two qubits using single-qubit rotations, Z, and a c-phase gate, CZ

The use of quantum buses allows for the implementation of multi-qubit gates and spin-boson interactions, coupling the electromagnetic field in the resonator with superconducting circuits [11, 21, 24, 32].

3.1 Ising Model with Transverse Field Via Capacitive Nearest-Neighbour Gates

One of the most studied spin models in condensed matter is the Ising model with a transverse field, which is used for describing the behavior of interacting nearest-neighbor dipoles in the presence of a transverse magnetic field. The Hamiltonian of N spins can be written as

$$H_{ITF} = J \sum_{\langle jk \rangle} \sigma_j^z \sigma_k^z + B \sum_j \sigma_j^x, \tag{3}$$

where σ_j^α is the Pauli operator acting over the jth spin with $j = 1, \ldots, N$, in the direction $\alpha = x, y, z$. J stands for the coupling between nearest-neighbor spins and B is the coupling between a spin and the transverse field. Depending on the sign of J the model is ferromagnetic ($J < 0$) or antiferromagnetic ($J > 0$). In order to reproduce this interaction in superconducting circuits, we make use of a high-fidelity set of gates, as introduced in Eq. (2): single-qubit rotations along the x direction, $X_j(\phi) = \exp(-i\phi\sigma_j^x)$, and two-qubit ZZ gates, $ZZ_{jk}(\theta) = \exp(-i\theta\sigma_j^z\sigma_k^z)$.

As shown in Sect. 2, it is possible to decompose a complex interaction into discrete series of gates through a Trotter expansion. In order to implement the spin-spin interaction, we need to execute $(N-1)$ two-qubit gates. In this case, there is no digital error because all the gates in this decomposition commute,

$$\exp(-i\,\theta \sum_{\langle jk \rangle} \sigma_j^z \sigma_k^z) = e^{-i\theta\sigma_1^z\sigma_2^z}\, e^{-i\theta\sigma_2^z\sigma_3^z} \,\cdots\, e^{-i\theta\sigma_{N-1}^z\sigma_N^z}, \tag{4}$$

with $\theta = Jt$, t being the simulation time of the experiment. The coupling among the spins and the transverse field can be simulated in a similar way using N single qubit rotations,

$$\exp(-i\,\phi \sum_j \sigma_j^x) = e^{-i\phi\sigma_1^x}\, e^{-i\phi\sigma_2^x} \,\cdots\, e^{-i\phi\sigma_N^x}, \tag{5}$$

with $\phi = Bt$. Given that the two interactions in Eqs. (4) and (5) do not commute, one has to implement them in sequential short-time Trotter steps to minimize the digital error. In Fig. 2, we show a scheme of the protocol for the quantum simulation of the transverse field Ising model for four spins. The recent achievement of high-fidelity single and two-qubit (CZ) gates with superconducting circuits will allow one to perform many Trotter steps for several qubits, using hundreds of gates.

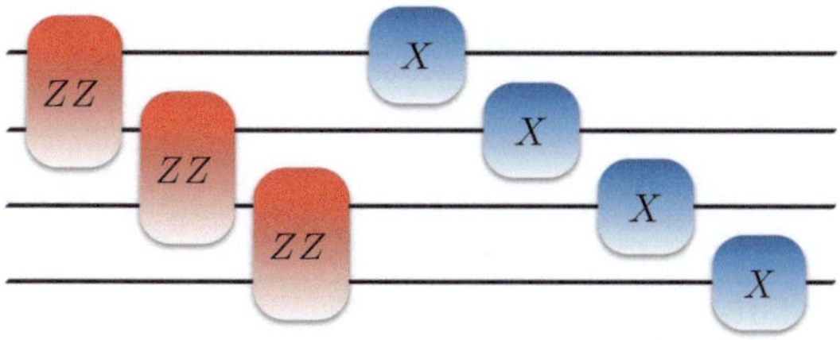

Fig. 2 Protocol for digital quantum simulation of the Ising model with transverse magnetic field in terms of ZZ two-qubit gates and single qubit rotations along x axis

In order to reduce the digital error, it is necessary to increase the number of Trotter steps. In Fig. 3, we plot a numerical simulation of the Ising model with transverse field for different digital steps. The simulated dynamics with digital decomposition is more accurate when compared with the exact dynamics when the number of Trotter steps is increased.

3.2 *Spin Chain Coupled to a Bosonic Field Mode Via Tavis-Cummings Interaction*

We now analyze a model consisting of a spin-chain with nearest-neighbour couplings interacting with a bosonic mode. In this sense, both free energies of the bosonic mode and spins are taken into account, as well as spin-spin and spin-boson interactions. The spin-spin evolution is modelled with the Ising dynamics, while the Tavis-Cummings model describes the interactions between spins and bosons. The resulting Hamiltonian is

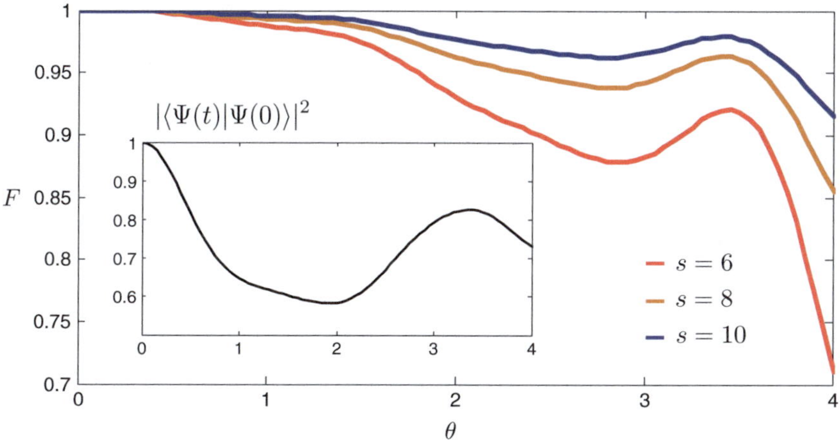

Fig. 3 Digital simulation of the ferromagnetic Ising model with a transverse field for four spins in superconducting circuits, up to a phase of $\theta = 4$, with $J/B = 2$. The plot shows the fidelity of the digitally evolved state versus the ideally evolved one for different number of Trotter steps, $s = 6, 8, 10$. The *inset* shows the overlap between the ideally evolved state with the initial state, that is, all qubits in $|0\rangle_z$

$$H_{ITC} = \omega\, a^\dagger a + \sum_j \frac{\Omega}{2}\sigma_j^z - J\sum_{\langle jk\rangle}\sigma_j^z\sigma_k^z + g\sum_j (a\sigma_j^+ + a^\dagger\sigma_j^-). \tag{6}$$

Following the notation presented above, σ_j^z is the Pauli operator along z direction, $\sigma_j^+(\sigma_j^-)$ is the creation(annihilation) spin excitation operator acting on the ith spin and $a(a^\dagger)$ is the annihilation(creation) operator of the bosonic mode. ω and Ω are the free energies of each boson and spin, respectively. Moreover, J is the coupling constant between nearest spins and g stands for the coupling among spins and bosonic field.

The implementation in circuit QED requires the simulation not only of the spin dynamics, as in the previous example, but also of the bosonic mode. To achieve this, the superconducting qubits play the role of spins while the photons in a transmission line resonator emulate the bosonic excitations in the simulation. In order to perform the interactions of the model, it is necessary to couple the resonator to all the superconducting qubits. The Tavis-Cummings interaction appears straightforwardly in circuit QED setups once the rotating wave approximation is performed,

$$H_1 = \omega_1\, a^\dagger a + \sum_j \frac{\Omega_1}{2}\sigma_j^z + g\sum_j (a\sigma_j^+ + a^\dagger\sigma_j^-), \tag{7}$$

being ω_1 the frequency of the photons in the resonator, Ω_1 the frequency associated with the excitation energy of the superconducting qubits, and g the qubit-resonator coupling constant. The spin-spin interaction for qubits j and k is achieved by means of the ZZ gate presented in Eq. (2). Detuning to high frequencies the qubit-resonator interaction we are able to reproduce the model

$$H(j,k) = \omega'\, a^\dagger a + \sum_j \frac{\Omega'}{2}\sigma_j^z - J\sigma_j^z\sigma_k^z. \tag{8}$$

Since $[H(j,k), H(j',k')] = 0\ \forall\ j, j', k, k'$, we can define and implement sequentially the interaction

$$H_2 = \sum_{\langle jk\rangle} H(j,k) = \omega_2\, a^\dagger a + \sum_j \frac{\Omega_2}{2}\sigma_j^z - J\sum_{\langle jk\rangle}\sigma_j^z\sigma_k^z, \tag{9}$$

where $\omega_2 = (N-1)\omega'$, $\Omega_2 = (N-1)\Omega'$ and N the number of simulated spins, and it fulfills the condition $\exp(-itH_2) = \prod_{\langle jk\rangle}\exp(-itH(j,k))$, being t the execution time.

Summing the interactions H_1 and H_2 we recover the model we wanted to reproduce, H_{ITC}. Nevertheless, $[H_1, H_2] \neq 0$, so we need to employ the Trotter method in order to make the digital error decrease, as shown in Fig. 4. Moreover, for considering the resonator photonic leakage, we have calculated the evolution of the system making use of the master equation,

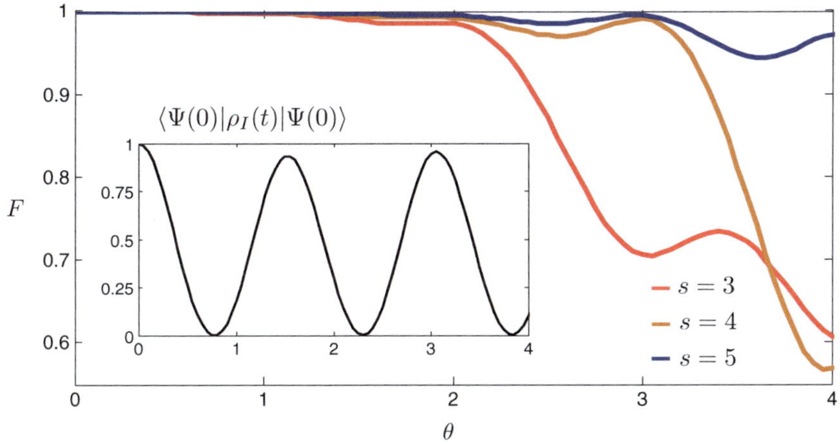

Fig. 4 Fidelity F of the simulation of a four-spin chain coupled to a bosonic mode with circuit QED for different Trotter steps, $s = 3, 4, 5$. The *upper curves* correspond to larger number of Trotter steps. Here, the parameters of Hamiltonians H_1 and H_2 are $\omega_1 = 2\pi \times 200$ MHz, $\Omega_1 = 2\pi \times 180$ MHz, $g = 2\pi \times 80$ MHz, $\omega_2 = 2\pi \times 600$ MHz, $\Omega_2 = 2\pi \times 18$ MHz, $J = 2\pi \times 200$ MHz and the decay rate of the resonator is given by $\kappa = 2\pi \times 10$ kHz. F is defined as the overlap between the ideally evolved density matrix and the digitally evolved one, $F(t) = \text{Tr}(\rho_I(t)\rho_T(t))$. The *inset* shows the overlap between the ideally evolved density matrix and the state of the system at $t = 0$, $1/\sqrt{2}(a^\dagger + (a^\dagger)^2/\sqrt{2})|0\rangle_p \otimes |1_1 0_2 0_3 0_4\rangle_z$, i.e., the same probability for having 1 and 2 photons in the resonator and all the superconducting qubits in the ground state of σ_i^z except the first, which is excited

$$\dot{\rho} = -i[H_t, \rho] + \kappa L(a)\rho, \tag{10}$$

where $L(a)\rho = (2a\rho a^\dagger - a^\dagger a\rho - \rho a^\dagger a)/2$ is the Lindblad superoperator acting on a, κ is the decay rate of the resonator, and $H_t = \{H_1, H_2\}$ is the Hamiltonian that governs the evolution. Notice that we have considered a coherence time much longer for the qubits than for the resonator. In Fig. 5, we plot the steps for implementing the protocol for four spins interacting with a bosonic mode.

Fig. 5 Protocol for the digital quantum simulation of a spin-chain coupled to a bosonic mode with superconducting circuits, in terms of unitary evolutions of Hamiltonians H_1, H_2 and $\tilde{H}$, being the interaction defined in Eq. (8)

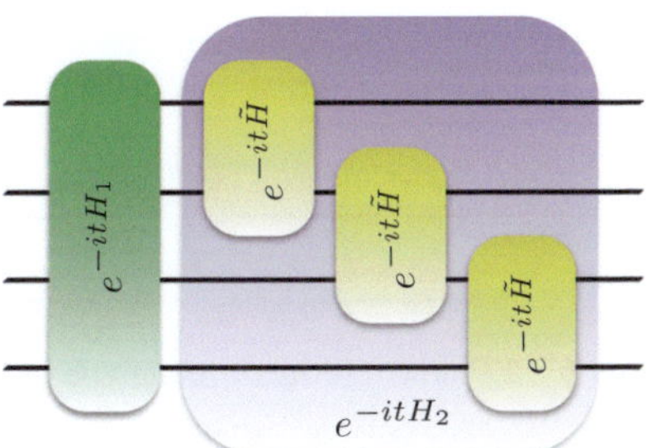

3.3 Collective Spin Coupling Mediated by Resonators

In this subsection, we extend the Ising model presented in Eq. (3) by adding three-body interactions. The method can be generalized to arbitrary interaction orders. This extension allows us to simulate problems such as quantum chemistry [1, 16, 28, 39], as well as fermionic lattice models [6, 17], by using the Jordan-Wigner mapping to map fermionic interactions into spin interactions. The Hamiltonian for N spins including three-body interactions can be written as

$$H = J\sum_{\langle jk\rangle}\sigma_j^z\sigma_k^z + G\sum_{\langle jkl\rangle}\sigma_j^z\sigma_k^z\sigma_l^z + B\sum_j \sigma_j^x, \qquad (11)$$

Here, we have added one collective interaction term with coupling constant G, which is the coupling among three nearest neighbour spins. This model can be simulated by enriching with additional gates the protocol for the Ising model in Sect. 3.1. That is, together with single-qubit rotations along the x direction, $X_j(\phi) = \exp(-i\phi\sigma_j^x)$, and two-qubit zz gates, $ZZ_{jk}(\theta) = \exp(-i\theta\sigma_j^z\sigma_k^z)$, we also consider the combination of collective gates shown in Fig. 6. This will allow us to couple three qubits, $ZZZ_{jkl}(\beta) = \exp(-i\beta\sigma_j^z\sigma_k^z\sigma_k^z)$.

The collective spin interaction of this model can be decomposed into $(N-1)$ two-qubit gates and $2(N-2)$ three-qubit gates, and the transverse field is mimicked by N single qubit rotations. Moreover, we notice that the digital error of the Trotter expansion in Eq. (1) is reduced due to the fact that the interaction summands of the Hamiltonian commute with each other. The Trotter expansion for this model reads

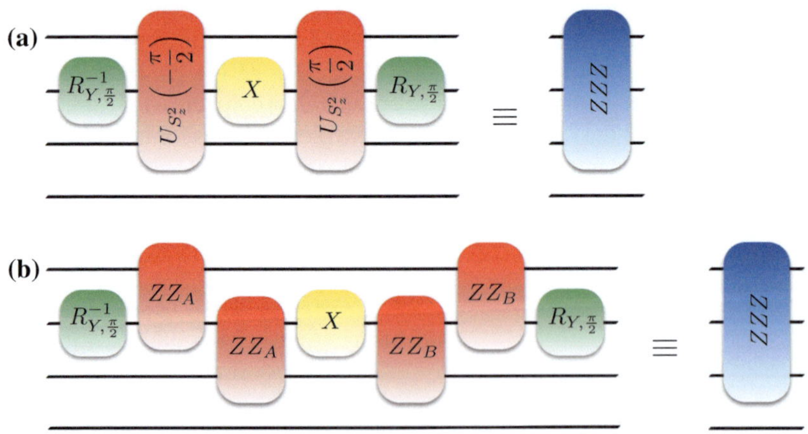

Fig. 6 **a** Protocol for performing one of the three-qubit interactions appearing in Eq. (11) with collective gates. Here $ZZZ_{123}(\alpha) = \exp(-i\alpha\sigma_1^z\sigma_2^z\sigma_3^z)$, $R_{Y,\theta} = \exp(-i\theta\sigma^y/2)$ is the rotation along the Y-axis of a qubit, $X = \exp(i\alpha\sigma^x)$, and $U_{S_z^2}(\theta) = \exp(-i\theta/2\sum_{i<j}\sigma_i^z\sigma_j^z)$. **b** The same interaction ZZZ can be realized with two-qubit gates where $ZZ_A = \exp(i\pi\sigma^z\otimes\sigma^z/4)$, and $ZZ_B = \exp(-i\pi\sigma^z\otimes\sigma^z/4)$

$$e^{-iHt} \simeq \left(e^{-i\,t/s\,J\sum_{\langle jk\rangle}\sigma_j^z\sigma_k^z} e^{-i\,t/s\,G\sum_{\langle jkl\rangle}\sigma_j^z\sigma_k^z\sigma_l^z} e^{-i\,t/s\,B\sum_j\sigma_j^x}\right)^s, \quad (12)$$

where

$$\exp(-i\theta\sum_{\langle jk\rangle}\sigma_j^z\sigma_k^z) = e^{-i\theta\sigma_1^z\sigma_2^z}\, e^{-i\theta\sigma_2^z\sigma_3^z}\,\ldots\, e^{-i\theta\sigma_{N-1}^z\sigma_N^z},$$
$$\exp(-i\beta\sum_{\langle jkl\rangle}\sigma_j^z\sigma_k^z\sigma_l^z) = e^{-i\beta\sigma_1^z\sigma_2^z\sigma_3^z}\, e^{-i\beta\sigma_2^z\sigma_3^z\sigma_4^z}\,\ldots\, e^{-i\beta\sigma_{N-2}^z\sigma_{N-1}^z\sigma_N^z},$$
$$\exp(-i\phi\sum_j\sigma_j^x) = e^{-i\phi\sigma_1^x}\, e^{-i\phi\sigma_2^x}\,\ldots\, e^{-i\phi\sigma_N^x}, \quad (13)$$

with $\theta = -Jt$, $\beta = Gt$ and $\phi = Bt$, t being the simulated execution time. The collective gate for three qubits can be decomposed into two-qubit gates, as in Fig. 6. Recently, the implementation of collective gates with a quantum bus has been proposed in superconducting circuits [23].

In Fig. 7, we plot a numerical simulation of the extended Ising model with higher-order terms and transverse field for several Trotter steps. The figure shows as in the previous examples how the simulated dynamics with digital methods becomes more accurate when compared with the exact one when the number of Trotter steps is increased.

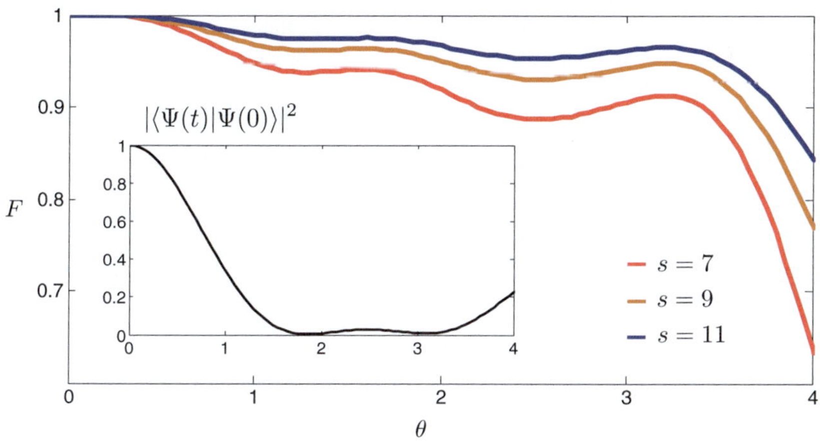

Fig. 7 Digital simulation of the extended Ising model with a transverse field and three-body terms for four spins in superconducting circuits, where $J = G = -2\pi \times 400$ MHz and $B = 2\pi \times 200$ MHz, for a phase of $\theta \equiv -Jt = 4$. The plot shows the fidelity of the digitally evolved state with the ideally evolved one for different Trotter steps, $s = 7, 9, 11$. The *inset* shows the overlap between the ideally evolved state with the initial state, that is, all qubits in $|0\rangle_z$ state

4 Conclusions

In this article, we have proposed a digital quantum simulation of spin chains coupled to bosonic modes by means of circuit quantum electrodynamics architectures. We have presented a method for decomposing spin interactions and implementing them stroboscopically with available single and two-qubit gates. Furthermore, we have considered both circuit QED setups implementing capacitive couplings between superconducting qubits and transmission line resonators acting as quantum buses. We have exemplified our method with the quantum simulation of the Ising model with transverse field, a spin chain coupled to a bosonic field mode, and a many-body spin model with three-body terms, which are realized through a bosonic quantum bus. These results show that spin chains coupled to bosonic field modes can be implemented efficiently with superconducting qubits.

We thank Rami Barends for useful discussions and acknowledge support from Spanish MINECO FIS2012-36673-C03-02; Ramón y Cajal Grant RYC-2012-11391; UPV/EHU Project EHUA14/04, and two UPV/EHU PhD grants, Basque Government IT472-10; PROMISCE, and SCALEQIT EU projects.

References

1. Babbush R., Love, P. J., Aspuru-Guzik, A., Adiabatic Quantum Simulation of Quantum Chemistry, Sci. Rep. **4**, 6603 (2014)
2. Ballester, D., Romero, G., García-Ripoll, J.J., Deppe, F., Solano, E.: Quantum simulation of the ultrastrong-coupling dynamics in circuit quantum electrodynamics. Phys. Rev. X **2**, 021007 (2012)
3. Barends, R., et al.: Superconducting quantum circuits at the surface code threshold for fault tolerance. Nature **508**, 500 (2014)
4. Barends, R., et al., Digital quantum simulation of fermionic models with a superconducting circuit, Nature Commun. **6**, 7654 (2015)
5. Bloch, I., Dalibard, J., Nascimbene, S.: Quantum simulations with ultracold quantum gases. Nat. Phys. **8**, 267 (2012)
6. Casanova, J., Mezzacapo, A., Lamata, L., Solano, E.: Quantum simulation of interacting fermion lattice models in trapped ions. Phys. Rev. Lett. **108**, 190502 (2012)
7. Casanova, J., Romero, G., Lizuain, I., García-Ripoll, J.J., Solano, E.: Deep strong coupling regime of the Jaynes-Cummings model. Phys. Rev. Lett. **105**, 263603 (2010)
8. Chiesa, A., Santini, P., Gerace, D., Raftery, J., Houck, A. A., Carretta, S., Digital quantum simulators in a scalable architecture of hybrid spin-photon qubits, arXiv:1504.05667 (2015)
9. Devoret, M.H., Schoelkopf, R.J.: Superconducting circuits for quantum information: an outlook. Science **339**, 1169 (2013)
10. Feynman, R.P.: Simulating physics with computers. Int. J. Theor. Phys. **21**, 467 (1982)
11. García-Álvarez, L., Casanova, J., Mezzacapo, A., Egusquiza, I.L., Lamata, L., Romero, G., Solano, E.: Fermion-Fermion scattering in quantum field theory with superconducting circuits. Phys. Rev. Lett. **114**, 070502 (2015)
12. Geller, M. R., Martinis, J. M., Sornborger, A. T., Stancil, P. C., Pritchett, E. J., Galiautdinov, A., Universal quantum simulation with pre-threshold superconducting qubits: Single-excitation subspace method, arXiv:1210.5260 (2012)
13. Georgescu, I.M., Ashhab, S., Nori, F.: Quantum simulation. Rev. Mod. Phys. **86**, 153 (2014)
14. Koch, J., Houck, A.A., Hur, K.L., Girvin, S.M.: Time-reversal-symmetry breaking in circuit-QED-based photon lattices. Phys. Rev. A **82**, 043811 (2010)

15. Lanyon, B.P., Hempel, C., Nigg, D., Müller, M., Gerritsma, R., Zähringer, F., Schindler, P., Barreiro, J.T., Rambach, M., Kirchmair, G., Hennrich, M., Zoller, P., Blatt, R., Roos, C.F.: Universal digital quantum simulation with trapped ions. Science **334**, 57 (2011)
16. Lanyon, B.P., Whitfield, J.D., Gillet, G.G., Goggin, M.E., Almeida, M.P., Kassal, I., Biamonte, J.D., Mohseni, M., Powell, B.J., Barbieri, M., Aspuru-Guzik, A., White, A.G.: Towards quantum chemistry on a quantum computer. Nat. Chem. **2**, 106 (2009)
17. Las Heras, U., García-Álvarez, L., Mezzacapo, A., Solano, E., Lamata, L.: Fermionic models with superconducting circuits. EPJ Quant. Technol. **2**, 8 (2015)
18. Las Heras, U., Mezzacapo, A., Lamata, A., Filipp, S., Wallraff, A., Solano, E.: Digital quantum simulation of spin systems in superconducting circuits. Phys. Rev. Lett. **112**, 200–501 (2014)
19. Lewenstein, M., Sanpera, A., Ahufinger, V.: Ultracold Atoms in Optical Lattices. Oxford University Press, USA (2012)
20. Lloyd, S.: Universal quantum simulators. Science **273**, 1073 (1996)
21. Mei, F., Stojanovic, V. M., Siddiqi, I., Tian, L., Analog superconducting quantum simulator for Holstein polarons, Phys. Rev. B **88**, 224502 (2013)
22. Mezzacapo, A., Casanova, J., Lamata, L., Solano, E.: Digital quantum simulation of the Holstein model in trapped ions. Phys. Rev. Lett. **109**, 200501 (2012)
23. Mezzacapo, A., Lamata, L., Filipp, S., Solano, E.: Many-body interactions with tunable-coupling transmon qubits. Phys. Rev. Lett. **113**, 050501 (2014)
24. Mezzacapo, A., Las Heras, U., Pedernales, J.S., DiCarlo, L., Solano, E., Lamata, L.: Digital quantum Rabi and Dicke models in superconducting circuits. Sci. Rep. **4**, 7482 (2014)
25. Mostame, S., et al., Towards Outperforming Classical Algorithms with Analog Quantum Simulators, arXiv:1502.00962 (2015)
26. Niemczyk, T., Deppe, F., Huebl, H., Menzel, E.P., Hocke, F., Schwarz, M.J., García-Ripoll, J.J., Zueco, D., Hümmer, T., Solano, E., Marx, A., Gross, R.: Circuit quantum electrodynamics in the ultrastrong-coupling regime. Nat. Phys. **6**, 772 (2010)
27. Pedernales, J.S., Di Candia, R., Ballester, D., Solano, E.: Quantum simulations of relativistic quantum physics in circuit QED. New. J. Phys. **15**, 055008 (2013)
28. Poulin, D., Hastings, M. B., Wecker, D., Wiebe, N., Doherty, A. C., Troyer, M., The Trotter Step Size Required for Accurate Quantum Simulation of Quantum Chemistry, arXiv:1406.4920 (2014)
29. Rabi, I.I.: On the process of space quantization. Phys. Rev. **49**, 324 (1936)
30. Rotondo, P., Cosentino Lagomarsino, M., Viola, G.: Dicke simulators with emergent collective quantum computational abilities, Phys. Rev. Lett. **114**, 143601 (2015)
31. Salathé. Y., et al., Digital Quantum Simulation of Spin Models with Circuit Quantum Electrodynamics, Phys. Rev. **X 5**, 021027 (2015)
32. Seo, K., Tian, L., Quantum phase transition in a multi-connected superconducting Jaynes-Cummings lattice, Phys. Rev. B **91**, 195439 (2015)
33. Suzuki, M.: Fractal decomposition of exponential operators with applications to many-body theories and Monte Carlo simulations. Phys. Lett. A **146**, 319 (1990)
34. Tavis, M., Cummings, F.W.: Exact solution for an N-moleculeradiation-field Hamiltonian. Phys. Rev. **170**, 379 (1968)
35. van Oudenaarden, A., Mooij, J.E.: One-dimensional mott insulator formed by quantum vortices in Josephson junction arrays. Phys. Rev. Lett. **76**, 4947 (1996)
36. Viehmann, O., von Delft, J., Marquardt, F.: Observing the nonequilibrium dynamics of the quantum transverse-field Ising chain in circuit QED. Phys. Rev. Lett. **110**, 030601 (2013)
37. Wallraff, A., Schuster, D.I., Blais, A., Frunzio, L., Huang, R.-S., Majer, J., Kumar, S., Girvin, S.M., Schoelkopf, R.J.: Strong coupling of a single photon to a superconducting qubit using circuit quantum electrodynamics. Nature **431**, 162 (2004)
38. You, J.Q., Shi, X.-F., Hu, X., Nori, F.: Quantum emulation of a spin system with topologically protected ground states using superconducting quantum circuits. Phys. Rev. B **81**, 014505 (2010)
39. Yung, M.-H., Casanova, J., Mezzacapo, A., McClean, J., Lamata, L., Aspuru-Guzik, A., Solano, E.: From transistor to trapped-ion computers for quantum chemistry. Sci. Rep. **4**, 3589 (2014)

Age Structures in Mathematical Models for Infectious Diseases, with a Case Study of Respiratory Syncytial Virus

Alexandra B. Hogan, Kathryn Glass, Hannah C. Moore and Robert S. Anderssen

Abstract Mathematical modelling plays an important role in understanding the dynamics of transmissible infections, as information about the drivers of infectious disease outbreaks can help inform health care planning and interventions. This paper provides some background about the mathematics of infectious disease modelling. Using a common childhood infection as a case study, age structures in compartmental differential equation models are explored. The qualitative characteristics of the numerical results for different models are discussed, and the benefits of incorporating age structures in these models are examined. This research demonstrates that, for the SIR-type model considered, the inclusion of age structures does not change the overall qualitative dynamics predicted by that model. Focussing on only a single age class then simplifies model analysis. However, age differentiation remains useful for simulating age-dependent intervention strategies such as vaccination.

Keywords Respiratory syncytial virus · Infectious disease · Mathematical modelling · Bifurcation analysis · Seasonality

---

The original version of this chapter was revised: The erratum to this chapter is available at DOI 10.1007/978-4-431-55342-7_23.

A.B. Hogan (✉) · K. Glass
The Australian National University, Canberra, Australia
e-mail: alexandra.hogan@anu.edu.au

K. Glass
e-mail: kathryn.glass@anu.edu.au

H.C. Moore
The University of Western Australia, Perth, Australia
e-mail: hannah.moore@telethonkids.org.au

R.S. Anderssen
The Commonwealth Scientific and Industrial Research Organisation (CSIRO), Acton, Australia
e-mail: bob.anderssen@csiro.au

R.S. Anderssen et al. (eds.), *Applications + Practical Conceptualization + Mathematics = fruitful Innovation*, Mathematics for Industry 11,
DOI 10.1007/978-4-431-55342-7_9

1 The Purpose and Applications of Infectious Disease models

In epidemiology, mathematical modelling is used to understand the dynamics of transmissible infections. This knowledge has important implications for health care planning and disease interventions. Mathematical models are tools that can help predict the impact of various control strategies on patterns of infection; can assist with understanding the groups of a population that are driving the transmission; and can allow control strategies to be tested and subsequently implemented effectively. For example, mathematical modelling techniques have been widely used in influenza pandemic planning [15, 16], in helping identify the drivers of seasonality in pre-vaccination measles epidemics [17, 18] and in understanding the dynamics of pertussis (whooping cough) outbreaks in young children [31].

This paper provides an overview of the mathematics of infectious disease modelling in the context of a common childhood respiratory infection. In particular, this work examines the extent to which age structures should be incorporated into the modelling, and shows how seasonality and waning immunity can be implemented in relatively simple compartmental models.

2 Introduction to the Mathematics of Infectious Disease Models

Mathematical models applied to infectious disease dynamics typically use deterministic, stochastic, or time series approaches. This paper will focus on a specific type of deterministic, ordinary differential equation model, employing the Susceptible-Infectious-Recovered (SIR) model approach. This model was first introduced by Kermack and McKendrick in 1927 [26] and has since been widely applied to model infectious disease dynamics.

The approach of the SIR framework is to divide a specified population into different compartments that correspond to the states of an infection. These compartments describe whether individuals are susceptible to an infection, infectious, or recovered. The basic SIR model is presented at Eqs. 1–4. Assuming a homogeneous, well-mixed population, S represents the number of individuals in a defined population who are susceptible, while I represents the number of individuals who are infectious and able to infect susceptible individuals. The class R represents the number of individuals who are 'removed' (recovered and immune to reinfection).

Demography is represented by the inclusion of a birth rate μ, which corresponds to an average life expectancy of $1/\mu$ years. In this example, the birth rate is assumed

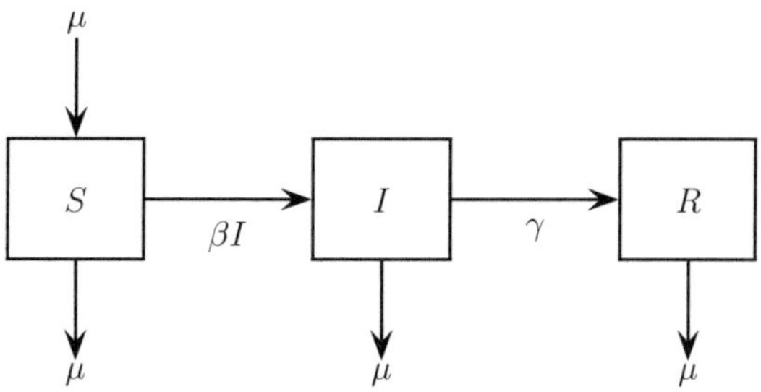

Fig. 1 Schematic diagram for the SIR deterministic ordinary differential equation model

to equal the death rate, such that the total population, represented by N, remains constant over time t. Such an assumption is suitable for infectious diseases where the infection life cycle is relatively short compared to the average individual lifespan and the death rate due to the disease is negligible. The recovery rate is represented by γ, where $1/\gamma$ is the average duration of infection in years. A schematic representation of this structure is shown in Fig. 1. The differential equation model corresponding to such dynamics is given by

$$\frac{dS}{dt} = \mu N - \beta S \frac{I}{N} - \mu S \tag{1}$$

$$\frac{dI}{dt} = \beta S \frac{I}{N} - \gamma I - \mu I \tag{2}$$

$$\frac{dR}{dt} = \gamma I - \mu R \tag{3}$$

$$N = S + I + R. \tag{4}$$

Though the SIR model is one of the simplest forms of a suite of compartmental models, it encapsulates the essence of many infectious disease situations with sufficient accuracy to be able to make useful predictions. Additional states may be included (such as a temporarily immune class) and complexities added (such as age structures). The SIR model and its variations are described extensively elsewhere [4, 11, 23, 25].

3 Respiratory Syncytial Virus

Respiratory syncytial virus (RSV) causes respiratory tract infections in young children. It is the most common pathogen found in children aged less than two years hospitalised with respiratory symptoms and studies indicate that almost all children will have been infected by the time they reach two years [19, 28, 32]. Because of the significant health care and economic burden of RSV (discussed for example in [12, 20, 38]), an improved understanding of its transmission dynamics is required to assist with health care planning. However, because the dynamics of RSV infection are poorly understood, there remains a need for representative models that are validated by the available data.

RSV dynamics have a clear age structure. RSV incidence is higher for children under 12 months than those between 12 and 24 months [29]. Peak incidence is observed in children between two and four months [32]. Newborn infants are typically protected from RSV infection by maternal antibodies until about six weeks of age (although infection can still occur in this early phase of life) [9, 13]. RSV infection data is usually collected from hospitalised cases, hence the cases observed are for severe infection only. However, the dynamics observed at the severe end of the disease spectrum may be representative of the dynamics in the broader community.

Few studies have been undertaken to examine the transmission of RSV among adults, but it is thought that repeated infection can occur throughout life [9, 22], and that in older children and adults, RSV symptoms present as those of a common cold [19]. Several studies have reported on outbreaks of RSV in aged care facilities and estimated the mortality caused by RSV in these older age groups [21, 35].

An important feature of RSV, in terms of understanding its transmission patterns and burden, is its seasonal behaviour. In temperate climates, RSV typically displays annual seasonal patterns, with high numbers of infections in winter and relatively low numbers in the summer months. In some temperate regions, biennial patterns of RSV infection have been detected. Such dynamics have been observed in Switzerland [14], Finland [36], Chile [7] and Australia [29].

Finally, immunity to RSV following recovery from infection is thought to be short-lived, averaging around 200 days [37]. Consequently, children can be infected in consecutive years.

Mathematical models of RSV must therefore take account of different patterns of severe illness with age, of seasonality in disease transmission, and must allow for the waning of disease immunity following recovery from infection.

4 An Age Structured Modelling Approach to RSV

Several models for RSV that implement the SIR approach have been published [10, 30, 37, 39, 40]. A time series approach has been examined by Spaeder et al. [33], and a network approach by Acedo et al. [2, 3]. Stochastic methods have been investigated by Arenas et al. [5].

In the work of Leecaster et al., an age-structured compartmental approach is used to distinguish between children less than two years old, and adults, with an additional 'Detected' class [27]. Acedo et al. divide the population into children under one year of age, and the remaining population, in order to model a vaccination strategy for RSV [1]. In recent work by Moore et al. [29], age structuring is used to fit a compartmental model to RSV detection data for children up to two years of age in Perth, Western Australia. The present paper builds upon the work presented published by Moore et al. [29], using parameters relevant to RSV dynamics in Western Australia.

In the following sections, compartmental models for RSV are presented. These models take into account the known clinical characteristics and epidemiological features of RSV, such as waning immunity, a latent period, and seasonal changes in the degree of transmission. The age-structured modelling approach is implemented for one, two and three age groups, in order to capture the transmission and susceptibility characteristics of different age groups. Numerical solutions are found and the resulting qualitative characteristics of the dynamics are discussed.

4.1 A Single Age Class Model for RSV

The simplest model for RSV transmission is that of a single age class. In this situation the age group was chosen to be the combined child and adult population, with no differentiation between age groups, and with the birth rate equal to the death rate. A latent disease class, represented by E, is included to reflect the state where an individual is infected with RSV, but not yet infectious. The disease states are presented as proportions of a population, such that $S + E + I + R = N = 1$. This model was first presented in [24] and the relevant equations are reproduced here as Eqs. 5–9.

$$\frac{dS}{dt} = \mu - \beta SI + \nu R - \mu S \tag{5}$$

$$\frac{dE}{dt} = \beta SI - \delta E - \mu E \tag{6}$$

$$\frac{dI}{dt} = \delta E - \gamma I - \mu I \tag{7}$$

$$\frac{dR}{dt} = \gamma I - \nu R - \mu R \tag{8}$$

$$\beta = b[1 + a\sin(2\pi t)] \tag{9}$$

To incorporate the effect of seasonal fluctuations in the number of infected cases, the transmission parameter β is replaced with a sinusoidal forcing function, shown at Eq. 9. The birth rate μ was chosen based on birth and population data from Western Australia [6] and corresponds to a life expectancy of 74 years. The infectious rate δ is based on previous studies and corresponds to a latent period of four days [37]. Similarly, the recovery rate γ is based on previous studies in the literature and corresponds to an infectious period of nine days [2, 27, 37].

The waning immunity parameter ν is less well understood and is therefore chosen by fitting models to data from Western Australia, as demonstrated in [29], and corresponds to an immunity period of 230 days. The amplitude of seasonal forcing a was selected based on the same fitting routine, and the parameter b was allowed to vary. Parameter definitions and values are summarised at Table 1.

Table 1 Parameter values for the compartmental models are estimated from the literature, from population data, and from fitting the two age class model to weekly detection data for metropolitan Western Australia as demonstrated in [29]

Parameter	Description	Value	References
μ	Birth rate for Perth, Western Australia	0.0135	[6]
b	Overall transmission		
a	Amplitude of seasonal forcing	0.522	Fitted value, as in [29]
δ	Infectious rate	91.479	[37]
γ	Recovery rate	40.110	[2, 27, 37]
ν	Waning immunity rate	1.585	Fitted value, as in [29]
η_i	Ageing rate	$\eta_1 = \eta_2 = 1$, $\eta_3 = 0.0139$	
σ_i	Transmission scaling factor	$\sigma_2 = 1$, $\sigma_3 = 0.6$	
α_i	Susceptibility scaling factor	$\alpha_2 = 0.228$, $\alpha_3 = 0.6$	Fitted value, as in [29]

The rates are given in years

4.2 Multiple Age Class Models for RSV

There are two main approaches employed in the literature to simulate the ageing process in models for disease transmission. One is the continuous approach, where each compartment in the model is assumed to be a function of both age and time (see [25] for a concise explanation of this method), and the model can be represented as a system of partial differential equations. While realistic, this approach is complicated and the equations are more difficult to solve numerically.

A simpler approach is to treat age groups as compartments in the model, replicating the susceptible, infectious and removed states for each age class. While increasing the number of ordinary differential equations, the system remains straightforward to solve numerically. In this paper, we concentrate only on this second approach to age structures. Detailed examples of where the continuous approach has been used can be found elsewhere [8, 34, 41], although not for RSV.

The multiple age class compartmental model for three age classes is shown in Eqs. 10–17. The age classes are children up to 12 months of age; children aged between 12 and 24 months; and the remaining population. The model may also be adjusted for two age classes only. The seasonal forcing term for each age class is shown in Eqs. 18–19. In the following model, the youngest age class (denoted '1') includes the birth term, with additional classes (denoted 'i') representing older age groups.

$$\frac{dS_1}{dt} = \mu - \beta_1 S_1(I_1 + \sum_{i=2}^{3} \sigma_i I_i) + \nu R_1 - \eta_1 S_1 \tag{10}$$

$$\frac{dE_1}{dt} = \beta_1 S_1(I_1 + \sum_{i=2}^{3} \sigma_i I_i) - \delta E_1 - \eta_1 E_1 \tag{11}$$

$$\frac{dI_1}{dt} = \delta E_1 - \gamma I_1 - \eta_1 I_1 \tag{12}$$

$$\frac{dR_1}{dt} = \gamma I_1 - \nu R_1 - \eta_1 R_1 \tag{13}$$

$$\frac{dS_i}{dt} = \eta_{i-1} S_{i-1} - \beta_i S_i(I_1 + \sum_{i=2}^{3} \sigma I_i) + \nu R_i - \eta_i S_i \tag{14}$$

$$\frac{dE_i}{dt} = \eta_{i-1} E_{i-1} + \beta_i S_i(I_1 + \sum_{i=2}^{3} \sigma I_i) - \delta E_i - \eta_i E_i \tag{15}$$

$$\frac{dI_i}{dt} = \eta_{i-1} I_{i-1} + \delta E_i - \gamma I_i - \eta_i I_i \tag{16}$$

$$\frac{dR_i}{dt} = \eta_{i-1} R_{i-1} + \gamma I_i - \nu R_i - \eta_i R_i \tag{17}$$

$$\beta_1 = b[1 + a \sin(2\pi t)] \tag{18}$$

$$\beta_i = \alpha_i \beta_{i-1} \tag{19}$$

$$i = 2, 3$$

In comparison to the single age class model of Eqs. 5–9, the additional parameters in the model with multiple age classes are σ_i, which represents the reduced transmissability for age class i, and α_i which represents the reduced susceptibility in age group i. The extent to which transmission is reduced for older age classes is not well understood. Hence, transmission was not scaled for the 12–24 month old age class, but was selected as 0.6 for the older age group as in [29]. Similarly for the susceptibility scaling parameter α, the value for reduced susceptibility was selected for the second age group based on Western Australian data, and chosen to be 0.6 for the older age group. The parameter η_i represents the rate of ageing out of age group i, where $1/\eta_i$ is the time spent in age group i. Infection-specific parameters are the same as those for the single age class model. The parameter values for the chosen age structure are presented at Table 1.

4.3 Numerical Solutions

The compartmental ordinary differential equation systems for one, two and three age classes, shown in Eqs. 5–19, were solved numerically using MATLAB's inbuilt *ode45* integrator. The values of the fixed parameters are given in Table 1.

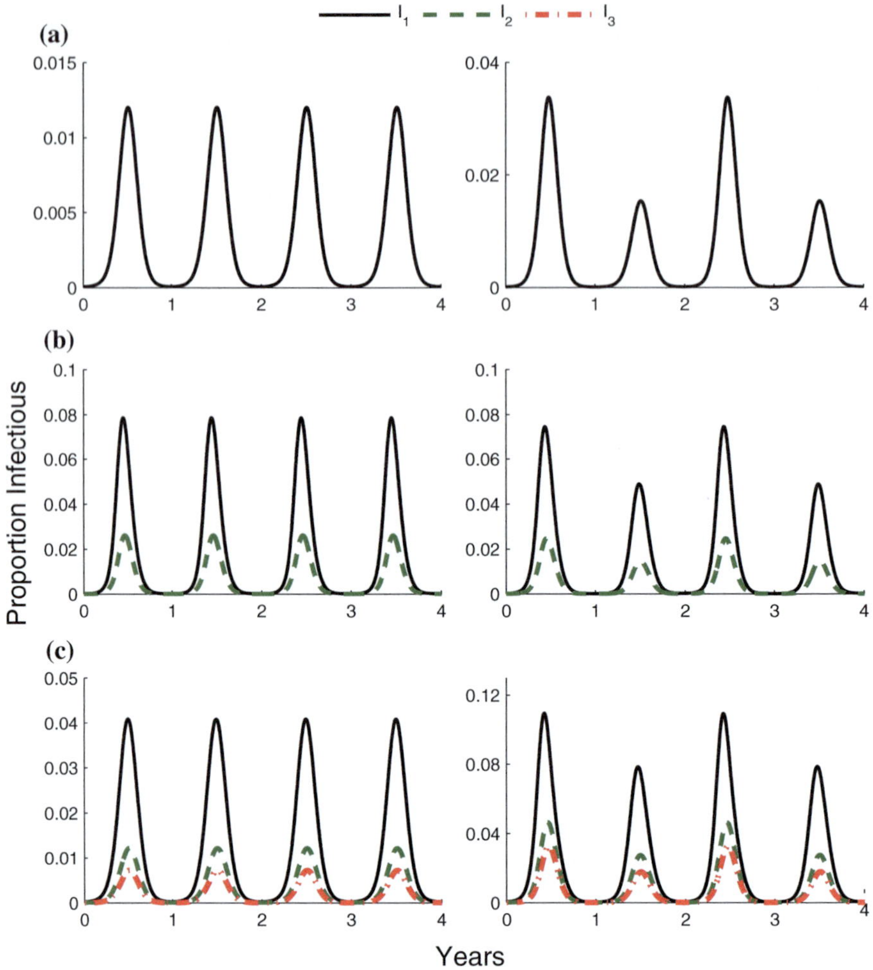

Fig. 2 This figure shows two numerical solutions for compartmental RSV models, for each of **a** one, **b** two and **c** three age classes. For each model, the numerical output demonstrates that either an annual or biennial pattern may be produced, depending on the value of the transmission parameter b. The values of b are as follows: **a** 45, 49; **b** 3400, 3200; **c** 460, 530. Other parameter values are provided in Table 1

The transmission parameter b was chosen to vary, in order to demonstrate different numerical solutions. The range of possible b values that produced plausible numerical solutions varied depending on the model age structure. This is a consequence of how the models were formulated. The model compartments (S, E, I, R), were assumed to be proportions of the chosen population, rather than numbers of individuals. For each age class, the compartments in that age classes summed to 1 at $t = 0$, and the total population did not remain constant as the birth rate and the 'ageing out' or death rate were not equal for all model structures; this changed the degree of transmission required to sustain annual epidemics. In practice, exact values of the transmission parameter b will vary according to the data the model is fitted to.

Depending on the value of b chosen, and holding other parameters constant, the model solutions produced either annual or biennial patterns. Examples of solutions for different values of the transmission parameter b are shown in Fig. 2. In this figure, the values of b were selected to be within a range that produced plausible solutions and so as to demonstrate markedly different dynamics.

In order to more clearly show the range of b values that produce either annual or biennial patterns of infection, a numerical bifurcation analysis was undertaken using XPP-AUT software. The analysis was conducted for each of the compartmental models, for one, two and three age classes, with bifurcation parameter b. The output is shown in Fig. 3. For each plot, the y-axis is the proportion of infectious individuals I at the seasonal peak, for the youngest age class. We can observe how the infectious peak changes as the value of transmission parameter increases, and whether the model dynamics are annual or biennial.

Figure 3 shows that for each model, there exists a region of solutions with biennial dynamics contained by two period doubling bifurcations. Either side of this region the solutions revert to annual seasonal dynamics. This analysis demonstrates the similar qualitative dynamics for the one, two and three age class models for RSV. A more

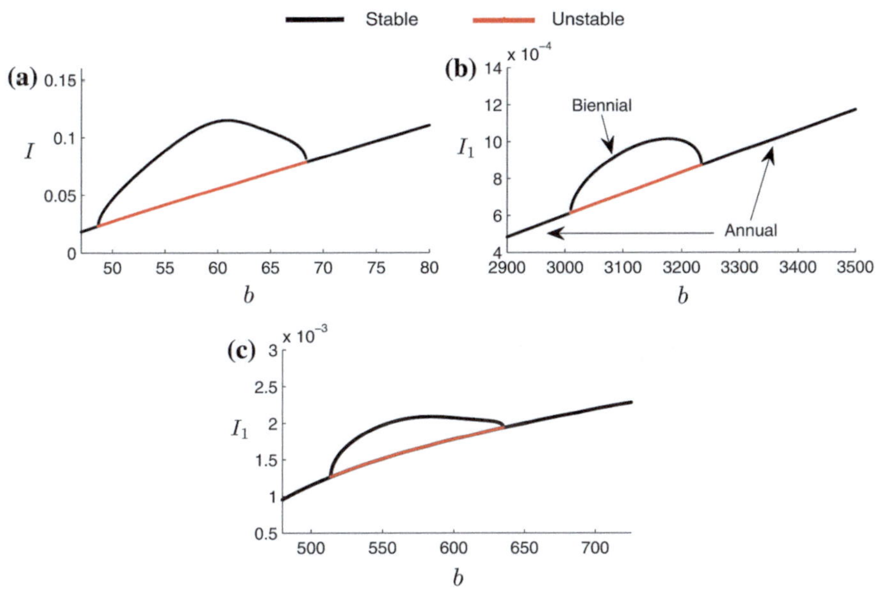

Fig. 3 This figure shows a bifurcation diagram for each of the compartmental RSV models, for **a** one, **b** two and **c** three age classes. The bifurcation parameter is the overall transmission b. For each model, the qualitative behaviour is similar, where there is a region of values of b that produces period two (biennial) solutions, and a region either side that produces period one (annual) solutions

detailed bifurcation analysis (for example, exploring other bifurcation parameters) is outside the scope of the present paper, but will be explored in a forthcoming publication.

5 Discussion

Mathematical modelling is an important tool for understanding the patterns of infectious disease transmission, and one use of these models is for simulation of disease-specific intervention strategies. Infectious disease interventions are often targeted for different age groups, particularly for children. The ability to implement age-structured modelling approaches is useful for studying the theoretical outcome of a vaccination strategy that is concentrated on specific age groups.

The purpose of this paper is to provide an overview of infectious disease modelling, with a focus on a common childhood respiratory infection. A series of ordinary differential equation models for RSV are presented, incorporating the known clinical characteristics of the infection. Age structures for one, two and three age classes are implemented using the compartmental approach, in order to examine the dynamics of the numerical solutions.

It is found that for the models considered, the qualitative dynamics of the numerical solutions are either annual or biennial, depending on the degree of transmission. Further, the dynamics of the solutions are qualitatively similar for the one, two and three age class systems. Despite adding the complexities of additional age classes, the overall patterns of disease transmission were the same, with a higher transmission rate in younger age groups producing a higher proportion of infectious cases.

This finding has useful implications for studying the dynamics of infectious disease models. It suggests that where the overall behaviour of a model is being investigated, and different possible patterns of disease transmission being considered, the age structuring may be ignored and the simplest single age class model considered. As the number of differential equations is reduced, numerical analyses (such as phase plane and bifurcation) are much simpler and often more readily interpreted. Age structures may of course be included at a later stage of model development, but considering only the single age class (here the younger population only) greatly simplifies analytical and numerical testing in situations where interventions are not being studied.

References

1. Acedo, L., Díez-Domingo, J., Moraño, J.-A., Villanueva, R.-J.: Mathematical modelling of respiratory syncytial virus (RSV): vaccination strategies and budget applications. Epidemiol. Infect. **138**(6), 853–60 (2010)
2. Acedo, L., Moraño, J.-A., Díez-Domingo, J.: Cost analysis of a vaccination strategy for respiratory syncytial virus (RSV) in a network model. Math. Comp. Model. **52**(7–8), 1016–1022 (2010)

3. Acedo, L., Moraño, J.-A., Villanueva, R.-J., Villanueva-Oller, J., Díez-Domingo, J.: Using random networks to study the dynamics of respiratory syncytial virus (RSV) in the Spanish region of Valencia. Math. Comp. Model **54**(7–8), 1650–1654 (2011)
4. Anderson, R.M., May, R.M.: Infectious Diseases of Humans: Dynamics and Control. Oxford University Press, Oxford (1991)
5. Arenas, A.J., González-Parra, G., Moraño, J.-A.: Stochastic modeling of the transmission of respiratory syncytial virus (RSV) in the region of Valencia, Spain. BioSystems **96**(3), 206–212 (2009)
6. Australian Bureau of Statistics.: Births, Australia, 2009, Table 2 Births, Summary, Statistical Divisions 2004 to 2009, time series spreadsheet, cat. no. 3301.0, viewed 20 September 2014. http://www.abs.gov.au/AUSSTATS/abs
7. Avendaño, L.F., Palomino, M.A., Larrañaga, C.: Surveillance for respiratory syncytial virus in infants hospitalized for acute lower respiratory infection in Chile (1989 to 2000). J. Clin. Microbiol. **41**(10), 4879–4882 (2003)
8. Busenberg, S., Cooke, K., Iannelli, M.: Endemic thresholds and stability in a class of age-structured epidemics. SIAM J. Appl. Math. **48**(6), 1379–1395 (1988)
9. Cane, P.A.: Molecular epidemiology of respiratory syncytial virus. Rev. Med. Virol. **11**(2), 103–16 (2001)
10. Capistrán, M., Moreles, M., Lara, B.: Parameter estimation of some epidemic models. The case of recurrent epidemics caused by respiratory syncytial virus. Bull. Math. Biol. **71**, 1890–1901 (2009)
11. Diekmann, O., Heesterbeek, H., Britton, T.: Mathematical Tools for Understanding Infectious Disease Dynamics. Princeton University Press, Princeton (2013)
12. Díez-Domingo, J., Pérez-Yarza, E.G., Melero, J.A., Sánchez-Luna, M., Aguilar, M.D., Blasco, A.J., Alfaro, N., Lázaro, P.: Social, economic, and health impact of the respiratory syncytial virus: a systematic search. BMC Infect. Dis. **14**(1), 544 (2014)
13. Domachowske, J.B., Rosenberg, H.F.: Respiratory syncytial virus infection: immune response, immunopathogenesis, and treatment. Clin. Microbiol. Rev. **12**(2), 298–309 (1999)
14. Duppenthaler, A., Gorgievski-Hrisoho, M., Frey, U., Aebi, C.: Two-year periodicity of respiratory syncytial virus epidemics in Switzerland. Infection **31**(11–12), 75–80 (2003)
15. Ferguson, N.M., Cummings, D.A.T., Cauchemez, S., Fraser, C., Riley, S., Meeyai, A., Iamsirithaworn, S., Burke, D.S.: Strategies for containing an emerging influenza pandemic in Southeast Asia. Nature **437**(7056), 209–214 (2005)
16. Germann, T.C., Kadau, K., Longini, I.M., Macken, C.A.: Mitigation strategies for pandemic influenza in the United States. Proc. Natl. Acad. Sci. USA **103**(15), 5935–5940 (2006)
17. Grenfell, B.T., Bjørnstad, O.N., Kappey, J.: Travelling waves and spatial hierarchies in measles epidemics. J. Nat. **414**, 716–723 (2001)
18. Grenfell, B.T., Bolker, B.M.: Population dynamics of measles. In: Scott, M.E., Smith, G. (eds.) Parasitic and Infectious Diseases: Epidemiology and Control, pp. 219–234. Academic Press, Orlando (1994)
19. Hall, C.B.: Respiratory syncytial virus. In: Feigin, R.D., Cherry, J.D. (eds.) Textbook of Paediatric Infectious Diseases, vol. II, pp. 1247–1267. W.B. Saunders Company, Philadelphia (1981)
20. Hall, C.B., Weinberg, G.A., Iwane, M.K., Blumkin, A.K., Edwards, K.M., Staat, M.A., Auinger, P., Griffin, M.R., Poehling, K.A., Erdman, D., Grijalva, C.G., Zhu, Y., Szilagyi, P.: The burden of respiratory syncytial virus infection in young children. N. Engl. J. Med. **360**(6), 588–598 (2009)
21. Hardelid, P., Pebody, R., Andrews, N.: Mortality caused by influenza and respiratory syncytial virus by age group in England and Wales 1999–2010. Influenza Other Respir. Viruses **7**(1), 35–45 (2013)
22. Henderson, F.W., Collier, A.M., Clyde Jr, W.A., Denny, F.W.: Respiratory-syncytial-virus infections, reinfection and immunity: a prospective, longitudinal study in young children. N. Engl. J. Med. **300**(10), 530–534 (1979)
23. Hethcote, H.W.: The mathematics of infectious diseases. SIAM Rev. **42**(4), 599–653 (2007)

24. Hogan, A.B., Mercer, G.N., Glass, K., Moore, H.C.: Modelling the seasonality of respiratory syncytial virus in young children. In: 20th International Congress on Modelling and Simulation, vol. 9, pp. 338–344. Adelaide, Australia (2013)
25. Keeling, M.J., Rohani, P.: Modeling Infectious Diseases in Humans and Animals. Princeton University Press, Princeton (2008)
26. Kermack, W.O., McKendrick, A.G.: A contribution to the mathematical theory of epidemics. Proc. R. Soc. Lond. A. **115**, 700–721 (1927)
27. Leecaster, M., Gesteland, P., Greene, T., Walton, N., Gundlapalli, A., Rolfs, R., Byington, C., Samore, M.: Modeling the variations in pediatric respiratory syncytial virus seasonal epidemics. BMC Infect. Dis. **11**(1), 105 (2011)
28. Moore, H.C., de Klerk, N., Keil, A.D., Smith, D.W., Blyth, C.C., Richmond, P., Lehmann, D.: Use of data linkage to investigate the aetiology of acute lower respiratory infection hospitalisations in children. J. Paediatr. Child Health **48**(6), 520–528 (2012)
29. Moore, H.C., Jacoby, P., Hogan, A.B., Blyth, C.C., Mercer, G.N: Modelling the seasonal epidemics of respiratory syncytial virus in young children. PLoS ONE **9**(6), e100422 (2014)
30. Paynter, S., Yakob, L., Simões, E.A.F., Lucero, M.G., Tallo, V., Nohynek, H., Ware, R.S., Weinstein, P., Williams, G., Sly, P.D.: Using mathematical transmission modelling to investigate drivers of respiratory syncytial virus seasonality in children in the Philippines. PLoS ONE **9**(2), e90094 (2014)
31. Rohani, P., Zhong, X., King, A.A.: Contact network structure explains the changing epidemiology of pertussis. Science **330**, 982–985 (2010)
32. Sorce, L.R.: Respiratory syncytial virus: from primary care to critical care. J. Pediatr. Health Care **23**(2), 101–108 (2009)
33. Spaeder, M.C., Fackler, J.C.: A multi-tiered time-series modelling approach to forecasting respiratory syncytial virus incidence at the local level. Epidemiol. Infect. **140**(4), 602–607 (2012)
34. Tudor, D.: An age-dependent epidemic model with application to measles. Math. Biosci. **147**, 131–147 (1985)
35. van Asten, L., van den Wijngaard, C., van Pelt, W., van de Kassteele, J., Meijer, A., van der Hoek, W., Kretzschmar, M., Koopmans, M.: Mortality attributable to 9 common infections: significant effect of influenza A, respiratory syncytial virus, influenza B, norovirus, and parainfluenza in elderly persons. J. Infect. Dis. **206**(5), 628–639 (2012)
36. Waris, M.: Pattern of respiratory syncytial virus epidemics in Finland: two-year cycles with alternating prevalence of groups A and B. J. Infect. Dis. **163**(3), 464–469 (1991)
37. Weber, A., Weber, M., Milligan, P.: Modeling epidemics caused by respiratory syncytial virus (RSV). Math. Biosci. **172**(2), 95–113 (2001)
38. Welliver, R.C.: Review of epidemiology and clinical risk factors for severe respiratory syncytial virus (RSV) infection. J. Pediatr. **143**(5 Suppl), S112–S117 (2003)
39. White, L.J., Mandl, J.N., Gomes, M.G.M., Bodley-Tickell, A.T., Cane, P.A., Perez-Brena, P., Aguilar, J.C., Siqueira, M.M., Portes, S.A., Straliotto, S.M., Waris, M., Nokes, D.J., Medley, G.F.: Understanding the transmission dynamics of respiratory syncytial virus using multiple time series and nested models. Math. Biosci. **209**(1), 222–239 (2007)
40. White, L.J., Waris, M., Cane, P.A., Nokes, D.J., Medley, G.F.: The transmission dynamics of groups A and B human respiratory syncytial virus (hRSV) in England & Wales and Finland: seasonality and cross-protection. Epidemiol. Infect. **133**(2), 279–289 (2005)
41. Zhu, G.: Threshold and stability results for an age-structured epidemic model. Comput. Math. Appl. **42**, 883–907 (2001)

Mathematical Theory to Compute Stochastic Cellular Processes

Keita Iida and Yoshitaka Kimura

Abstract A central challenge of gene expression analysis during the last few decades has been the characterization of the expression patterns experimentally and theoretically. Modern techniques on single-cell and -molecule resolution reveal that transcriptions and translations are stochastic in time and that clonal population of cells displays heterogeneity in the abundance of a given RNA and protein per cell. Hence, to take into account a cell-to-cell variability, we consider a stochastic model of transcription and the chemical master equation. Our stochastic analysis and Monte-Carlo simulation show that the limiting distribution of mRNA copy number can be expressed by a Poisson-beta distribution. The distribution represents the four different types of expression patters, which are typically found in various experimental profiles.

Keywords Gene expression · Single-cell analysis · Biological noise · Markov process · Master equation · Poisson-beta distribution

1 Introduction

Gene expression analysis is being used to investigate the functions of gene products (RNA and protein), to improve our understanding of various aspects of cellular function and disease, and to facilitate drug development [1]. Expression analysis has ever revealed the key regulators for various cell differentiations, which may help scientists establish novel cells [2]. However, little is known about the regulatory mechanisms of dynamic gene expressions. Although many biological processes, such as transcription factors binding, chromatin remodeling and cell cycle, have been reported as the important factors, a systematic understanding of unidirectional

K. Iida · Y. Kimura (✉)
Tohoku University, 1-1 Seiryo-machi, Aoba-ku, Sendai, Miyagi 980-8574, Japan
e-mail: ykimura@med.tohoku.ac.jp

K. Iida
e-mail: kiida1984@med.tohoku.ac.jp

R.S. Anderssen et al. (eds.), *Applications + Practical Conceptualization + Mathematics = fruitful Innovation*, Mathematics for Industry 11,
DOI 10.1007/978-4-431-55342-7_10

cell differentiations remains to be acquired. Today systems biology gives us a novel methodology to systematically understand the complex intracellular dynamics [3].

Modern techniques on single-cell and -molecule resolution reveal that transcriptions and translations are stochastic in time and that clonal population of cells displays heterogeneity in the abundance of a given RNA and protein per cell [4–7]. Thus, expression analysis based on probability statistics becomes an indispensable tool today [8], which may shed light on the classical biological knowledge [9]. In the present article, we mathematically investigate a variety of expression patterns by analyzing a simple model of transcription. Then, we discuss a reduction of the model equation, which is the key step to make gene regulatory networks [10].

2 Mathematical Model

Based on previous papers [11], we consider the following mathematical model for a single gene expression induced by a transcription factor:

$$\begin{aligned} \mathrm{G} &\overset{a}{\longrightarrow} \mathrm{G}^*, & \mathrm{G}^* &\overset{b}{\longrightarrow} \mathrm{G}, \\ \mathrm{G}^* &\overset{c}{\longrightarrow} \mathrm{G}^* + \mathrm{mRNA}, & \mathrm{mRNA} &\overset{d}{\longrightarrow} \phi, \end{aligned} \tag{1}$$

where G and G^* denote the genes being 'off' and 'on' states, respectively, and ϕ the degraded mRNA. Here, we assume that the transcription event can only occur under the 'on' state [12]. The parameters a and b are the probabilities per unit time of the promoter switching from inactive to active and active to inactive, respectively, c and d are the probabilities per unit time of transcription and mRNA degradation, respectively.

We assume that the time evolution of the mRNA copy number is modeled by a simple Markov process in continuous time, of which the state space is defined as

$$S = \{(i, n) \mid i \in \{0, 1\}, n \in \mathbb{Z}_{\geq 0}\},$$

where $i = 0$ and 1 are for 'off' and 'on' states, respectively, and n denotes the mRNA copy number in the system. Let $P_n^{(i)}(t)$ be the probability of having (i, n) state at time t, which obeys the following master equation:

$$\frac{\mathrm{d}}{\mathrm{d}t}\begin{bmatrix} P_n^{(0)} \\ P_n^{(1)} \end{bmatrix} = \boldsymbol{A}\begin{bmatrix} P_n^{(0)} \\ P_n^{(1)} \end{bmatrix} + \boldsymbol{C}\begin{bmatrix} P_{n-1}^{(0)} - P_n^{(0)} \\ P_{n-1}^{(1)} - P_n^{(1)} \end{bmatrix} + \boldsymbol{D}\begin{bmatrix} (n+1)P_{n+1}^{(0)} - nP_n^{(0)} \\ (n+1)P_{n+1}^{(1)} - nP_n^{(1)} \end{bmatrix}, \tag{2}$$

where $A = \begin{bmatrix} -a & b \\ a & -b \end{bmatrix}$, $C = \begin{bmatrix} 0 & 0 \\ 0 & c \end{bmatrix}$ and $D = \begin{bmatrix} d & 0 \\ 0 & d \end{bmatrix}$. The initial condition is

$$(P_n^{(0)}(0), P_n^{(1)}(0)) = (\delta_{0,n}, 0), \tag{3}$$

where δ is the Kronecker delta.

3 Analysis

The limiting distribution $\overline{P}_n$ of the system (2) and (3) becomes as follows:

$$\overline{P}_n = \frac{\gamma^n}{n!} e^{-\gamma} \frac{(\alpha)_n}{(\alpha+\beta)_n} {}_1F_1(\beta, \alpha+\beta+n; \gamma), \tag{4}$$

where we define $\alpha = a/d$, $\beta = b/d$ and $\gamma = c/d$. Here, $()_n$ is the Pochhammer symbol and ${}_1F_1$ is the Kummer function. As indicated in an earlier paper [13], (4) can be further simplified as follows:

$$\overline{P}_n = \int_0^1 \frac{(\gamma x)^n}{n!} e^{-\gamma x} \frac{x^{\alpha-1}(1-x)^{\beta-1}}{B(\alpha, \beta)} dx, \tag{5}$$

where B is the beta function. The distribution (5), which is called a Poisson-beta distribution, shows that the transcription rate can be regarded as γx in which x follows the beta distribution with parameters α and β. Hence, in the long-time limit, the model (1) can be approximated by the following scheme:

$$G^* \xrightarrow{cX} G^* + \text{mRNA}, \qquad \text{mRNA} \xrightarrow{d} \phi, \tag{6}$$

where the stochastic variable X follows the beta distribution $B(x; \alpha, \beta)$.

Figure 1a–d shows the limiting distribution (4) with various parameter sets. As one can see in Fig. 1, the expression patterns widely change depending on the para-

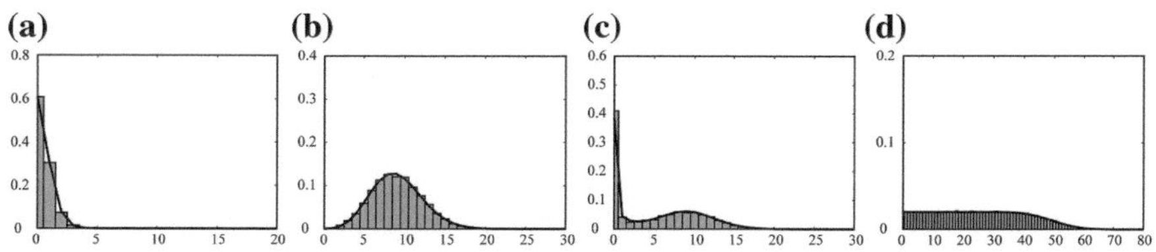

Fig. 1 The limiting distribution with respect to the mRNA copy number obtained from (2) and (3). The exact solutions (*bold line*) are obtained from (4) and numerical solutions (*filled bar graph*) from Monte-Carlo simulation with $\Delta t = 0.1$. The parameters (α, β, γ) are **a** (50, 50, 1), **b** (1, 10, 10), **c** (0.1, 0.1, 10), **d** (1, 1, 50)

meters α, β and γ. From the analytical result (5), we found that the beta distribution produces the variation and the Poisson distribution guarantees the discreteness of mRNA molecules.

4 Conclusion

Mathematical models of gene regulation have been studied since 1960s [14–16]. However, the classical deterministic approaches based on the population-wide average methods, such as the statistical procedure and the modeling with ordinary differential equations, are not enough to understand cell-to-cell variability. To understand the mechanisms of cell-to-cell variation in gene expressions, we should consider intrinsic and extrinsic noises ('biological noise') when constructing a mathematical model [4]. In the present article, we considered a simple model of transcription with only two gene states ('on' and 'off') and investigated the probability distribution of mRNA copy number. We found that the limiting distribution can be described by the Poisson-beta distribution, which represents four different types of expression patterns Fig. 1. Thus, the classical model (1) can be approximated by the scheme (6) in the long-time limit.

Acknowledgments This work was supported by Tohoku University's Juten Senryaku Program.

References

1. Holstege, F.C.P., Jennings, E.G., Wyrick, J.J., Lee, T.I., Hengartner, C.J., Green, M.R., Golub, T.R., Lander, E.S., Young, R.A.: Cell **95**, 717–728 (1998)
2. Takahashi, K., Yamanaka, S.: Cell **126**, 663–676 (2006)
3. Kitano, H.: Science **295**, 1662–1664 (2002)
4. Elowitz, M.B., Levine, A.J., Siggia, E.D., Swain, P.S.: Science **297**, 1183–1186 (2002)
5. Golding, I., Paulsson, J., Zawilski, S.M., Cox, E.C.: Cell **123**, 1025–1036 (2005)
6. Chubb, J.R., Trcek, T., Shenoy, S.M., Singer, R.H.: Curr. Biol. **16**, 1018–1025 (2006)
7. Sanchez, A., Golding, I.: Science **342**, 1188–1193 (2013)
8. Burrage, K., Hegland, M., Macnamara, S., Sidje, R.B.: Markov Anniversary Meeting: an International Conference to Celebrate the 150th Anniversary of the Birth of A.A. Markov, Boson Books, pp. 21–38 (2006)
9. Selvarajoo, K.: Wiley Interdiscip. Rev. Syst. Biol. Med. **4**, 385–399 (2012)
10. Hegland, M., Burden, C., Santoso, L., MacNamara, S., Booth, H.: J. Comput. Appl. Math. **205**, 708–724 (2007)
11. Peccoud, J., Ycart, B.: Theor. Popul. Biol. **48**, 222–234 (1995)
12. Golding, I., Cox, E.C.: Curr. Biol. **16**, R371–R373 (2006)
13. Kim, J.K., Marioni, J.C.: Genome Biol. **14**, 1–12 (2013)
14. Monod, J., Jacob, F.: Cold Spring Harb. Symp. Quant. Biol. **26**, 389–401 (1961)
15. Simon, Z.: J. Theor. Biol. **8**, 258–263 (1965)
16. Griffith, J.S.: J. Theor. Biol. **20**, 202–208 (1968)

Mathematical Model of Epidermal Structure

Yasuaki Kobayashi and Masaharu Nagayama

Abstract Using a mathematical model of the epidermis, we investigate how the structure of the dermis affects the spatio-temporal pattern of the upper structure, especially the stratum corneum (SC). We find that, while large scale undulations greatly affects the upper structure, small scale undulations do not propagate into the upper structure, which is consistent with experimentally observed cross sections of the epidermis.

Keywords Off-lattice model · Calcium dynamics · Dermatology

1 Introduction

Skin is an important organ. It serves as barrier by blocking chemical and physical attacks from outside, and at the same time keeping body fluid inside. Skin is composed of two layers: the inner part is called *dermis*, where a network of blood vessels can be found; and the outer part is *epidermis*, where epidermal cells (keratinocytes) form a layered structure. The barrier function is maintained by the outermost layers in the epidermis, called the stratum corneum (SC). Several experiments have shown that, in healthy normal epidermis, there exists a layer of high concentration region of Ca^{2+} just beneath the SC. It is also known that the Ca^{2+} distribution in this layer is closely related to the status of the layered structure of the SC.

In order to fully understand the role of the barrier function, a model of the organized structure of the SC is required. In fact, cross sections of skin show that, while

Y. Kobayashi (✉) · M. Nagayama
RIES, Hokkaido University, Sapporo 060-0811, Japan
e-mail: yasuaki.kobayashi@es.hokudai.ac.jp

M. Nagayama
e-mail: nagayama@es.hokudai.ac.jp

Y. Kobayashi · M. Nagayama
CREST, Japan Science and Technology Agency, 7, Gobancho, Chiyoda-ku, Tokyo 102-0076, Japan

R.S. Anderssen et al. (eds.), *Applications + Practical Conceptualization + Mathematics = fruitful Innovation*, Mathematics for Industry 11,
DOI 10.1007/978-4-431-55342-7_11

the dermis is highly undulated, the SC has a flat structure. Considering the fact that keratinocytes are supplied from the basal layer, which also has undulations matching the dermis beneath it, how a flat structure of the SC is created from the undulated basal layer is not clear.

We have recently proposed a mathematical model for epidermis, which is based on the Ca^{2+} localization phenomenon [1, 2]. This model could successfully reproduce the layered structure of the epidermis, namely the low-Ca^{2+} keratinocytes on the dermis, the SC on top, and the Ca^{2+}-localized monolayer between them. This model has also revealed that Ca^{2+} localization can reduce the spatio-temporal fluctuations of the SC. Using this model, in this paper, we investigate how the dermal structure influences the upper structure, especially the SC.

2 Particle Dynamics Model

Let us first briefly introduce our mathematical model of epidermis. Details about the model will be reported elsewhere [3].

Each epidermal cell is represented by a spherical particle, which is continuously supplied from the basal layer. Cells interact with each other through a short-range repulsive interaction: The cell at $\boldsymbol{x} = \boldsymbol{x}_i$ with the radius r_i feels the force $-\frac{\partial}{\partial \boldsymbol{x}_i} V(|\boldsymbol{x}_i - \boldsymbol{x}_j|)$ from the other cell at $\boldsymbol{x} = \boldsymbol{x}_j$ with the radius r_j, where we assume that V has a Lennard-Jones type interaction structure:

$$V(|\boldsymbol{x}_i - \boldsymbol{x}_j|) = 2\left(\frac{r_i + r_j}{|\boldsymbol{x}_i - \boldsymbol{x}_j|}\right)^6 - \left(\frac{r_i + r_j}{|\boldsymbol{x}_i - \boldsymbol{x}_j|}\right)^{12}. \tag{1}$$

The i-th cell is assigned a degree of differentiation, denoted by $S_i(t)$, which is a non-decreasing function of t. When $S_i(t) = S^*$, the cell is considered to be cornified, becoming a part of the SC. We assume that this cornification process is accelerated when Ca^{2+} inside the cell is high:

$$\dot{S}_i = \omega' + \alpha'(c_i - \bar{c})_+, \tag{2}$$

where $(x)_+$ equals x if x is positive and otherwise 0, c_i is the Ca^{2+} concentration of the i-th cell, and ω' and α' are constants.

The Ca^{2+} dynamics are modelled using our recently proposed mathematical model of intra-cellular Ca^{2+} in the epidermis [2]. This model is in essence a dicretized version of a reaction-diffusion equation, with the source term coming from the cornified cell. Here it is assumed that, when a cell undergoes cornification, it releases a stimulant, which induces Ca^{2+} excitation in the neighbouring cells. There is thus an interdependence between the Ca^{2+} dynamics and the cell differentiation.

This model has successfully reproduced the experimentally observed Ca^{2+} localization phenomenon beneath the SC. Also it has been shown that when we neglect Ca^{2+} dynamics, the spatio-temporal structure of the Ca^{2+} localization is greatly affected, leading to instability in the SC layers [3].

3 Numerical Simulations

In this paper, new numerical results are presented about the effect of the dermal structure. We consider a three dimensional space $[0, L] \times [0, L] \times [f(x, y), L_z]$, where $f(x, y)$ determines the dermis. Here the following sinasoidal undulations, uniform in y, are assumed:

$$f(x, y) = \frac{A}{2} \sin\left(\frac{2\pi n x}{L}\right). \tag{3}$$

Periodic boundary conditions are assigned in the x and y directions. Particles on the dermis reproduce themselves at a constant rate, and thus supply new particles to the upper layers. Particles on the top layer disappear when a certain time elapses after cornification.

Figure 1 shows two examples of numerical simulations with different wavelengths of the dermal undulations. In both cases, we find that organized structures of epidermis are reproduced, where a localized layer of high-Ca^{2+} keratinocytes forms between the low-Ca^{2+} layers and the SC, which is in accordance with our previous results in the case of a flat dermis, as mentioned above [3]. However, in the case of undulated dermis, the upper structure shows significant differences depending on the wavelength of the dermis.

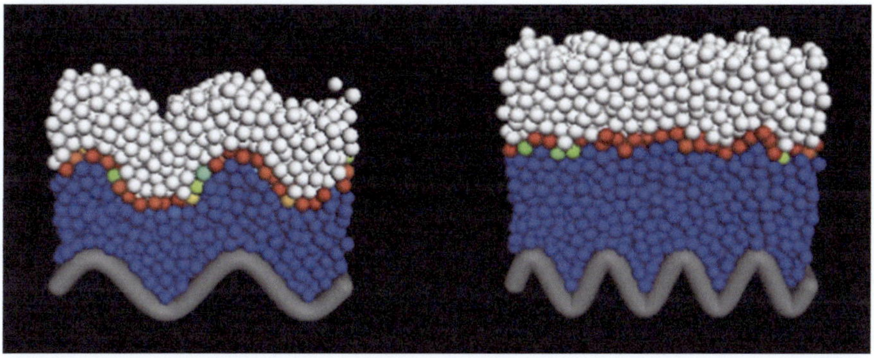

Fig. 1 Numerical simulation of the epidermis. *Left* $n = 2$, $A = 10$; *right* $n = 4$, $A = 10$. The dermal boundary is *colored gray*, and the SC is *colored white*. In the *middle* between the dermis and the SC, *colors* indicate the Ca^{2+} level: *blue* for *low* Ca^{2+}, *red* for high Ca^{2+}

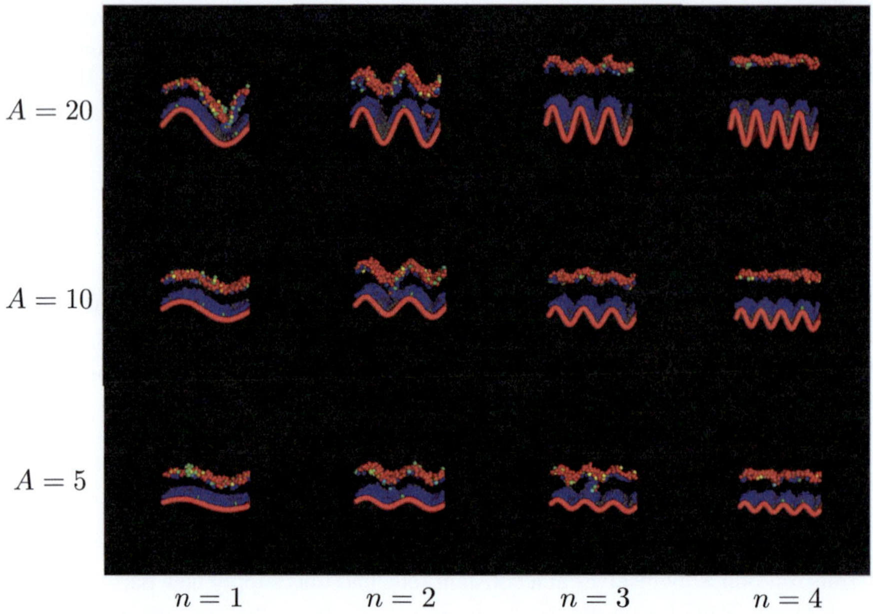

Fig. 2 Comparison of the spatial structure of Ca^{2+} localization layers for different values of n and A. For each panel, the *bottom red* represents the dermis. *Blue* and *gray* particles on the dermis represent the reproducible particles; *blue* ones are in the reproduction process. For the other particles, only the Ca^{2+} localization layer, the particles which are in contact with the SC, are shown

For different values of wavenumber n and amplitude A, we performed numerical simulations, as shown in Fig. 2. Here, apart from the dermis and the basal layer, we visualized only the cells which form Ca^{2+} localization layers: the cells which are in contact with the cornified cells. We found that, as n increases, which means the decreased wavelength of dermal undulations, the Ca^{2+} localization layer becomes flatter and flatter. In the case $n = 1$, for example, when the amplitude of the dermal undulation increases, the undulation of the Ca^{2+} localization layer becomes amplified. On the other hand, in the case $n = 5$, the undulation of the Ca^{2+} localization layer remains flat even for larger amplitudes.

To quantify these observations, we introduce the following evaluation function:

$$E(t) = \frac{1}{N_c(t)} \sum_{j \in \Omega} \frac{(z_j(t) - \bar{z}(t))^2}{D^2}, \tag{4}$$

where Ω is the set of particles which are in contact with the SC, $N_c(t)$ is the number of particles in Ω, $z_j(t)$ is the height of the j-th particle, D is the diameter of the particles, and $\bar{z}(t) \equiv \frac{1}{N_c(t)} \sum_{j \in \Omega} z_j(t)$. Note that all these values are time dependent. $E(t)$ can be used as a measure of deformation of the Ca^{2+} localization layer: $E(t) \geq 0$ by definition, and in particular $E(t) = 0$ if the layer is completely flat.

Figure 3 shows the time evolution of $E(t)$ for different values of n, with $A = 10$. It is clear that, when the wavelength of the dermal undulation is large (small n), the structure of the Ca^{2+} localization layer not only deforms in space, but also fluctuates in time. On the other hand, as the wavelength becomes shorter (larger n), this spatio-temporal flucation becomes reduced.

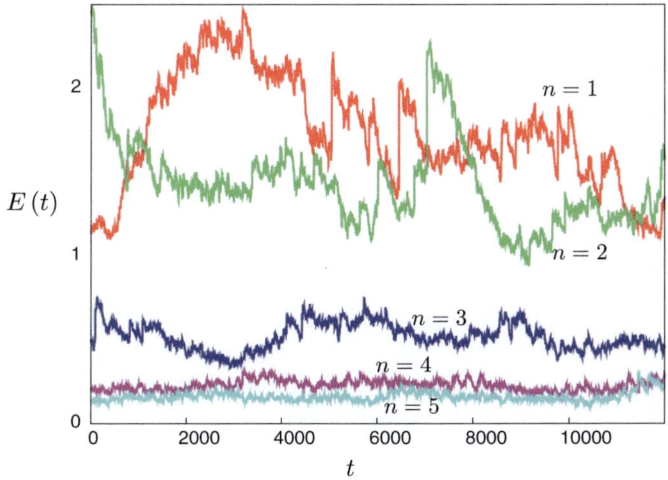

Fig. 3 Time evolution of $E(t)$ for different values of n, with $A = 10$

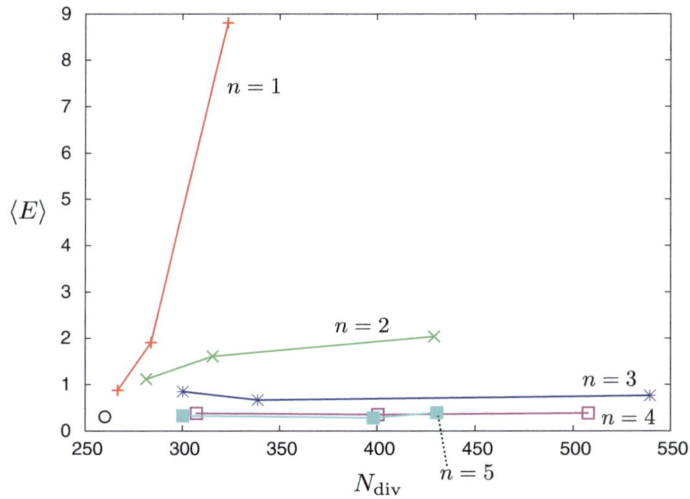

Fig. 4 Time average of $E(t)$ against the number of reproducible particles N_{div} for different values of n and A. For each n, three different values of A (5, 10, 20) are plotted, for each of which N_{div} is counted. The *open circle* near the *left bottom* corresponds to the case of the flat dermis ($n = 0$, $A = 0$)

Since different values n and A lead to different surface area of the dermis, the number of cells reproduced from the bottom layer, denoted by $N_{\rm div}$, which is proportional to the surface area, also depends on n and A. In order to compare the deformation with the same supply rate of particles, we plotted the time-average of $E(t)$, $\langle E \rangle$, for different values of n and A, as a function of $N_{\rm div}$, as shown in Fig. 4. This shows that the wavenumber dependence becomes larger as increasing the number of reproducible particles, i.e., as increasing the supply rate of particles. It is also indicated in Fig. 4 that there is a critical wavenumber between $n = 2$ and $n = 3$, above which the upper structure is not affected by the dermis.

4 Concluding Remarks

We have investigated the effect of dermal undulations on the epidermal structure, and found that, while large scale undulations are directly reflected in the upper structure, small scale undulations affect the upper structure in a limited way. This can be considered as a result of Ca^{2+} dynamics: since the stimulant of Ca^{2+} excitation diffuses into space, the high-stimulant-concentration surface tends to become flat, and particles which are about to undergo cornification also tends to follow this surface. Although a thorough analysis of our particle-based model beyond numerical investigation is difficult, a reduced model based on a reaction-diffusion system might be constructed, which can grasp the nature of reduced short-wavelength fluctuations observed here.

References

1. Denda, M., et al.: Frontiers in epidermal barrier homeostasis—an approach to mathematical modelling of epidermal calcium dynamics. Exp. Dermatol. **23**, 79–82 (2014)
2. Kobayashi, Y., et al.: Mathematical modeling of calcium waves induced by mechnical stimulation in keratinocytes. PLOS ONE **9**(3), e92650 (2014)
3. Kobayashi, Y., Sawabu, Y., Kitahata, H., Nagayama, M.: Mathematical model for epidermal homeostasis, in preparation

Beyond Adiabatic Elimination: Effective Hamiltonians and Singular Perturbation

Mikel Sanz, Enrique Solano and Íñigo L. Egusquiza

Abstract Adiabatic elimination is a standard tool in quantum optics, that produces an effective Hamiltonian for a relevant subspace of states, incorporating effects of its coupling to states with much higher unperturbed energy. It shares with techniques from other fields the emphasis on the existence of widely separated scales. Given this fact, the question arises whether it is feasible to improve on the adiabatic approximation, similarly to some of those other approaches. A number of authors have addressed the issue from the quantum optics/atomic physics perspective, and have run into the issue of non-hermiticity of the effective Hamiltonian improved beyond the adiabatic approximation. Even though non-hermitian Hamiltonians are interesting in their own right, this poses conceptual and practical problems. Here, we first briefly survey methods present in the physics literature. Next we rewrite the problems addressed by the adiabatic elimination technique to make apparent the fact that they are singular perturbation problems from the point of view of dynamical systems. We apply the invariant manifold method for singular perturbation problems to this case, and show that this method produces the equation named after Bloch in nuclear physics. Given the wide separation of scales, it becomes intuitive that the Bloch's equation admits iterative/perturbative solutions. We show, using a fixed point theorem, that indeed the iteration converges to a perturbative solution that produces in turn an exact Hamiltonian for the relevant subspace. We propose thus several sequences of effective Hamiltonians, starting with the adiabatic elimination and improving on it. We show the origin of the non-hermiticity, and that it is inessential given the isospectrality of the effective non-hermitian operator and a corresponding effective

M. Sanz (✉) · E. Solano
Department of Physical Chemistry, University of the Basque Country UPV/EHU, Apartado 644, 48080 Bilbao, Spain
e-mail: mikel.sanz@ehu.eus

E. Solano
IKERBASQUE, Basque Foundation for Science, Maria Diaz de Haro 3, 48013 Bilbao, Spain
e-mail: enr.solano@gmail.com

Í.L. Egusquiza
Department of Theoretical Physics and History of Science, University of the Basque Country UPV/EHU, Apartado 644, 48080 Bilbao, Spain
e-mail: inigo.egusquiza@ehu.eus

R.S. Anderssen et al. (eds.), *Applications + Practical Conceptualization + Mathematics = fruitful Innovation*, Mathematics for Industry 11,
DOI 10.1007/978-4-431-55342-7_12

hermitian operator, which we build. We propose an application of the introduced techniques to periodic Hamiltonians.

Keywords Adiabatic elimination · Singular perturbation · Effective Hamiltonians · Schrieffer-Wolff expansion · Born-Oppenheimer approximation · Bloch's equation · Lambda system · Floquet's Hamiltonians

1 Adiabatic Elimination

Situations in which there is a wide separation among energy or time scales present in a physical system pervade all of Physics. It is therefore of recurrent interest to develop techniques to obtain approximate, effective descriptions of the low energy or slow sector of the system, since they are the most likely accessible to experimentation and control. In general, a naïve perturbation expansion will not provide us with the required effective description; depending on the scheme and approach, the obstacle to do so will turn up as secular terms in a time evolution, or as zero denominators in state expansions. Therefore, the approximation schemes valid for these situations will not be directly perturbative; they will be asymptotic, or resummation-based, or cumulant, or combinations thereof. Often, we have heuristic arguments for the construction of an effective description of the system, which do not naturally lead to improvements, at least not systematically.

A very common and useful approximation in quantum optics, normally based on a heuristic argument, is the so-called *adiabatic elimination* technique [1–3]. There are several ways of introducing this approximation. For this first presentation, consider the Schrödinger equation as a dynamical system, evolving with a Hamiltonian H as $i\partial_t\psi = H\psi$ (we set here and henceforth $\hbar = 1$). Let state ψ be partitioned into $\alpha = P\psi$ and $\gamma = Q\psi$, with P and $Q = 1 - P$ projectors, in such a manner that the eigenvalues of PHP are widely separated from those of QHQ. Let it be the case in which the coupling between the P and Q subspaces is very small when compared to the eigenvalues of QHQ. To be more specific, let τ be a characteristic scale, for instance, the norm of the restricted inverse $(QHQ)^{-1}$. Then, the Schrödinger equation may be rewritten as

$$\mathrm{i}\partial_t\alpha = PHP\alpha + PHQ\gamma, \tag{1a}$$

$$\mathrm{i}\partial_t\gamma = QHP\alpha + QHQ\gamma, \tag{1b}$$

where the last line of the system is to be multiplied by τ. Heuristically, we are asking that $\tau QHP \ll 1$, while $\tau QHQ = O(1)$. This can be achieved if γ is small, and if we neglect its time evolution. In other words, we slave γ to α in the approximation $QHQ\gamma = -QHP\alpha$, or, formally, $\gamma = -(QHQ)^{-1}QHP\alpha$. By substituting this approximation for γ in the first component of the system, we obtain the effective evolution for the *slow sector* α as

$$\mathrm{i}\partial_t \alpha = \left(PHP - PHQ\frac{1}{QHQ}QHP \right) \alpha. \tag{2}$$

By eliminating the fast component γ, we obtain an effective Hamiltonian for the slow sector

$$H_{\text{adiabatic}} = PHP - PHQ\frac{1}{QHQ}QHP. \tag{3}$$

This adiabatic elimination process has proven to be extremely useful in isolating the effective slow evolution for a number of systems. However, the aforementioned heuristic presentation does not lend itself readily to a systematic improvement. Thus, the question of how to transform this approximation into a controlled expansion is recurrently posed in quantum optics literature [4, 5]. In essence, both papers present an expansion of an integral kernel, either in the energy domain [4] or the time domain [5].

As a matter of fact, this problem of finding a systematic expansion when there is a wide divergence in energy scales has also appeared in many other contexts: nuclear physics, condensed matter, and atomic and molecular physics, among others. In each case, there are expansion techniques, under different guises and names: Bloch's equation [6, 7], Schrieffer–Wolff expansion [8], or Born–Oppenheimer approximation [9]. In this paper, we want to merge these techniques by presenting the system of equations as a singular perturbation problem, and by using the invariant manifold scheme for resummation of secular terms. For this purpose, we shall derive systematic expansions which improve on the adiabatic elimination technique and furnish us with an effective Hamiltonian for the low energy sector. Furthermore, we show the equivalence of this expansion and the Schrieffer–Wolff method. We apply this method to some simple examples in quantum optics, and suggest its future applicability for time-dependent periodic Hamiltonians.

2 A Singular Perturbation Problem and Bloch's Equations

In Physics, there are often problems depending on a small parameter which can be solved by applying perturbation theory. The singular perturbation theory approach has been developed for those cases for which no uniform regular expansion is possible, the so-called *singular problems*. Heuristically, one can identify a problem as being singular when it is qualitatively different at the zero order of the expansion, and immediately out of it. For example, the algebraic problem $\varepsilon x^2 + x - 1 = 0$ is singular for the small parameter ε, since it is second order if $\varepsilon \neq 0$, but first order if $\varepsilon = 0$. Similarly, singular perturbation problems in systems of ordinary differential equations [10] appear frequently by reducing the order of the differential equation when the small parameter is set to be 0.

Let us now consider the Λ system depicted in Fig. 1, which is ubiquitous in atomic physics. This is governed by the system of ordinary differential equations

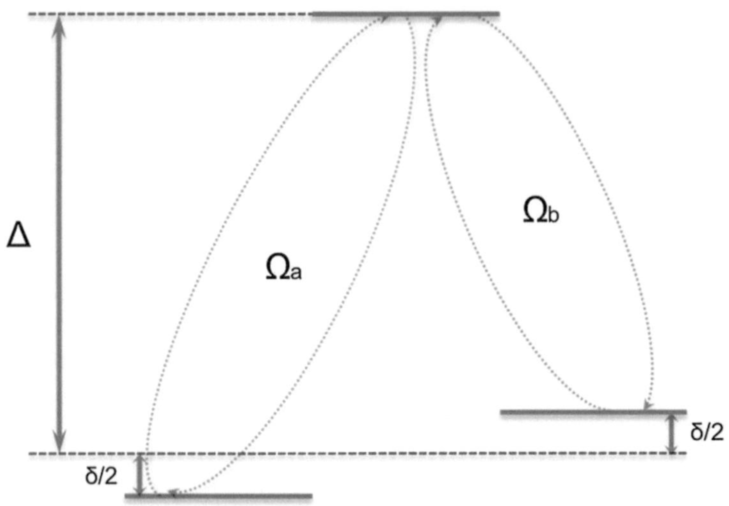

Fig. 1 The Λ system is a paradigmatic model in atomic physics and quantum optics, where a three-level atomic system may be coupled to classical or quantum fields

$$\begin{aligned} \mathrm{i}\dot{\alpha} &= -\frac{\delta}{2}\alpha + \frac{\tilde{\Omega}_a^*}{2}\gamma, \\ \mathrm{i}\dot{\beta} &= \frac{\delta}{2}\beta + \frac{\tilde{\Omega}_b^*}{2}\gamma, \\ \mathrm{i}\dot{\gamma} &= \frac{\tilde{\Omega}_a}{2}\alpha + \frac{\tilde{\Omega}_b}{2}\beta + \Delta\gamma. \end{aligned} \tag{4}$$

We are interested in the regime in which $\Delta \gg \delta, |\Omega_i|$, with δ real. Under the change of variable $t \to \delta t$ and the definitions $\Omega_i = \tilde{\Omega}_i/\delta$ and $\varepsilon = \delta/\Delta$, the system transforms into

$$\begin{aligned} \mathrm{i}\dot{\alpha} &= -\frac{1}{2}\alpha + \frac{\Omega_a^*}{2}\gamma, \\ \mathrm{i}\dot{\beta} &= \frac{1}{2}\beta + \frac{\Omega_b^*}{2}\gamma, \\ \mathrm{i}\varepsilon\dot{\gamma} &= \gamma + \frac{\varepsilon}{2}\left(\Omega_a\alpha + \Omega_b\beta\right). \end{aligned} \tag{5}$$

This is heuristically identifiable as a singular problem, since the last differential equation changes its character into an algebraic one as $\varepsilon \to 0$. Indeed, if one were to attempt a naïve perturbation expansion in the small parameter ε, one would obtain secular terms already at the first order, rendering the expansion invalid already for times t of the order of $\Delta/|\tilde{\Omega}_i|^2$ (in terms of the original variables).

The issue at hand is therefore how to identify a time-uniform scheme which provides us with approximations to the slowly varying variables. Among the numerous existing proposals, we shall concentrate here on the invariant manifold method [10]. This consists in constructing, among the submanifolds invariant under the flow, those that are perturbative in the small parameter. In the case of interest, as the system is linear, it is pertinent to only examine linear subspaces.

Let us write system (1) in the matrix form

$$\mathrm{i}\partial_t \begin{pmatrix} \alpha \\ \gamma \end{pmatrix} = \begin{pmatrix} \omega & \Omega^\dagger \\ \Omega & \Delta \end{pmatrix} \begin{pmatrix} \alpha \\ \gamma \end{pmatrix}. \tag{6}$$

For clarity, let us notice that if we were to write the Λ system in this more compact notation, we would be using the isomorphism $\mathbb{C}^3 = \mathbb{C}^2 \oplus \mathbb{C}$. The three component vector $(\alpha, \beta, \gamma)^T$ would be written (again for the Λ system) as a two component object $(\alpha, \gamma)^T$, where the first component is in turn a vector in $\mathbb{C}^2$. It follows that, in this case, $\omega \in M_2(\mathbb{C})$ (in fact, hermitian), $\Delta \in M_1(\mathbb{C})$ and hermitian, so in fact $\Delta \in \mathbb{R}$ for the Λ system case. Finally, Ω is a row two component vector for the Λ system.

For the general case, consider that the full Hamiltonian of system (1) acts on the Hilbert space $\mathscr{H}$. Then the projectors P and Q give the Hilbert subspaces $P\mathscr{H}$ and $Q\mathscr{H}$ respectively, with $P\mathscr{H} \oplus Q\mathscr{H} = \mathscr{H}$. We have introduce new notation in (6) for the operators $P_i H P_j$, where P_i stand for either P or Q. Namely, $PHP \to \omega \in \mathscr{B}(P\mathscr{H})$, $QHQ \to \Delta \in \mathscr{B}(Q\mathscr{H})$, while $QHP \to \Omega \in \mathscr{B}(P\mathscr{H}, Q\mathscr{H})$. Now, α stands for an element of $P\mathscr{H}$, while $\gamma \in Q\mathscr{H}$.

As we are looking for linearly invariant subspaces, let us define them by the relation $\gamma = B\alpha$. Notice that the embedding operator B belongs to $\mathscr{B}(P\mathscr{H}, Q\mathscr{H})$, as Ω. The invariance condition or embedding equation then reads

$$\Omega + \Delta B = B\omega + B\Omega^\dagger B. \tag{7}$$

By itself, this equation, known in the literature as *Bloch's equation* [11], is not advantageous with respect to the direct analysis of the Hamiltonian. However, in the case of interest to us, the spectrum of Δ is energetically very separated from the spectrum of ω, and we shall carry out a perturbative expansion or an iterative procedure to determine B. Once B has been obtained, the evolution in the subspace is determined by $\mathrm{i}\partial_t\alpha = \left(\omega + \Omega^\dagger B\right)\alpha$. This also provides us with approximate evolutions in the subspace, given an approximate solution to Bloch's equation. Notice the existence of an, in general, non-hermitian linear operator $h_{\mathrm{eff}} = \omega + \Omega^\dagger B$, which plays the role of an effective Hamiltonian.

3 Expansion Beyond Adiabatic Elimination

Let us assume that Δ has a bounded inverse, and that, for some definition of the operator norm, $\varepsilon = \|\Delta^{-1}\|\|\omega\| \ll 1$ and $\varepsilon' = \|\Delta^{-1}\|\|\Omega\| \ll 1$. We define the nonlinear transformation of operators

$$T(A) = -\Delta^{-1}\Omega + \Delta^{-1}A\omega + \Delta^{-1}A\Omega^{\dagger}A. \tag{8}$$

The invariance condition (7) may be now written as a fixed point equation, namely $B = T(B)$. We shall now prove that the nonlinear transformation T has a fixed point which is the required solution to the problem at hand. Observe that, due to the definition of T and the properties of operator norms, we have

$$\begin{aligned} \|T(A)\| &\leq \|\Delta^{-1}\| \, \|\Omega\| + \|\Delta^{-1}\| \, \|\omega\| \, \|A\| + \|\Delta^{-1}\| \, \|\Omega\| \, \|A\|^2 \\ &\leq \varepsilon' \left(1 + \|A\|^2\right) + \varepsilon\|A\|. \end{aligned} \tag{9}$$

From this observation we obtain the following central proposition:

Theorem 1 *Assume* $\varepsilon, \varepsilon' \geq 0$ *and* $\varepsilon' \leq (1-\varepsilon)/2$. *Let us define*

$$r(\varepsilon, \varepsilon') = \frac{1-\varepsilon}{2\varepsilon'} + \sqrt{\left(\frac{1-\varepsilon}{2\varepsilon'}\right)^2 - 1}. \tag{10}$$

Then, the fixed point equation $T(A) = A$ *has at least one solution* A_* *such that*

$$\|A_*\| \leq r(\varepsilon, \varepsilon'). \tag{11}$$

Proof By direct analysis of the function $g(x) = \varepsilon'(1 + x^2) + \varepsilon x$, and using inequality (9) we conclude that, for every A such that

$$\|A\| \leq r(\varepsilon, \varepsilon') \tag{12}$$

it holds that

$$\|T(A)\| \leq r(\varepsilon, \varepsilon'). \tag{13}$$

As T maps a bounded closed convex set of the corresponding Banach space of operators into itself, there exists at least one fixed point $A_* = T(A_*)$ in the set, by Schauder's fixed point theorem.

Furthermore, choosing the operator $-\Delta^{-1}\Omega$ as the initial point of the iteration, one can readily see that

$$\|-\Delta^{-1}\Omega\| \leq \|\Delta^{-1}\| \, \|\Omega\| = \varepsilon' \leq r(\varepsilon, \varepsilon') \tag{14}$$

under the stated conditions. Thus, we know that there is at least one fixed point in its vicinity.

To sum up, under those conditions, we are assured of the existence of a solution of Bloch's equation which is small in the sense that $\|B\| \leq r(\varepsilon, \varepsilon')$. It is therefore natural to attempt either an iterative process or a perturbative expansion to compute approximations for that solution.

More concretely, we define

$$\begin{aligned} B^{(0)} &= -\Delta^{-1}\Omega\,, \\ B^{(k+1)} &= T\left[B^{(k)}\right]. \end{aligned} \tag{15}$$

The sequence of operators $B^{(k)}$ is thus defined by iteration, and lies in the region of applicability of the proposition. Although we have not proven the convergence of this sequence, the mapping of problem (7) to the Schrieffer–Wolff expansion presented in Sect. 5 allows us to conclude the uniqueness of the fixed point in some circumstances, given the results presented in [12].

Alternatively, we can define a perturbative expansion as

$$B = \sum_{k=1}^{\infty} B_{(k)} \tag{16}$$

and insert it into the embedding or Bloch's equation to obtain the recursive equation

$$B_{(k+1)} = \Delta^{-1}B_{(k)}\omega + \Delta^{-1}\sum_{l=1}^{k-1} B_{(k-l)}\Omega^{\dagger}B_{(l)}, \tag{17}$$

with the initial conditions $B_{(1)} = -\Delta^{-1}\Omega$ and $B_{(2)} = -\Delta^{-2}\Omega\omega$. Notice that, formally,

$$B^{(k)} - \sum_{l=1}^{k+1} B_{(l)} = O\left(\Delta^{-(k+2)}\right),$$

which relates the perturbative and the recurrence results to a given order. Explicit formulae relating the iteration approximations $B^{(k)}$ and the perturbative contributions $B_{(k)}$ are hardly illuminating beyond this relation to a given order.

4 The Effective Hamiltonian

As pointed out above, the evolution in the subspace fixed by B is determined by $\omega + \Omega^\dagger B$. In the first order, this provides us with a hermitian operator

$$h_{(1)} = h^{(1)} = \omega - \Omega^\dagger \Delta^{-1} \Omega \,, \tag{18}$$

either by iteration or perturbatively. This is, in fact, what we would obtain from direct adiabatic elimination, namely, by setting $\partial_t \gamma$ to zero, solving γ as $-\Delta^{-1}\Omega\alpha$, and replacing this by α in the differential equation. Furthermore, this operator is hermitian by construction.

Hermiticity, however, is not maintained in higher recurrence or perturbation orders [13]. This has been a sticking point in the literature, and the source of some confusion. It is clear that, by construction, the linear generator of evolution for the α part of the full $\psi = (\alpha, \gamma)^T$ does not need to be hermitian, even if the total Hamiltonian is hermitian. In any case, its spectrum must be real. In order to prove it, let us consider an eigenvector of the total Hamiltonian belonging to the linear subspace determined by $\gamma = B\alpha$. Then, the restriction to its α part will be an eigenvector of the effective Hamiltonian with the same eigenvalue as that for the total Hamiltonian. Conversely, let us assume that α_* is an eigenstate of the effective Hamiltonian. Then, $\psi_* = (\alpha_*, B\alpha_*)^T$ is also an eigenstate of the total Hamiltonian with the same eigenvalue, and hence the latter is real.

In the finite-dimensional case of interest, it follows that the full effective Hamiltonian must be similar to a hermitian Hamiltonian. Let us construct this similarity transformation assuming that we already have a solution B for Eq. (7) at hand. As a first step, notice that the total conserved norm $\langle \psi, \psi \rangle$ can be expressed as

$$\begin{aligned} \langle \psi, \psi \rangle &= \langle \alpha, \alpha \rangle + \langle \gamma, \gamma \rangle \\ &= \left\langle \alpha, \left(1 + B^\dagger B\right) \alpha \right\rangle . \end{aligned}$$

One is immediately led to examine $h_1 = \left(1 + B^\dagger B\right)\left(\omega + \Omega^\dagger B\right)$. By using Eq. (7) and its conjugate, one concludes that h_1 is indeed hermitian, if B is a solution to Bloch's equation. Its spectrum, however, is not the one corresponding to the time evolution in the low energy sector $P\mathscr{H}$. Let us define

$$S_B = \sqrt{1 + B^\dagger B}, \tag{19}$$

which is always possible, since $1 + B^\dagger B$ is positive. Then, for any constant unitary V acting on $P\mathscr{H}$, we obtain a hermitian Hamiltonian

$$h_V = V S_B \left(\omega + \Omega^\dagger B\right) S_B^{-1} V^\dagger . \tag{20}$$

Hermiticity is easily proven by noticing that

$$\begin{aligned} h_V &= V S_B \left(\omega + \Omega^\dagger B\right) S_B^{-1} V^\dagger \\ &= V S_B^{-1} h_1 S_B^{-1} V^\dagger . \end{aligned}$$

The unitary V can be subsumed in the choice of the square root defining S_B in (19). In what follows, the omission of the subscript $_V$ is associated to the assumption of a choice for S_B.

In point of fact, an exact solution for (7) is as hard to come by as an exact diagonalisation of the initial Hamiltonian, so we have to use approximate methods. However, if we truncated B to some approximation, h_1 (and thus h_V) would no longer be automatically hermitian. This can be mended by introducing in

$$h = \left(1 + B^\dagger B\right)^{-1/2} \left(\omega + \Omega^\dagger B + B^\dagger \Omega + B^\dagger \Delta B\right) \left(1 + B^\dagger B\right)^{-1/2} \tag{21}$$

an approximate solution B_a to Bloch's equation. This expression would be equivalent (up to unitary transformations V) to h_V, if B_a were an exact solution of Bloch's equation (7); on the other hand, it is explicitly hermitian for any B_a, which provides us with a hermitian approximate Hamiltonian encoding the effective evolution. In the case of a perturbative expansion, one can compute (21) to second order, for example, giving

$$h^{(2)} = \omega - \Omega^\dagger \Delta^{-1} \Omega - \frac{1}{2} \left(\Omega^\dagger \Delta^{-2} \Omega \omega + \omega \Omega^\dagger \Delta^{-2} \Omega\right). \tag{22}$$

5 Relation to Schrieffer–Wolff Expansion

In 1966, J.R. Schrieffer and P.A. Wolff introduced a perturbatively-built canonical transformation, with an antihermitian generator S, in order to eliminate small denominators in the perturbative expansion of the Anderson Hamiltonian [8]. However, this technique is nowadays a staple in condensed matter physics.

Let us define an antihermitian operator S by

$$\tanh(S) = \begin{pmatrix} 0 & B^\dagger \\ -B & 0 \end{pmatrix}, \tag{23}$$

where S is called Schrieffer–Wolff's operator (see, for instance, [12]). Then, Bloch's equation (7) and its conjugate can be written together in the form

$$\begin{pmatrix} 0 & \Omega^\dagger \\ \Omega & 0 \end{pmatrix} - \tanh(S) \begin{pmatrix} 0 & \Omega^\dagger \\ \Omega & 0 \end{pmatrix} \tanh(S) = \left[\begin{pmatrix} \omega & 0 \\ 0 & \Delta \end{pmatrix}, \tanh(S) \right].$$

Under the assumption that $\cosh(S)$, with antihermitian and block anti-diagonal S), exists and is invertible, this equation is equivalent to $\exp(S)H\exp(-S)$ being block-diagonal.

Some further formal manipulations give us the additional identification

$$
\begin{aligned}
e^{-S} &= [1-\tanh(S)]\left[1-\tanh^2(S)\right]^{-1/2} \\
&= \begin{pmatrix} 1 & -B^\dagger \\ B & 1 \end{pmatrix}\begin{pmatrix} \left(1+B^\dagger B\right)^{-1/2} & 0 \\ 0 & \left(1+BB^\dagger\right)^{-1/2} \end{pmatrix}.
\end{aligned} \tag{24}
$$

This expression, together with the equivalence between Bloch's equation and that $\exp(S)H\exp(-S)$ is block diagonal, give us the equivalence between the two methods. The hermitian effective Hamiltonian h_V is, in fact, the block obtained from the Schrieffer–Wolff method corresponding to the low energy sector, and an approximate S gives us an approximate B, and viceversa.

In the Schrieffer–Wolff method, the approximations are built by expanding S in powers of the interaction, and imposing that, to that order, the transformed Hamiltonian must be block-diagonal. To establish the connection between both methods, let us denote $S = \begin{pmatrix} 0 & -\mathscr{S}^\dagger \\ \mathscr{S} & 0 \end{pmatrix}$. Then, by using our notation, the first order Schrieffer–Wolff condition reads

$$
\mathscr{S}_0\omega - \Delta\mathscr{S}_0 = \Omega\,. \tag{25}
$$

Notice that this is a Sylvester equation, which will have a unique solution if the spectra of ω and Δ are disjoint. It should be pointed out that, in the context we are interested in, it might be necessary to compute approximate solutions for Eq. (25). Alternatively, if a solution for (25) can be computed, it will be a *resummation* of our perturbative or iterative expansion.

To the same expansion order, the effective Hamiltonian for the low energy sector reads

$$
h_{SW}^{(1)} = \omega + \frac{1}{2}\left(\mathscr{S}_0^\dagger\Omega + \Omega^\dagger\mathscr{S}_0\right). \tag{26}
$$

6 The Λ-system Example

The Λ-system in quantum optics described in the system of Eq. (4) is amenable to both the approach based on Bloch's equation and Schrieffer–Wolff's approach if $\Delta \gg \delta, \tilde{\Omega}_i$. Direct application of expressions (22) and (26) yields

$$
\begin{aligned}
h_{SW}^{(1)} &= -\frac{\Delta}{2}\sigma^z - \frac{1}{\Delta}\frac{1}{1-(\delta/2\Delta)^2}\tilde{\Omega}^\dagger\tilde{\Omega} \\
&+ \frac{\delta}{4\Delta^2}\frac{1}{1-(\delta/2\Delta)^2}\left(\sigma^z\tilde{\Omega}^\dagger\tilde{\Omega} + \tilde{\Omega}^\dagger\tilde{\Omega}\sigma^z\right),
\end{aligned} \tag{27}
$$

$$h^{(2)} = -\frac{\Delta}{2}\sigma^z - \frac{1}{\Delta}\tilde{\Omega}^\dagger\tilde{\Omega} + \frac{\delta}{4\Delta^2}\left(\sigma^z\tilde{\Omega}^\dagger\tilde{\Omega} + \tilde{\Omega}^\dagger\tilde{\Omega}\sigma^z\right). \quad (28)$$

In this simple example one can already see some features of the Schrieffer–Wolff expansion as compared to the Bloch expansion; namely that the coefficients are not purely perturbative, but involve a resummation of perturbative terms. Additionally, this resummation presents with a pole that is not seen in the first perturbation terms. One should not expect the location of the pole to this order to be exact, and, in fact, working out specific exactly solvable examples (such as $\tilde{\Omega} \to (1, 0)$) it is easy to see that it is not located at $\delta = 2\Delta$.

Alternatively, the iteration expansion readily lends itself to numerical implementations. In Figs. 2 and 3, we show the presence of a secular shift for the adiabatic elimination approximation, the lack of normalisation with the non-hermitian Hamiltonian to some orders, and that the numerical solution of Bloch's equation matches perfectly to the real one already at fourth order of iteration, when ignoring large frequency oscillations.

In particular, in Fig. 2, we show the exact evolution of the populations of the lowest lying states in the Λ system. In the context of atomic physics, "populations" translates into the square modulus of the coefficients of the state in an orthogonal basis.

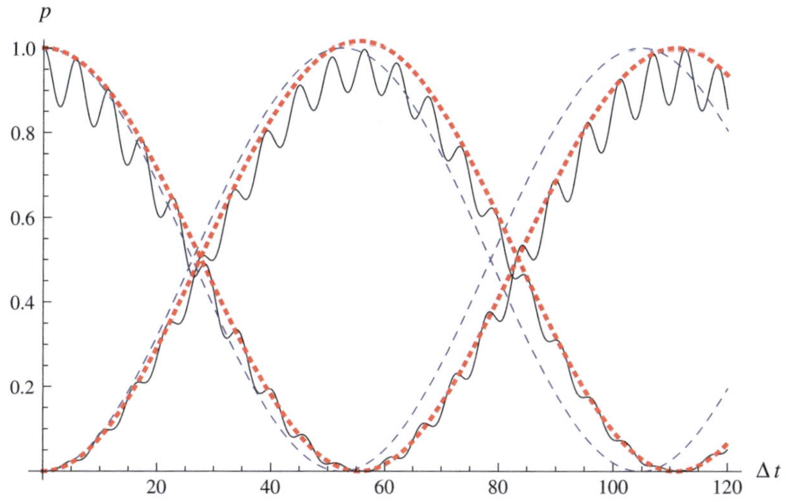

Fig. 2 Evolution of the population of the ground and excited states, with initial state $(1, 0, 0)^T$, under **a** the exact Hamiltonian (continuous *black line*), **b** zeroth order effective Hamiltonian (*dashed blue line*) and **c** fourth iteration of T (*dotted red line*). The parameters are $\delta = -0.0175\Delta$, $\tilde{\Omega}_a = 0.4\Delta$, $\tilde{\Omega}_b = 0.3\Delta$, for direct comparison with Ref. [5]

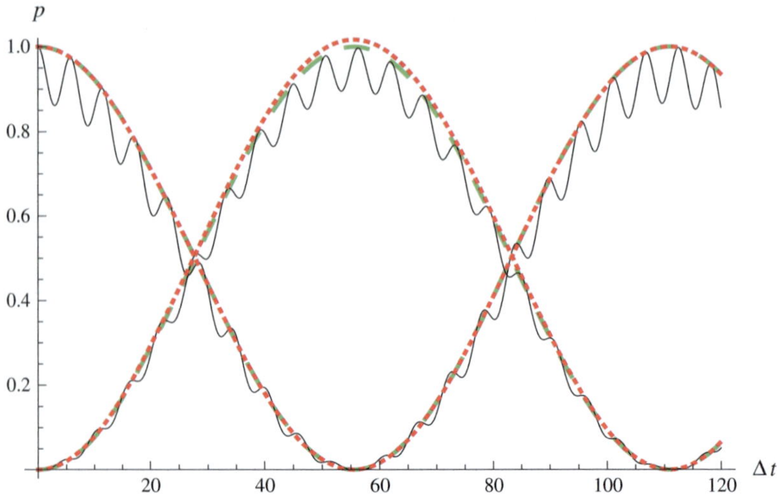

Fig. 3 Evolution of the population of the ground and excited states, with initial state $(1, 0, 0)^T$, under **a** the exact Hamiltonian (*continuous black line*), **b** $h_{\text{eff}}^{(4)}$ (*dotted red line*) and **c** $h_V^{(10)}$ with $V = 1$ (*dashed green line*). The parameters are $\delta = -0.0175\Delta$, $\tilde{\Omega}_a = 0.4\Delta$, $\tilde{\Omega}_b = 0.3\Delta$, as before

Thus, for a state $\psi = (\alpha, \beta, \gamma)^T$ we are depicting $|\alpha|^2$ and $|\beta|^2$ as a function of normalised time Δt, and these two quantities are compared with the evolution of the same populations with the first order of Bloch's approximation, i.e. adiabatic elimination approximation, and with the fourth iteration of the recurrence. One should observe the secular shift in the adiabatic elimination: the maxima of the relevant population under the evolution dictated by the adiabatic elimination are recurrently advanced with respect to the maxima given by the exact evolution. We also depict evolution under a non-hermitian effective Hamiltonian, and its effect is reflected in the fact that population maxima can be larger than one, as shown in the central maximum depicted in red. Since we are looking at the low energy effective evolution, the fast oscillations do not appear in the evolution under the effective Hamiltonian.

Analogously, in Fig. 3, one notices that the hermitian Hamiltonian to 10th order (obtained from computing the perturbative expansion of B to 10th order and substituting in (21)) perfectly matches the evolution of the exact populations, when one discards the rapid fluctuations that populate temporarily the highest energy state.

7 Large Frequency Expansions for Periodic Hamiltonians

Let us consider a time-dependent periodic Hamiltonian with period $2\pi/\omega$, so that $H(t) = H(t + 2\pi/\omega)$. Let $U(t)$ be the operator solution of $i\partial_t U(t) = H(t)U(t)$, with $U(0) = \mathbb{1}$. Then, by Floquet's theorem, the eigenvalues of $U(2\pi/\omega)$, when written as $\exp(-i2\pi\varepsilon/\omega)$, provide us with the so-called *quasi-energies* ε. It is well-known that the quasi-energies can be identified with the eigenvalues of the corresponding Floquet's Hamiltonian [14]. In order to understand it, let us assume that the original periodic Hamiltonian can be expanded as

$$H(t) = \sum_k H_k e^{-ik\omega t} \tag{29}$$

with integer k and $H_k^\dagger = H_{-k}$. Let us define the operators K and T acting on a different Hilbert space (the canonical example is $L^2\left(S^1\right)$), and such that (a) K is hermitian and its spectrum consists of all integer numbers, i.e. if we denote the eigenbasis of K as $|k\rangle$, then $K|k\rangle = k|k\rangle$; (b) T is unitary; (c) $[K, T] = T$. The Floquet's Hamiltonian associated to the original one is hence

$$H_F = \sum_k H_k T^k - \omega K. \tag{30}$$

In order to avoid cluttering formulae unnecessarily we frequently omit the tensor product sign; by $H_k T^k$ we actually denote $H_k \otimes T^k$, which acts on $\mathscr{H} \otimes L^2(S^1)$ if $\mathscr{H}$ is the original Hilbert space on which $H(t)$ acts. Similarly, K stands for $1 \otimes K$.

We shall now apply the formalism of adiabatic expansions to this Floquet's Hamiltonian to obtain approximate values for the quasi-energies. The relevant projectors are $P = \mathbb{1} \otimes |0\rangle\langle 0|$ and $Q = \mathbb{1} - P$. Therefore,

$$PH_FP = H_0, \tag{31}$$

$$PH_FQ = \sum_k H_k P T^k Q = \sum_{k\neq 0} H_k \otimes |0\rangle\langle -k|, \tag{32}$$

$$QH_FP = \sum_k H_k Q T^k P = \sum_{k\neq 0} H_k \otimes |k\rangle\langle 0|, \tag{33}$$

$$QH_FQ = -\omega QKQ + \sum_k H_k Q T^k Q. \tag{34}$$

The problem in the direct application of the formalism presented above lies in the computation of the inverse restricted to the Q subspace. However, in the limit of ω much larger than H_k in norm, the inverse may be approximated by

$$\begin{aligned} Q(QH_FQ)^{-1}Q &= -\frac{1}{\omega}\left(1 - \frac{1}{\omega}K_Q^{-1}\sum_k H_k Q T^k Q\right)^{-1} K_Q^{-1} \\ &= -\sum_{l=0}^{\infty} \frac{1}{(\omega)^{l+1}} \left[K_Q^{-1}\sum_k H_k Q T^k Q\right]^l K_Q^{-1}. \end{aligned} \tag{35}$$

where K_Q^{-1} is the inverse of K restricted to the complement of $|0\rangle$. By using now this approximation and truncating, we obtain a first approximation of the effective hermitian Hamiltonian in the constant sector as

$$H_0 - \frac{1}{\omega}\sum_{k\neq 0}\frac{1}{k}H_{-k}H_k\,. \tag{36}$$

As an example admitting an exact solution, let us consider

$$H(t) = g\left(\sigma^+ e^{i\omega t} + \sigma^- e^{-i\omega t}\right). \tag{37}$$

The effective Hamiltonian reads

$$H_{\text{eff}} = -\frac{g^2}{\omega}\left(1 - \frac{g^2}{\omega^2}\right)\sigma^z\,, \tag{38}$$

up to $O(g^6/\omega^5)$. It should be stressed at this point that the crucial information carried by the effective Hamiltonian is the set of eigenvalues. As stated, this example is exactly solvable, but it is also an interesting toy model to test the power of the techniques shown in this paper. To solve it, it is sensible to consider the evolution of a time-dependent Hamiltonian, which is, in physics language, the interaction picture Hamiltonian obtained from $H_S = (\Delta + \omega)\sigma^z/2 + g\sigma^x$ with respect to a free part given by $H_{\text{free}} = \omega\sigma^z/2$. For the sake of clarity, let us define precisely what the interaction picture is. Let $U_S(t)$ be the unitary solution of the initial value problem $i\partial_t U_S = H_S U_S$ with $U_S(0) = 1$. Similarly, let $U_{\text{free}}(t)$ be the unitary solution of the initial value problem $i\partial_t U_{\text{free}} = H_{\text{free}} U_{\text{free}}$ with $U_{\text{free}}(0) = 1$. The interaction picture Hamiltonian with respect to the free part H_{free} is defined as

$$H_I(t) = U_{\text{free}}(t)^\dagger\,(H_S - H_{\text{free}})\,U_{\text{free}}.$$

Now, let us consider the initial value problem $i\partial_t U_I(t) = H_I(t)U_I(t)$, with $U_I(0) = 1$ and $U_I(t)$ unitary. Then one readily sees that the solution for this initial value problem can be written as

$$U_I(t) = U_{\text{free}}(t)^\dagger U_S(t). \tag{39}$$

Alternatively, if we are presented with a time-dependent Hamiltonian, the corresponding evolution operator can be computed if the Hamiltonian is identified as the interaction picture Hamiltonian with respect to some free part.

For the specific H_S and H_{free} above we have

$$H_I(t) = e^{i\omega t\sigma^z/2}\left(\frac{\Delta}{2}\sigma^z + g\sigma^x\right)e^{i\omega t\sigma^z/2}$$

$$= \frac{\Delta}{2}\sigma^z + g\left(\cos(\omega t)\sigma^x - \sin(\omega t)\sigma^y\right)$$
$$= \frac{\Delta}{2}\sigma^z + g\left(\sigma^+ e^{i\omega t} + \sigma^- e^{-i\omega t}\right).$$

Obviously, the time dependent Hamiltonian $H(t)$ of (37) corresponds to this $H_I(t)$ in the case $\Delta = 0$. Particularising to the example the solution (39) we have

$$U_I(t) = \exp\left(\frac{i\omega t}{2}\sigma^z\right)\exp\left(-\frac{i(\omega+\Delta)t}{2}\sigma^z - igt\sigma^x\right). \tag{40}$$

Hence, by computing $U_I(2\pi/\omega)$, its eigenvalues read $(2\pi/\omega) \times (\omega/2 \pm \sqrt{(\omega+\Delta)^2/4 + g^2})$. Thus, the quasi-energies, which are defined modulo ω, can be fixed as

$$\mp\left[\frac{\omega}{2} - \sqrt{\left(\frac{\omega+\Delta}{2}\right)^2 + g^2}\right]. \tag{41}$$

If $\omega \gg \Delta, g$, the quasi-energies may be expanded to obtain

$$\pm\left(\frac{\Delta}{2} + \frac{g^2}{\omega} - \frac{g^2\Delta}{\omega^2} + \frac{g^2(\Delta^2 - g^2)}{\omega^3} + \cdots\right),$$

matching (38) in the case $\Delta = 0$.

One can also apply Schrieffer–Wolff's method in this case; no resummation is gained, however, since $\mathscr{S}_0$ is computed exactly as

$$\mathscr{S}_0 = -\frac{g}{\omega}\left(\sigma^+ \otimes |-1\rangle\langle 0| - \sigma^- \otimes |1\rangle\langle 0|\right), \tag{42}$$

giving $h_{SW}^{(1)} = \frac{g^2}{\omega}\sigma^z$ for the resonant case $\Delta = 0$. The case $\Delta \neq 0$ can also be solved exactly for $\mathscr{S}_0$, resulting in

$$\mathscr{S}_0 = \frac{2g}{\Delta - 2\omega}\left(\sigma^+ \otimes |-1\rangle\langle 0| - \sigma^- \otimes |1\rangle\langle 0|\right). \tag{43}$$

Therefore, in this case $\Delta \neq 0$, the corresponding Hamiltonian is

$$h_{SW}^{(1)} = \frac{\Delta}{2}\sigma^z - \frac{2g^2}{\Delta - 2\omega}\sigma^z. \tag{44}$$

To summarise, we have shown that both Bloch expansion and Schrieffer–Wolff's provide us with large frequency expansions for the quasi-energies in periodic Hamiltonians. As a further scope, it would be interesting to consider the application of these novel techniques to the quantum Rabi model beyond rotating-wave approximation [15, 16].

8 Conclusions

The well established adiabatic elimination procedure in quantum optics has produced some controversy in that scientific community on its meaning and on the feasibility of systematic improvements on that approximation. We have shown that, in fact, adiabatic elimination is the first term of a systematic expansion (be it perturbative or iterative) using Bloch's equation from nuclear optics. We have further shown the existence of relevant solutions. We next connected this approach to Schrieffer–Wolff's method, showing in which sense it can be said that Schrieffer–Wolff is a resummation. After the presented example, we have looked into the case of periodic Hamiltonians, employing the previous techniques on Floquet's Hamiltonian, to obtain high frequency expansions for quasi-energies. We expect these novel methods to be useful in current models of interest as is the case of the quantum Rabi model.

We acknowledge financial support from Basque Government Grants IT472-10 and IT559-10, UPV/EHU UFI 11/55, Spanish MINECO FIS2012-36673-C03-02 and FPA2009-10612, PROMISCE and SCALEQIT European projects.

References

1. Walls, D., Milburn, G.: Quantum Optics. Springer (2008)
2. Shore, B.: Acta Phys. Slovaca Rev. Tutorials **58**(3), 243 (2010)
3. Yoo, H.I., Eberly, J.: Phys. Rep. **118**(5), 239 (1985)
4. Brion, E., Pedersen, L.H., Mølmer, K.: J. Phys. A: Math. Theoret. **40**(5), 1033 (2007)
5. Paulisch, V., Rui, H., Ng, H.K., Englert, B.-G.: Eur. Phys. J. Plus **129**(1), 12 (2014)
6. Eden, R.J., Francis, N.C.: Phys. Rev. **97**, 1366 (1955)
7. Ellis, P.J., Osnes, E.: Rev. Mod. Phys. **49**, 777 (1977)
8. Schrieffer, J.R., Wolff, P.A.: Phys. Rev. **149**, 491 (1966)
9. Born, M., Oppenheimer, R.: Ann. Phys. **389**(20), 457 (1927)
10. Fenichel, N.: J. Differ. Equ. **31**(1), 53 (1979)
11. Killingbeck, J.P., Jolicard, G.: J. Phys. A: Math. Gen. **36**(20), R105 (2003)
12. Bravyi, S., DiVincenzo, D.P., Loss, D.: Ann. Phys. **326**(10), 2793 (2011)
13. Muga, G., Palao, J.P., Navarro, B., Egusquiza, I.L.: Phys. Rep. **395**, 357 (2004)
14. Shirley, J.H.: Phys. Rev. **138**, B979 (1965)
15. Braak, D.: Phys. Rev. Lett. **107**, 100401 (2011)
16. We would like to cite here the contribution to this Proceedings of Daniel Braak

The Formation and Launch of the Asia Pacific Consortium of Mathematics for Industry (APCMfI)

Masato Wakayama, Alexandra B. Hogan and Robert S. Anderssen

The evolving dynamics of the interaction between society and industry is recorded in the language of mathematics.

Abstract The Forum "Math-for-Industry" 2014 (FMfI 2014) represented the first opportunity to formally showcase the concept and formation of the Asia Pacific Consortium of Mathematics for Industry (APCMfI). This new initiative is intended to support the development of mathematics and its applications, and to enhance innovation and technology, in order to explore new research fields and improve the quality of life. A primary goal is to develop industrial mathematical research in the common Asia Pacific time zone of the East Asia and Oceania countries and to stimulate the two-way interaction between mathematics in academia and industry.

Keywords Mathematics-for-industry · Industrial problems · Two-way interaction

1 The Formation of APCMfI

The initial meetings for the planning of the Asia Pacific Consortium of Mathematics for Industry (APCMfI) took place in Canberra, Australia, from 31 March

M. Wakayama (✉)
Kyushu University, Fukuoka, Japan
e-mail: wakayama@imi.kyushu-u.ac.jp

A.B. Hogan
The Australian National University, Canberra, Australia
e-mail: alexandra.hogan@anu.edu.au

R.S. Anderssen
The Commonwealth Scientific and Industrial Research Organisation, Acton, Australia
e-mail: bob.anderssen@csiro.au

R.S. Anderssen et al. (eds.), *Applications + Practical Conceptualization + Mathematics = fruitful Innovation*, Mathematics for Industry 11,
DOI 10.1007/978-4-431-55342-7_13

to 2 April, 2014. Those present were a small group of colleagues from Japan, Malaysia, New Zealand and Australia, in order to provide representation across the Asia Pacific region. The meetings formalized a planning structure for the formation of APCMfI: formulated a management structure; discussed opportunities for seed funding; planned a membership base; and considered future activities that would benefit APCMfI members.

A central idea that emerged from the meetings was the need to emphasis the 'for' in Mathematics ***for*** Industry, so as to reflect how, in the solution of industrial problems, one must turn to the utilization of mathematics in the answering of the questions under consideration. It is of crucial importance to recognize the impact that industrial applications are having and can have on the development of fundamental mathematical concepts, and how solving practical complex problems can stimulate ideas for new mathematics.

2 Earlier History

Before the Canberra meeting, there were various discussions about the need to have an Asia Mathematics-for-Industry (MfI) consortium similar to the European Consortium for Mathematics in Industry (ECMI). The first meeting to test the water about the merits of such a collaboration occurred in Fukuoka prior to the Forum "Math-for-Industry" (FMfI) 2009. There was strong support for the idea with different colleagues agreeing to assist. The possibility of having such a consortium was again discussed and supported among the participants at the FMfI 2010 Forum in Hawaii. The next discussion occurred at Joint Workshop of the Institute for Mathematical Sciences (IMS) in the National University of Singapore and Institute of Mathematics for Industry (IMI) in Kyushu University on Mathematics for Industry "Biological and Climatic Prospects", held at the IMS meeting in 2012. It was on this occasion that it was agreed that the consortium, when formalized, should be called the Asia Pacific Consortium of Mathematics *for* Industry (APCMfI). During these various discussions, there was always strong support for the concept from colleagues across the Asia Pacific region including Australia, China, Indonesia, Japan, New Zealand, Singapore, South Korea and Vietnam.

Even though the support was strong that such a consortium should be formed, it was not until March of 2014 that the first formal organizational meeting occurred. The subsequent announcement about the decisions made at that meeting to have an APCMfI has since been circulated [1].

3 Management Structure and Membership

An interim management committee will provide direction for APCMfI for a two year establishment phase (starting April 2014). Following this phase, APCMfI management will consist of an elected board, as well as a council of representatives made up from APCMfI members. The operations of APCMfI will be in accordance with our constitution, which is currently being developed.

Current planning is that there will be three membership categories: full members (which include university departments or centers, and learned societies); industry and agency members; and individual members.

4 Planned Activities

A feature of APCMfI will be the practical activities undertaken to benefit its members. These planned activities include the following: participation in FMfI; student participation in the student poster session at FMfI; APCMfI prizes at the student poster session; internships for graduate students; regular Mathematics-for-Industry Study Groups (MISG), Math-for-Industry workshops and forums; the exchange of information and new research via a regular newsletter; and the organization of joint lectures and programs. Annual meetings will be held at a different Asia Pacific location.

The overall aim is to foster a fruitful two-way interaction between mathematical and statistical individuals and institutions on one side, and the needs of industry on the other. A special focus will be on encouraging and supporting students involvement in MfI activities, such as the FMfI students' poster session and internships.

Announcements about the establishment of APCMfI have recently been published in several international mathematical bulletins and we are pleased with the strong support and encouragement from our colleagues in China, Hawaii, South Korea, Malaysia and Singapore as well as Australia, New Zealand and Japan.

An official website for APCMfI has also been launched [2]. Here, interested parties can find information about the Consortium including information about how to become an industry, academic, or individual member. In addition, the plan is to make the Website available to colleagues throughout the Asia Pacific region to circulate information about industrial mathematics opportunities. This can be done by submitting information to the APCMfI website administrator.

5 The Importance of APCMfI

A key goal for APCMfI is to become a central body in the Asia Pacific region that not only connects people, but strengthens and highlights the excellent research being done in mathematics for industry in this region. There are many benefits in taking an international approach in building this new network. There are opportunities in sharing international experiences and benefits in linking junior and senior mathematicians. The broader geographical focus that APCMfI will stimulate will enhance industrial mathematics innovation in the Asia Pacific region (Fig. 1)

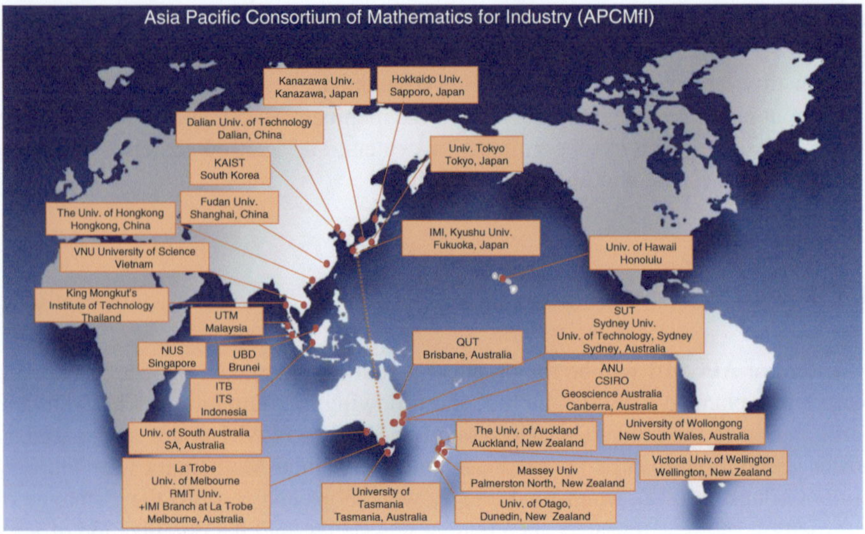

Fig. 1 The map shows the geographical region envisaged for APCMfI members, and highlights the opportunity provided by adjacent time zones

One way of stressing the future importance of APCMfI is to look at reinterpretations of the APCMfI acronym from various mathematical perspectives. The theme for FMfI 2014 was:

$$Applications + Perceptive\,Conceptualizations + Mathematics = fruitful\,Innovation.$$

As motivation for the study of mathematics, one possibility is:

$$Active\,Participation\,(in)\,Challenging\,Mathematics = future\,Intelligence.$$

6 APCMfI and Industrial Mathematics

The annual FMfI events have illustrated the growing importance of industrial mathematics for mathematics and innovation, and has fostered strong Asia Pacific collaboration and involvement, as well as international. The annual FMfI events started in Fukuoka in Japan, attracting guests from many countries. In order to support the APCMfI initiative, the plan is to extend this event throughout the Asia Pacific region, with plans for FMfI to be held in Brisbane, Australia, in 2016 and in Hawaii in 2017.

The different ways in which industrial mathematics performs a fundamental linking of the two way interaction between applications and mathematics have been the themes for the Forums in recent years. The details can be found using the links on the APCMfI website.

Acknowledgments The authors wish to thank the individuals who contributed to the first APCMfI planning meeting: Yasuhide Fukumoto (Japan), Graeme Wake (New Zealand), Zainal Aziz (Malaysia), Frank de Hoog (Australia) and the late Geoff Mercer (Australia). The authors also thank a secretary of the Institute of Mathematics for Industry at Kyushu University, Kazuko Ito, for her efficient assistance to APCMfI.

References

1. Announcement—Motivation and Planning for the Formation and Launch of APCMfI (2014). http://www.apcmfi.org/about/outline.html
2. APCMfI website. http://www.apcmfi.org/

Data Value Estimation for Privacy-Preserving Big/Personal Data Businesses

Shinsaku Kiyomoto

Abstract "Value Proposition" is a key factor when designing a business model. In personalized services, data value should be estimated using a certain model of data valuation. Generally, the data value depends on the size and precision of data, and it is expected to reflect the parameter k in the size of k-anonymized data sets and its data precision. A data set is said to have *k-anonymity* if each record is indistinguishable from at least $k - 1$ other records with respect to certain identifying attributes called *quasi-identifiers*. The parameter k influences not only the re-identification risk of the published data but also its value. When k is large, many attributes in the published data are replaced with uncharacteristic values in order to satisfy k-anonymity. On the other hand, a small k involves a serious risk of re-identification. There is a trade-off between privacy level and data value in the generation of k-anonymized data sets. Based only on the privacy requirement for reducing the re-identification risk, many people may assent to distribution of their private data in the form of a k-anonymized table when k is large enough or where as large k as possible is chosen. In this paper, we present a model for finding an appropriate k in k-anonymization. The model suggests that an optimal k exists that is appropriate to achieve a balance between value and anonymity when personal data are published.

Keywords Privacy · k-anonymity · Data value · Personal data · Data publishing · Anonymization

1 Introduction

Personalization has been implemented in a variety of services such as targeted advertisements, personalized searches, collaborative filtering, and location-based services. Breaches of privacy have been a major concern for users of personalized services, not only when using online web services but also with offline real services. O2O

S. Kiyomoto (✉)
KDDI R & D Laboratories Inc., 2-1-15, Ohara, Fujimino, Saitama 356-8502, Japan
e-mail: kiyomoto@kddilabs.jp

R.S. Anderssen et al. (eds.), *Applications + Practical Conceptualization + Mathematics = fruitful Innovation*, Mathematics for Industry 11,
DOI 10.1007/978-4-431-55342-7_14

(Online to Offline) is a new direction for commercial services; however, privacy concerns have become a serious issue due to the expansion of service collaborations [4, 22]. Users become very concerned when diverted to services; they were unaware of having any relationship with. In fact, some studies [14, 16] have suggested that Internet ads personalized by means of private data leak users' private information. Thus, privacy is an increasingly important aspect of data publishing. Sensitive data, such as medical records in public databases, are recognized as a valuable source of information for the allocation of public funds, medical research and statistical trend analysis [2]. However, if personal confidential information is leaked from the database, the service will be regarded as unacceptable by the original owners of the data [9]. Thus, anonymization methods represent a possible answer to the problem of securely protecting the privacy of individuals whose information is published [6]. One class of models, called *global-recoding*, maps the values of attributes to other values [28] in order to generate an anonymized dataset.

k-Anonymity [20, 21, 25] is a well-established notion for securing the privacy required for data publishing services. Attempts to quantify anonymity in statistical databases have led to a number of metrics, mostly k-anonymity-based [3, 17, 18, 24, 27, 29]. This concept is widely used for defining privacy requirements in several data publishing services. However, it is very difficult to choose an appropriate k in real services. Currently, the problem of choosing an appropriate k is usually discussed based on the underlying assumption that the parameter k should be chosen according to the essential privacy requirement of each service. However, this approach is problematic due to the difficulty of defining the common privacy requirement for all data purveyors. People's concerns may be directly related to the size of k, but the level of concern is generally different for each data purveyor. Thus, another approach is required to find a method for choosing an appropriate k for data publishing services.

"Value Proposition" is a key factor when designing a business model. In personalized services, the data value should be estimated using a certain model of data valuation. Generally, the data value depends on the size and precision of data, and it is expected to reflect the parameter k in the size of k-anonymized data sets and its data precision. The parameter k influences not only the re-identification risk of the published data but also its value. When k is large, many attributes in the published data are replaced with uncharacteristic values in order to satisfy k-anonymity. On the other hand, a small k involves a serious risk of re-identification. There is a trade-off between privacy level and data value in the generation of k-anonymized data sets. Based only on the privacy requirement for reducing the re-identification risk, many people may assent to distribution of their private data in the form of a k-anonymized table when k is large enough or where as large k as possible is chosen.

In this paper, we consider the problem of how to conure k in k-anonymization. We model values of k-anonymized data sets, and discuss a method for finding an appropriate k with regard to the trade-off between value and anonymity. This paper is the first step towards realizing an optimal configuration for the publication of k-anonymized data sets. The rest of this paper is organized as follows: related work and preliminary information are presented in Sect. 2. We present our data valuation model and experimental results in Sect. 3. Some remaining issues are presented in Sect. 4. We conclude this paper in Sect. 5.

2 Related Work

Valuation of private information is an interesting academic challenge but it also has beneficial applications in commercial services. Examples of pricing sensitive data in buyer-seller relationships have been presented in [1, 19]. Empirical measurements of the monetary premium such as in [15] could be linked to information-theoretic quantities. A question that is inextricably associated with all considerations of privacy is how to define "privacy". Generally, more sensitive data would be more valuable, but the user may not be willing to provide such sensitive private data. *Differential Privacy* [10, 11] is a notion of privacy using perturbative methods and is based on the statistical distance between two database tables differing by at most one element. The basic idea is that, regardless of background knowledge, an adversary with access to the data set draws the same conclusions, whether or not a particular person's data is included in the data set. That is, any person's data has an insignificant effect on the processing of a query. This notion is theoretically sound, but is not applicable to all data publishing services due to the limitation of the model. The model does not include off-line anonymization and publication of the anonymized data sets.

Fisher-Hubner [13] uses the entropy of attributes in a database to measure the risk of re-identification as the reciprocal of the number of records. Serjantov and Danezis [23] as well as Diaz et al. [8] propose Shannon entropy to measure the uncertainty of an outside observer about the assignment of usersl roles in a communication system. Toth et al. [26] define an upper bound for the probability of identification as the "degree of anonymity" which must not be exceeded for any user. *Unlinkability* is another approach to privacy. Clauß provided unlinkability measurements in which each data subject's identity is defined by a set of finite discrete attributes. Fan et al. [12] presented an abstract model called Privacy Petri Net (PPN) for privacy leak analysis.

All these studies assume that the value of personal data is measured at a single point in time. Such measurements do not consider its value under a possible future scenario. Furthermore, no method has been suggested for finding an appropriate k for k-anonymized data sets. Berthold and Bohme [5] first presented a privacy valuation model for attribute values using option pricing theory. Their model assumed uncertainty of the value in a future situation. We apply their model for risk analysis of re-identification and design a new valuation model for k-anonymized data sets.

k-anonymity: In this paper, we use k-anonymity as a privacy concept for published data sets. A data set is said to have *k-anonymity* if each record is indistinguishable from at least $k-1$ other records with respect to certain identifying attributes called *quasi-identifiers* [7]. In other words, at least k records must exist in the data set for each combination of the identifying attributes. Clearly any generalization algorithm that converts a database into one with k-anonymity involves a loss of information in that database. A database table T in which the attributes of each user are denoted in one record is in the public domain, and an attacker obtains the table and tries to distinguish the record of an individual. Suppose that a database table T has m records and n attributes $\{A_1, \ldots, A_n\}$. Each record $\mathbf{a^i} = (a_1^i, \ldots, a_n^i)$ can thus be considered to be an n-tuple of attribute values, where a^i_j is the value of attribute A_j in record $\mathbf{a^i}$. The database table T itself can thus be regarded as the set of records $T = \{\mathbf{a^i} : 1 \leq i \leq m\}$. The definition of k-anonymity is as follows:

k-Anonymity [20]. A table T is said to have *k-anonymity* if and only if each n-tuple of attribute values $\mathbf{a} \in T$ appears at least k times in T.

Our objective is to find an appropriate k for generation of the k-anonymized data sets.

3 Data Valuation

In this paper, we discuss how to find an appropriate k for the k-anonymization of data to be published. We consider a trade-off between data value and the risk of re-identification. The basic concept of our approach is to choose the k that produces the maximum benefit from published data. The benefit is calculated as the positive value that is the value of the published data and a negative value that quantifies the re-identification risk. We define a valuation model for published data in this section.

3.1 Model of Data Distribution Services

A quasi-identifier is an attribute that can be linked with external information to re-identify individual records with sufficiently high probability [7]. Generally, an anonymization table $T = (T^q|T^s)$ consists of two types of information: a sub-table of quasi-identifiers T^q and a sub-table of other attributes T^s. Since the other attributes represent the essential information with regard to database queries, a generalization method is used to modify (anonymize) T^q in order to prevent the identification of the owners of the other attributes while retaining the full information in T^s. That is, only T^q is modified by an anonymization algorithm and T^s is held. We assume that a published data set is an anonymized table T satisfying k-anonymity. A data holder has an original table including sensitive private information, and provides a service that generates an anonymization table. The anonymization mechanism generates a table T for data users and the data holder distributes the k-anonymized table. The data holder receives remuneration as a reasonable reward for distribution of the table from data users.

The data holder discloses a privacy requirement that is defined as a value of k and gathers personal data from data purveyors. A potential data purveyor checks the privacy requirement (the value of k), and sends personal data to the data holder if the privacy requirement fits his/her own privacy policy. The data holder hopes to choose an appropriate k to maximize the total value of the k-anonymized table: that is related not only to the number of data purveyors but also to the granularity of data sets. The granularity of data sets is generally determined by the value of k.

3.2 Model for Data Valuation

In this subsection, we define a data value model and show the experimental results. Generally, the value of data depends on two factors: the amount of data and the granularity of the data.

The precision of data depends on k in cases of k-anonymized data sets. Thus, we can define the granularity g of data as a function of k; that is $\mathsf{g} = f_g(k)$. The granularity is directly related to the value of the data. Generally, more precise data is more valuable to data users.

On the other hand, the amount of data depends on the number of users who agree to provide their personal data. It is assumed that the number of users providing their personal data increases when the level of privacy protection is higher. The level of privacy is defined as k in k-anonymized data sets. Therefore, the amount of data h is also defined as a function of k such that $\mathsf{h} = f_h(k)$. Thus, the value of data v is calculated as $\mathsf{v} = \mathsf{gh} = f_g(k) f_h(k)$.

As an illustrative example, we assume that the distribution of the number of users who accept k as appropriate is a normal distribution. The normal distribution is a simple model that many users accept a value of k' or values around k'. The number of users is defined as:

$$f_h(k) \approx N \times \int_{-\infty}^{\frac{k-k'}{k'}} \frac{1}{\sigma\sqrt{2\pi}} e^{-\frac{1}{2}(\frac{x-k'}{\sigma})^2} dx,$$

where N is the total number of users, k' is the value of k that the largest number of users see as appropriate for protecting privacy (average of the normal distribution), and ρ is the variance of the normal distribution, respectively.

We define $f_g(k)$ based on Shannon's entropy definition. When k is selected, at least k people have the same set of attributes in the published data. If we randomly choose a person, the probability q that the person has a certain set of attributes is calculated as $q \geq k/N$, where N is the total number of people. Thus, Shannon's entropy is calculated as $-\frac{k}{N} log(k/N)$. The number of groups can be estimated to be $\frac{N}{k}$. Thus, we define $f_g(k)$ as follows:

$$f_g(k) = -\alpha \frac{N}{k} \times \frac{k}{N} \times log_2(k/N) = -\alpha log_2(k/N),$$

where α is a constant value. We assume that the number of data users is constant or reduced with the passage of time, and thus, the value is proportional to time $t^{1/\gamma}$, where γ is a constant value. The function $t^{1/\gamma}$ models the decrease in the number of data users. The number of data users will be reduced dramatically when γ is large. In this model, we can calculate the value of the data as follows:

$$V(k, t) = -\alpha t^{\frac{1}{\gamma}} log_2(k/N) \times N \times \int_{-\infty}^{\frac{k-k'}{k'}} \frac{1}{\sigma\sqrt{2\pi}} e^{-\frac{1}{2}(\frac{x-\mu}{\sigma})^2} dx,$$

where t is the time period during which the data will be used.

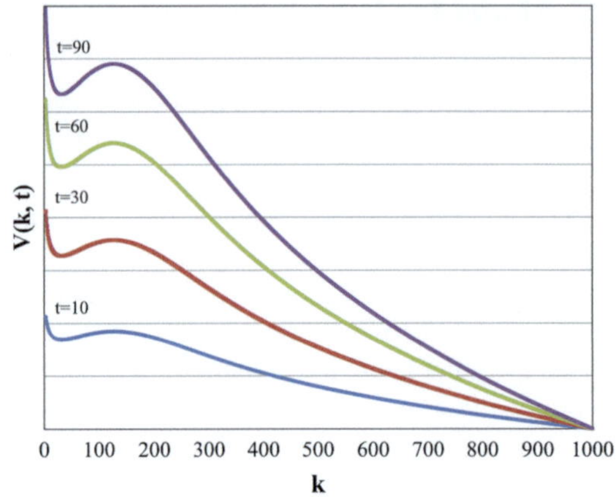

Fig. 1 The value of $V(k, t)$ for each k (normal distribution), when $\gamma = 0.6$

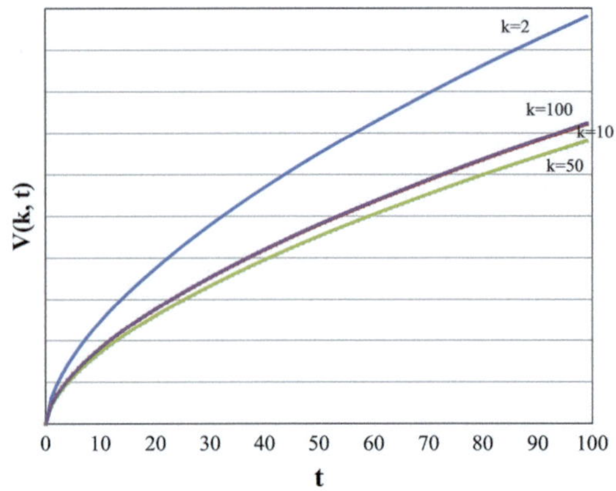

Fig. 2 The value of $V(k, t)$ for each t, when $\gamma = 0.6$

We examine the cases where $\gamma = 0.2, 0.4, 0.6, 0.8, 1.0$. The graphs of $V(k, t)$ have similar properties in all these cases. The case where $\gamma = 0.6$ is shown in Figs. 1 and 2. In Fig. 1, $V(k, t)$ has the maximum value where k is smallest, and has local maximum for particular values of k (between 100 and 200 in this case). Thus, we have two choices for appropriate k: the k that produces the local maximum of $V(k, t)$, or a possibly smaller k according to a privacy policy. In Fig. 2, the value of $V(k, t)$ monotonically increases with increasing t. The result suggests that the time period of data distribution should be as long as possible if we do not consider the risk of re-identification.

We can choose other models for the distribution of the number of users. For example, the value function $V(k, t)$ is shown in Fig. 3, where we use a χ^2-distribution as follows:

$$f_h(k) \approx N \times \int_0^k \frac{(\frac{1}{2})^{\frac{n}{2}} x^{\frac{n}{2}-1} e^{-\frac{x}{2}}}{\Gamma(\frac{n}{2})} dx$$

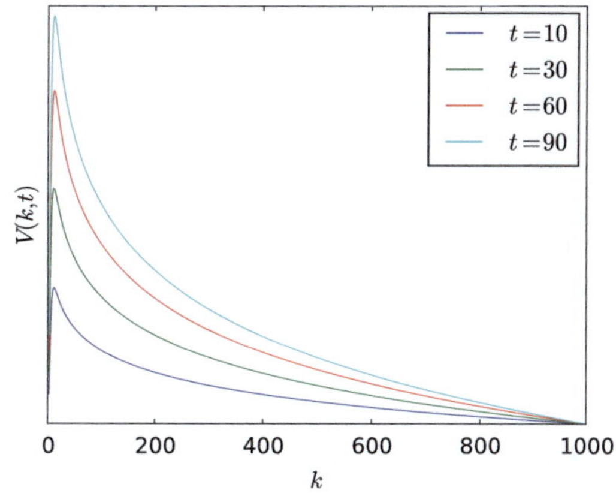

Fig. 3 The value of $V(k, t)$ for each k (χ^2 distribution case)

$$\Gamma(x) = \int_0^\infty t^{x-1} e^{-t} dt.$$

The above case means that the number of users concerned about privacy for the data distribution is not large. The χ^2-square distribution fits the case that the user distributes non-sensitive attributes, because the distribution has a peak around a low k' such as $k' = 2$ or 3, and an appropriate k for almost all users are around the low k.

Another case that uses a distribution model defined by the β- as follows is described in Fig. 4.

$$f_h(k) \approx N \times \int_0^{\frac{k}{N-1}} \frac{x^{\alpha-1}(1-x)^{\beta-1}}{B(\alpha, \beta)} dx,$$

where $\alpha = 0.5$, $\beta = 0.5$, and $B(\alpha, \beta) = \int_0^1 t^{\alpha-1}(1-t)^{\beta-1} dt$. The user distribution in this case is separated into two groups: a group of users unconcerned about privacy

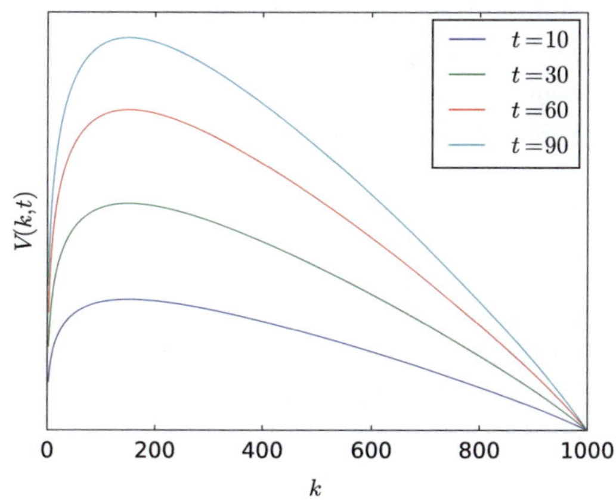

Fig. 4 The value of $V(k, t)$ for each k (β-function distribution case)

and another group of users very concerned about privacy. That is, the attributes in the data sets seem to include both sensitive and non-sensitive data. A user group that accepts a small value of k has non-sensitive data; thus, they can provide their personal data even though the value of k is low. On the other hand, a user group that requires a large value of k has sensitive data or their data is potentially sensitive and their privacy may be violated when the attributes are combined with other attributes. In this situation, a data collector should choose the target group of users and then decide the value of k. Figure 4 suggests that the target group should be the former group accepting a low value of k.

4 Remaining Issues

In this paper, we have presented a model for finding an appropriate k in personal data distribution services. The model suggested that we can find the optimal k for each service: however, before this model can be applied to real services some remaining issues need to be solved. The following issues still need to be addressed:

- **Choice of a Value Model and Parameter Setting**. Feasibility of the distribution of users was discussed in relation to the data valuation model. We presented several distribution models for the number of users such as a normal distribution. These distributions seem to be reasonable, but we have to choose an appropriate value model for each service and each set of personal data, and we need to validate it based on experimental evidence such as that obtained by a questionnaire. The method for choosing parameters such as α and β is another issue. Our model is both a concrete and abstract model. If we apply it to a real service, we have to select values for the corresponding parameters. However, the methods for choosing the reasonable parameters is an open issue; therefore the experimental results obtained from trials with a real service should be used for this purpose.
- **Risk Analysis**. There still remains a risk that a certain user is re-identified from a published data set, even though the published data set was k-anonymized. Thus, we should consider how to remove the remaining risk in a real service. In a business model, the risk is regarded as a negative factor and consists of the damaging impact of re-identification (which implies that payment of a certain amount of compensation would be necessary) and the probability that such re-identification is attributable to an adversary. If we design a model for calculating the negative factor, we can estimate the risk for publication of the k-anonymized data sets. The model may be designed according to data types and services that use k-anonymized data sets. This issue could also be examined in more detail by conducting a trial with a real service. If the model is established, insurance could be devised that covers the damage and thus the risk would be removed from the business model.

5 Conclusion

In this paper, we have presented a concrete model for analyzing the relationship between a parameter k and the value of published personal data. The model suggests that an optimal k exists, depending on the balance between the value of the data and the risk of its publication. We also considered some remaining issues related to the design of a business model for personal/big data publishing. This is the first step towards realizing an optimal configuration for the publication of personal data. We will continue to improve the model based on experiments with real services.

Acknowledgments This work was partially supported by the SURFBIRD project, CREST, JST.

References

1. Acquisti, A., Varian, H.R.: Conditioning prices on purchase history. Mark. Sci. **24**, 1–15 (2005)
2. Adam, N.R., Wortmann, J.C.: Security-control methods for statistical database: a comparative study. ACM Comp. Surv. **21**(4), 515–556 (1989)
3. Al-Fedaghi, S.S.: Balanced k-anonymity. Proc. of WASET **6**, 179–182 (2005)
4. Basu, A., Vaidya, J., Kikuchi, H.: Efficient privacy-preserving collaborative filtering based on the weighted slope one predictor. J. Internet Serv. Inf. Secur. **1**(4), 26–46 (2011)
5. Berthold, S., Bohme, R.: Valuating privacy with option pricing theory. In: Proceedings of the Eighth Workshop on the Economics of Information Security (WEIS 2009) (2009)
6. Ciriani, V., De Capitani di Vimercati, S., Foresti, S., Samarati, P.: k-anonymous data mining: a survey, models and algorithms. In: Privacy-Preserving Data Mining. Springer (2008)
7. Dalenius, T.: Finding a needle in a haystack—or identifying anonymous census record. J. Official Stat. **2**(3), 329–336 (1986)
8. Díaz, C., Seys, S., Claessens, J., Preneel, B.: Towards measuring anonymity. In: Proceedings of the 2nd International Conference on Privacy Enhancing Technologies, of PET'02. LNCS, vol. 2482, pp. 54–68 (2003)
9. Duncan, G., Lambert, D.: The risk of disclosure for microdata. J. Business Econ. Stat. **7**, 207–217 (1989)
10. Dwork, C.: Differential privacy. In: Proceedings of ICALP 2006. LNCS, vol. 4052, pp. 1–12 (2006)
11. Dwork, C.: Differential privacy: a survey of results. In: Proceedings of TAMC 2008. LNCS, vol. 4978, pp. 1–19 (2008)
12. Fan, L., Wang, Y., Cheng, X., Jin, S.: Quantitative analysis for privacy leak software with privacy petri net. In: Proceedings of the ACM SIGKDD Workshop on Intelligence and Security Informatics, ISI-KDD '12, pp. 7:1–7:9 (2012)
13. Fischer-Hubner, S.: It-security and privacy. IT-security and privacy design and use of privacy-enhancing security mechanisms series. LNCS **1958**, (2001)
14. Guha, S., Cheng, B., Francis, P.: Challenges in measuring online advertising systems. In: Proceedings of the 10th ACM SIGCOMM Conference On Internet Measurement, IMC '10, pp. 81–87 (2010)
15. Huberman, B.A., Adar, E., Fine, L.R.: Valuating privacy. Security privacy, IEEE **3**(5), 22–25 (2005)
16. Korolova, A.: Privacy violations using microtargeted ads: a case study. In: Proceedings of the 2010 IEEE International Conference on Data Mining Workshops, ICDMW '10, pp. 474–482 (2010)

17. Machanavajjhala, A., Gehrke, J., Kifer, D.: l-anonymity. In: Proceedings of ICDE'06, pp. 24–35 (2006)
18. Machanavajjhala, A., Gehrke, J., Kifer, D.: t-diversity. In: Proceedings of ICDE'07, pp. 106–115 (2007)
19. Odlyzko, A.: Privacy, economics, and price discrimination on the internet. In: Proceedings of the 5th International Conference on Electronic Commerce, ICEC '03, pp. 355–366 (2003)
20. Samarati, P.: Protecting respondents' identities in microdata release. IEEE Trans. Knowl. Data Eng. **13**(6), 1010–1027 (2001)
21. Samarati, P., Sweeney, L.: Generalizing data to provide anonymity when disclosing information. In: Proceedings of the 17th ACM SIGACT-SIGMOD-SIGART Symposium on Principles of Database Systems (PODS'98), p. 188 (1998)
22. Scipioni, M.P., Langheinrich, M.: Towards a new privacy-aware location sharing platform. J. Internet Serv. Inf. Secur. **1**(4), 47–59 (2011)
23. Serjantov, A., Danezis, G.: Towards an information theoretic metric for anonymity. In: Proceedings of the 2nd International Conference on Privacy Enhancing Technologies of PET'02. LNCS, vol. 2482, pp. 41–53 (2003)
24. Sun, X., Wang, H., Li, J., Truta, T.M., Li, P.: (p^{+}, α)-anonymity: a new enhanced privacy protection model. In: Proceedings of CIT'08, pp. 59–64 (2008)
25. Sweeney, L.: Achieving k-anonymity privacy protection using generalization and suppression. J. Uncertainty Fuzziness Knowl.-Based Syst. **10**(5), 571–588 (2002)
26. Toth, G., Hornak, Z., Vajda, F.: Measuring anonymity revisited. In: Proceedings of the 9th Nordic Workshop on Secure IT Systems, pp. 85–90 (2004)
27. Truta, T.M., Vinay, B.: Privacy protection: p-anonymity property. In: Proceedings of ICDE'06, pp. 94–103 (2006)
28. Willenborg, L., de Waal, T.: Elements of Statistical Disclosure Control, vol. 155. Springer, LNS (2001)
29. Wong, R.C.W., Li, J., Fu, A.W.C., Wang, K.: (α, k)-anonymity model for privacy preserving data publishing. In: Proceedings of ACM SIGKDD'06, pp. 754–759 (2006)

Australian Space Research Program—Platform Technologies for Space, Atmosphere and Climate Project: Selected Innovations

Robert Norman, Brett Carter, James Bennett, John Le Marshall, John Hearne and Kefei Zhang

Abstract In 2010 an international consortium led by RMIT University was awarded a multi-million-dollar Australian Space Research Program (ASRP) Project—"Platform Technologies for Space Atmosphere and Climate". This project is part of the Australian Government's recent space-related initiative to support national strategic, economic and social objectives. The aim of the project was to develop new algorithms, new approaches, software and system platforms for enhancing Australias capability in space-related research and promoting innovative applications of space-related cutting-edge technologies in Australia. The main project objectives and a selection of innovative research findings are presented, highlighting the applications, practical conceptualization and mathematics required.

Keywords Radio waves · Radio occultation · Ionosphere · Space debris · Orbital decay · Plasma bubbles

R. Norman (✉) · B. Carter · J. Bennett · J. Hearne · K. Zhang
School of Mathematical and Geospatial Sciences, RMIT University, Melbourne, Australia
e-mail: robert.norman@rmit.edu.au

B. Carter
e-mail: brett.carter@rmit.edu.au

J. Hearne
e-mail: john.hearne@rmit.edu.au

K. Zhang
e-mail: kefei.zhang@rmit.edu.au

J. Le Marshall
Australian Government Bureau of Meteorology, Melbourne, Australia
e-mail: j.lemarshall@bom.gov.au

R.S. Anderssen et al. (eds.), *Applications + Practical Conceptualization + Mathematics = fruitful Innovation*, Mathematics for Industry 11,
DOI 10.1007/978-4-431-55342-7_15

1 Introduction

An Australian Space Research Program (ASRP) project was awarded to an international consortium led by the SPACE Research Centre at RMIT University in 2010. This research formed an important part of the Australian Government's space-related initiatives to support national strategic, economic and social objectives, and to enhance Australia's space capabilities by developing integrated and advanced space-based platform technologies through a multi-sensor satellite remote sensing approach. The primary research tasks and a selection of fruitful innovations from this project are presented; particularly those research tasks in the context of Global Navigation Satellite System (GNSS) Radio Occultation (RO) and its innovative applications in the areas of space weather, geometrical optics, space debris tracking and weather prediction over the Australian region. The applications, concepts and mathematics used to develop these innovative models and important research findings are presented.

The GNSS RO is a robust Earth observation technique that provides important information for atmospheric parameter profiling and meteorological applications. The GNSS satellites are located at altitudes of $\sim$20,000 km above the Earth and they orbit the Earth twice daily. The GNSS satellites continuously transmit L-band frequency signals which are widely used for positioning, navigation and timing. The GNSS RO technique uses GNSS receivers on-board Low Earth Orbit (LEO) satellites (orbiting at altitudes of 200–2,000 km) to measure the phase delay in the received radio signals from GNSS satellites. The received GNSS signals are refracted by the Earth's atmosphere (Fig. 1) and using sophisticated atmospheric parameter retrieval processes profiles of atmospheric temperature, pressure, water vapour and electron concentrations in the ionosphere can be determined.

The international consortium consisted of RMIT University, the University of New South Wales (UNSW), Curtin University of Technology (CUT), the Australian government Bureau of Meteorology, Electro Optic Systems (EOS), GPSat Systems Australia (GSA) and the Centre of Space and Remote Sensing Research, National

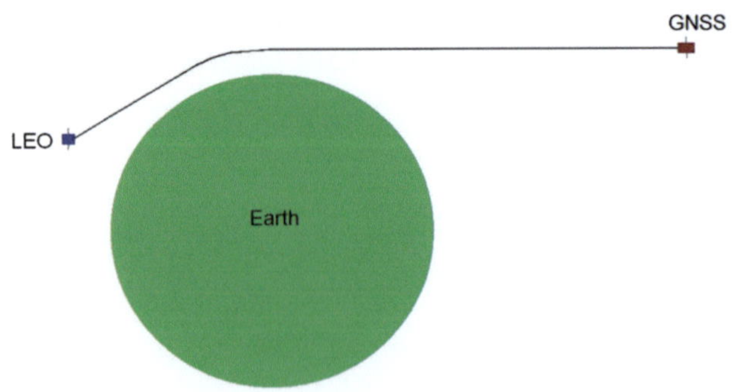

Fig. 1 An artist impression of the Earth's atmosphere occulting the GNSS to LEO signal path

Central University (NCU), Taiwan with support from the National Space Organization of Taiwan and the World Data Center for Meteorology (WDCM) from USA. The consortium is a mixture of government and academia research organizations and industry.

2 Aim and Objectives

The aim of the project was to enhance Australia's space capabilities by developing integrated and advanced space-based platform technologies through a multi-sensor satellite remote sensing approach. Theoretical, technical and practical application issues related to new generation navigation and geo-environmental remote sensing satellite systems were investigated, modeled and assessed.

The main objectives of this ASRP project were to:

- develop advanced algorithms for precise real-time or near real-time in-space tracking and navigation, and Precise Orbit Determination (POD) for current and future geo-environmental satellites
- investigate atmospheric mass density models in order to improve the accuracy, reliability and efficiency of the determination of space objects and space surveillance systems
- develop new algorithms and optimization for precise ubiquitous positioning and mapping in the context of new generation GNSS
- investigate the effects of the Earth's magnetic field, the troposphere, the stratosphere and the ionosphere on the electro-magnetic L-band frequency ray paths, including the development of comprehensive 3-D ray tracing application software packages
- study the atmosphere, the ionosphere and space weather using GNSS RO techniques
- evaluate and assimilate remote sensing data from multi-sensor satellites and to develop space-based platform technologies for investigating climate change and climatic hazards
- improve the characterization of climate in the Australian region based on the new models, algorithms, methodologies to be developed and implemented in the applications software for this project.

3 Project Outline

The project covered four main research themes with eight work packages (WPs). The relationship among these four main areas and associated eight WPs as well as the specific research is shown in Fig. 2.

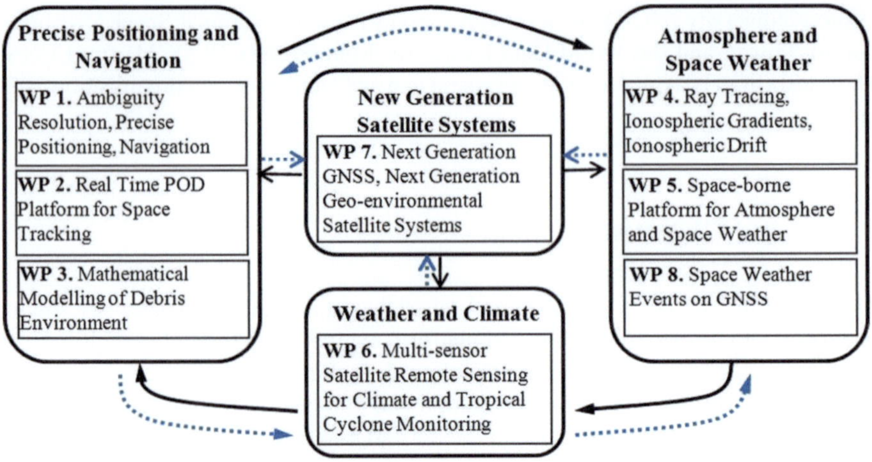

Fig. 2 A schematic identifying the four key research themes and their associated eight work packages and the road map displaying their relationships

4 Selection of Innovative Applications from the Project

Research in the aforementioned areas has been performed and substantial new novel and innovative models, algorithms and applications for the eight work packages has been achieved. More than 60 scientific research publications came from this 3 year project.

This paper will focus on just three of the innovative applications achieved, each on a very different aspect related to GNSS:

1. Innovative new geometrical ray tracing techniques for simulating GNSS propagation paths
2. Mathematical modeling of the low-Earth orbit environment
3. The effects of space weather on GNSS signals—Predicting equatorial plasma bubbles and the occurrence of GNSS scintillation.

5 Innovative New Geometrical Ray Tracing Techniques for Simulating GNSS Propagation Paths

GNSS RO is a relatively new meteorological observational data type that provides important information on the thermodynamic state of the atmosphere, improving atmospheric analyses and prognoses. One significant outcome as a result of this research project is the improvement in the accuracy of short to mid-term (3–5 days) weather forecasts by up to 10 h in the Australian region [16, 17]. GNSS RO data is now assimilated into the Australian government Bureau of Meteorology's numerical weather prediction (NWP) system and is currently considered to be one of the top 5 of the 30+ data sources used in reducing forecast error.

Although, GNSS RO data has improved weather forecasting and climate trend analysis, current GNSS RO retrieval techniques assume spherical stratification of the refractivity in the atmosphere. There is no accounting for GNSS signal paths bending via downrange and transverse refractive gradients in the atmosphere. These refractive gradients in the ionosphere and troposphere can be quite large [23]. In order to understand the effects that these refractive gradients have on GNSS signal paths we developed novel numerical and analytical ray tracing techniques. The potential benefits of improving GNSS RO atmospheric parameter retrieval techniques is more accurate weather forecasting, improved monitoring of short term climate trends, improved GNSS positioning, navigation and timing and a deeper understanding of the Earth system.

The new ray tracing techniques are based on geometrical optics and not wave theory, or full wave solutions, which are generally of far greater complexity. The applications, concepts and mathematics used to develop the innovative ray tracing techniques are described below.

Numerical and analytical ray tracing methods are commonly used for calculating the path of an electromagnetic signal in a medium specified by a refractive index, such as the Earth's atmosphere. We developed a new 3-D numerical ray tracing technique [22], based on Hamilton's equations [13], which traces finite flux or ray tubes traversing the refractive medium. This novel technique consists of eighteen differential equations representing the position and direction of the principle ray path as well two linearly independent variational ray paths. These eighteen differential equations are integrated simultaneously at each point along the ray path to determine ray parameters such as the group path, phase path, angle of arrival and signal strength. An extremely accurate homing-in algorithm capable of determining the simulated GNSS signal propagation paths reaching the desired target location, e.g., LEO satellite was developed.

Ray tracing is important for High Frequency (HF) communications and radar systems; such as Over The Horizon Radars (OTHR) and more recently for GNSS positioning and navigation and GNSS RO. These applications often require accurate near-real-time results to aid in the identification of appropriate communication channels or target locations. Analytic ray tracing techniques use explicit equations to define the ionosphere and to determine the ray path and ray parameters. Thus analytic ray tracing techniques are computationally more desirable than numerical ray tracing techniques providing near real time solutions however they are typically restricted to simple spherically stratified atmospheric models. More realistic models have been developed by Chen et al. [7] and Norman et al. [19] who developed the respective Quasi-Parabolic Segment (QPS) and Quasi-Cubic Segment (QCS) ionospheric models. These models automatically smoothly fit, using the method of least squares, QPS or QCS segments to electron density profiles of the ionosphere. Ray tracing using the QCS model involves solving elliptic integrals of the 1st, 2nd and 3rd kind in calculating the ray parameters such as phase path, group path and ground range.

The refractive index, μ, in the ionosphere for the isotropic case depends upon the plasma frequency, f_N, and signal frequency, f, and is given by:

$$\mu = 1 - \frac{f_N^2}{f^2}$$

For the QCS model each segment is of the form:

$$y = f_N^2 = \frac{A}{r^3} + \frac{B}{r^2} + \frac{C}{r} + D$$

where y, y' and y'' are smooth and continuous and using the method of least squares the technique attempts to fit a QCS to the electron density profile using all the data points to within a given root mean square error. If unsuccessful the program automatically discards the last electron density data point (lowest in altitude) and the process begins again. Once the QCS has been fitted a new QCS which is smoothly attached is fitted to the electron density profile. The process continues until the whole profile has been smoothly fitted with QCSs, Norman et al. [19].

The phase path P for the ray path traversing the ionosphere can be written as:

$$P = \int \frac{r^2\mu^2 dr}{r\sqrt{r^2\mu^2 - r_o^2 \cos\beta_o}}$$

where the term under the radical can be expressed as:

$$r^2\mu^2 - r_o^2\cos^2\beta_o = ar^2 + br + c + d/r$$

and a, b, c and d typically differ for each QCS segment and β_o represents the initial elevation angle. The Phase path in the ionosphere can be written as:

$$P = \int \frac{(1 - D/f^2)r^2 dr}{\sqrt{ar^4 + br^3 + cr^2 + dr}} - \int \frac{C/f^2 r dr}{\sqrt{ar^4 + br^3 + cr^2 + dr}} - \int \frac{B/f^2 dr}{\sqrt{ar^4 + br^3 + cr^2 + dr}} - \int \frac{A/f^2 dr}{\sqrt{ar^4 + br^3 + cr^2 + dr}}$$

where the integration is between the upper and lower heights of each QCS.

The QCS ionospheric model is a spherically stratified model and does not take into account the downrange atmospheric refractive gradients. A 2-D Segment Method for Analytic Ray Tracing (SMART) was introduced by Norman and Cannon [20] which approximates the horizontal gradients along and in the direction of the ray path. This technique was quite revolutionary, as at the time the research directions were focused on models "tilting" the ionosphere. The SMART technique allowed 2-D analytic ray tracing through complicated ionospheric models, still using the QPS or QCS formulism. The SMART technique allowed the ray path to pass smoothly from one downrange path segment to the next. This was achieved by determining a new initial elevation in the next downrange segment.

More recently and as an objective of this project we develop a new pseudo 3-D SMART technique, Norman et al. [21]. The 3-D SMART technique uses all that the 2-D SMART technique has to offer as well as taking into account the transverse horizontal refractive gradients acting on the propagated ray path. The transverse horizontal refractive gradients are typically much smaller than the vertical refractive gradients. The 3-D SMART technique follows a form of Hamilton's equations, requiring no numerical integration, where at the end of each downrange atmospheric segment the directional component of the ray path in the transverse, or azimuthal direction is calculated. The atmospheric refractive gradients in the new direction are used in the next downrange segment, so the computational time for ray tracing simulations in going from 2-D SMART to 3-D SMART is almost negligible. Ray simulations from the 3-D SMART technique compared well to corresponding ray paths using the 3-D numerical ray tracing program, Norman et al. [21].

The equations for the change in azimuth direction, δp_ϕ and position, $\delta\phi$, are:

$$\delta p_{\phi_i} = -\frac{1}{2f^2}\frac{\partial f_N^2}{\partial\phi}\Delta P_i' \quad \text{and} \quad \delta\phi_i = \delta p_\phi \Delta P_i'$$

where $\Delta P_i'$ represents a step in group path in the ith segment.

6 Mathematical Modeling of the Low-Earth Orbit Environment

One of the main objectives of the ASRP project was to create platform technologies to enhance Australia's capabilities in space research. This has been achieved and has led to subsequent successful research grants. This section briefly outlines part of this research on mathematical modeling of the evolution of the space debris population.

The LEO space environment, considered here as the region 200–2,000 km above the Earth, has become progressively cluttered from years of space operations. The prediction of a cascading of collisions in orbit leading to an exponential increase in debris [15], motivated several modeling studies and potential methods to curb the debris growth such as mitigation measures (e.g. end of life satellite deorbiting, explosion suppression, limiting the release of launch related debris, etc.) and remediation measures such as active debris removal. An excellent study, comparing results from six well-known models, showed that mitigation alone is insufficient and remediation measures should be considered to stabilize the environment (Liou et al. [18]).

A simple source-sink model has been constructed as part of the ASRP, similar to the earlier work by Rossi et al. [25], and is designed to be a platform for extension to a comprehensive tool for modeling the change in the space debris population and mitigation and remediation scenarios.

The following is a brief description of the model that was constructed and more details may be found in Bennett and Sang [1]. The LEO region is discretized into 18 altitudinal shells, each 100 Km wide. The model is coded in C++ and considers launches and release of operational debris, fragmentation generation from explosion and collision events, as well as natural atmospheric decay. Following [25], the objects or "species" were discretized into 10 logarithmic mass bins spanning the mass range 0.042 g to 14,278 kg. The deterministic model comprises a coupled system of non-linear differential equations for the change in population $N_{ij}(t)$ for species j in shell i, given by

$$\frac{dN_{ij}(t)}{dt} = L_{ij} + D_{ij}(t) + E_{ij} + C_{ij}(t), \qquad i = 1 \ldots 17, j = 1 \ldots 10,$$

where L_{ij} is the number of objects injected into shell i of species j resulting from launches and the release of operational debris. Similarly, the terms D_{ij}, E_{ij}, and C_{ij} represent the number of objects of species j injected into (or removed from) shell i from drag-induced orbital decay, explosions, and collisions, respectively.

The term L_{ij} is considered constant and was parameterised using historical data and information from the literature. The drag-induced orbital decay of an object in a shell at height h was estimated using a circular orbital decay rate given by

$$\frac{dh(t)}{dt} = -B_c \rho(t) \sqrt{h(t)\mu}$$

where B_c is the object's ballistic coefficient (assumed constant), μ is the gravitational parameter, ρ is the atmospheric density at height h, given by a piecewise exponential relationship:

$$\rho(t) = \rho_0 e^{-(h(t)-h_0)/H}$$

where ρ_0 is the atmospheric density at reference height h_0, and H is the scale height, which varies with height and values are given in Vallado [32]. The characteristic time taken for each species to decay from each shell, τ_{ij}, was calculated by numerically integrating the above two equations to determine D_{ij} as:

$$D_{ij}(t) = \frac{N_{(i+1)j}(t)}{\tau_{(i+1)j}} - \frac{N_{ij}(t)}{\tau_{ij}}$$

The number of objects of species j created/destroyed in shell i resulting from on-orbit explosions is given by:

$$E_{ij} = E_i \times g_j$$

where E_i is the number of explosions per year in each shell, determined by averaging historical events contained in Johnson et al. [14] between 1997–2006, and g_j is the number species j created from the explosion found using the formula contained in Su and Kessler [30].

The average number of collisions $C_i(k, l, t)$ in a time step Δt in shell i is approximated by:

$$\frac{C_i(k, l, t)}{\Delta t} = \frac{v_r \sigma_{kl}(N_{ik}(t) - \delta_{kl})N_{il}(t)}{V_i(1 + \delta_{kl})}$$

where v_r is the relative velocity, σ_{kl} is the collisional cross-section, $N_{ik}(t)$ and $N_{il}(t)$ are the populations of species k and l in shell i at time t, V_i is the volume of shell i, and δ_{kl} is the Kronecker delta. The number of objects of species j created/destroyed in shell i is given by:

$$C_{ij}(t) = \sum_{\substack{k,l=1 \\ k \leq l}}^{n} f_j(k, l) \times C_i(k, l, t)$$

where $f_j(k, l)$ is the number of objects created/destroyed of species j from a collision between species k and l, and n is the number of species. The number of objects created from a collision follows a power law function and is dependent on the impact strength of the collision. Although the model is simple it still shows the essential nature present in the more sophisticated models. More details may be found in Bennett and Sang [1].

Due to the necessity to improve the orbit determination and prediction accuracy for debris objects for reliable conjunction assessments, future revisions to the model were postponed. These advancements are continuing and the debris population model will undergo a needed upgrade to form part of a concerted effort to provide a reliable service to the international space community as part of a recently announced Cooperative Research Centre for Space Environment Management.[1] The overall goal of the Centre is to remotely manoeuvre debris objects for collision avoidance using photon pressure delivered from ground-based lasers to provide a cost-effective means of remediating the debris problem. A schematic of this concept is shown in Fig. 3.

When two objects are deemed to be on a collision course, photon pressure delivered from ground-based laser stations will be used to alter the orbits of one or both of the objects. Multiple engagements will be necessary to manoeuvre the object sufficiently to overcome the orbit prediction errors. The perturbations delivered on the object are small relative to the force required to lower an object for re-entry. More information on the research programs may be found at www.serc.org.au/research/.

The model described in this section has a few shortcomings which will be addressed in future revisions. Firstly, the model only considers the LEO region. This will be extended to cover the LEO to Geostationary Orbit (GEO) regions. The

[1] http://www.minister.industry.gov.au/ministers/macfarlane/media-releases/driving-research-and-delivering-results-australia, Department of Industry, Australian Government, accessed 16-Oct-2014.

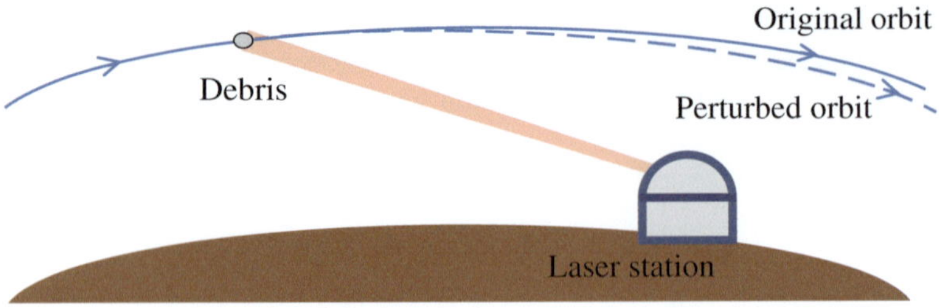

Fig. 3 Laser manoeuvre concept using photon pressure

model employs a simplified discretisation of the volumetric region into altitudinal shells, which will be relaxed in future to provide higher fidelity with object binning based on more orbital information. The sources and sinks will be extended and refined for better modeling of the population change and stochastic processes will be introduced. This will include a variety of mitigation and remediation scenarios such as active debris removal studies and ground-based laser manoeuvre for collision avoidance. The effects of solar radiation will be included which is important to model the effects at higher altitudes. Multiple fragmentation models will be included in the model for selection in debris growth studies.

Recently, our focus has been on better characterizations of the on-orbit objects and improving our orbit prediction capabilities where only sparse tracking data is available to fit in an orbit determination process [2–5, 26–28]. This is imperative for the laser manoeuvre described above. Capabilities in space situational awareness are growing for Australia to assist with curbing the debris problem.

The thrust of the modeling analysis is to determine the change in the debris population and the effects of mitigation and remediation scenarios—in particular laser manoeuvre—and assess their effectiveness in curbing the growth of space debris objects so that future space operations may continue.

7 The Effects of Space Weather on GNSS Signals—Predicting Equatorial Plasma Bubbles and the Occurrence of GNSS Scintillation

Mathematical modeling of the near-Earth space environment was another prime focus of the ASRP project with emphasis on investigating the impact of space weather on GNSS. Space weather is normally associated with the rare extreme events, when vast quantities of solar material are ejected out into space—a phenomenon called a Coronal Mass Ejection (CME)—that impacts satellite operations, power grids and many other modern technologies. Our research was primarily focused on the very common ionospheric plasma bubbles that cause problems on trans-ionospheric radio systems, such as GNSS and satellite communications, near the magnetic equator. Armed with the knowledge of what exactly causes these disturbances, one can accurately predict such disturbances before they occur, providing the increasing number of people and industries that rely on trans-ionospheric radio waves the necessary forewarning of disruptions to their systems.

Equatorial Plasma Bubbles (EPBs) are low-electron density regions that rise into the high-density plasma at higher altitudes during the local nighttime hours. The spectrum of plasma waves/irregularities associated with the EPBs effectively act as a diffraction grating on the radio waves that propagate through them, causing random fluctuations in the signal amplitude and phase called scintillations (e.g. [12, 33]). As a result, the coherent radio wave transmitted from space (e.g., from GNSS satellites) becomes unrecognizable by the receiver, and sometimes causes a break in the signal lock, called *a loss-of-lock.*

The physical process responsible can be described by the generalized Rayleigh-Taylor (R-T) plasma instability. This process is analogous to the traditional Rayleigh-Taylor fluid instability, in which a heavy fluid sits on the top of a light fluid, and gravity causes the light fluid to rise up into the heavy fluid by means of a series of localized "bubbles". The primary difference between the traditional R-T instability and the generalized R-T instability is the role of electric and magnetic fields and charged particles in the latter.

In the ionosphere, the Earth's magnetic field lines are highly conducting, meaning that both the ions and electrons can freely travel along them, provided that the collisions with the background neutrals (i.e., the thermospheric particles) are relatively infrequent. The highly conducting nature of the field lines means that the EPB-producing environment is non-local and that the R-T plasma instability growth rate must be field-line (or "flux-tube") integrated; i.e., integrating the effects of all necessary parameters along the flux tubes. The most recent formulation of the R-T plasma instability includes flux-tube integrated quantities, taking into account potential inter-hemispherical asymmetries in neutral winds and plasma distributions [31].

The flux-tube integrated generalized R-T plasma instability growth rate γ was derived by Sultan [31]:

$$\gamma = \frac{\Sigma_P^F}{\Sigma_P^F + \Sigma_P^E}\left(V_p - U_L^P - \frac{g_e}{v_{eff}}\right)K_F - R_T$$

where Σ_P^F and Σ_P^E are the flux-tube integrated E- and F-region Pederson conductivities, V_p is the upward plasma drift speed, U_L^P is the Pederson conductivity-weighted neutral wind speed perpendicular to the magnetic field line in the magnetic meridian plane, g_e is the altitude-corrected gravity, v_{eff} is the flux-tube integrated effective ion-neutral collision frequency that has been weighted by the electron density n_e, N_e is the flux-tube electron content, $K_F = \frac{1}{N_e}\frac{\partial N_e}{\partial h}$ is the F region flux-tube electron content height gradient and R_T is the electron density-weighted flux tube recombination rate. The flux-tube integrated quantities above take the form of $X = \int x \cdot dS$, where S is the distance along the field line, and those that are weighted by another parameter, such as electron density, take the form $Y = 1/N_e \int n_e y \cdot dS$ [31].

Despite the overall completeness of this theoretical formulation, ionospheric physicists have had difficulties in recent decades understanding the observed EPB occurrence rates observed around the world. The long-term seasonal/longitudinal climatology is rather well understood to be due to the strength of the V_p term in the growth rate expression above (e.g., [8]). However, why EPBs were generated on one day and not the next—i.e., the short-term EPB occurrence variability—has remained a significant issue (e.g., [9], and references therein). From an observational perspective, it is clear why such problems have dogged ionospheric physics for so long. The R-T growth rate above is a flux-tube integrated quantity, and observations of all the physical quantities that control the growth rate over an entire flux tube are virtually impossible with current ground-based and space-based instrumentation.

To deal with this difficult reality, researchers have either had to use a large EPB observational database to compile a statistical EPB occurrence model (e.g., ground-based data by Secan et al. [29] and space-based GPS RO data by Carter et al. [8] or use various ionosphere-thermosphere models in order to fill in the observational gaps. The first of these approaches doesn't require any knowledge of the physics associated with the generation of EPBs and instead uses simple occurrence statistics with the appropriate variables (e.g., solar activity, geomagnetic activity, season and location). Such empirical-type models are common and have been shown to be quite reliable in ionospheric physics; e.g. the International Reference Ionosphere (IRI) model [6]. However, as our knowledge of the ionosphere and how it couples to its surroundings advances, the physics-based models have increased in complexity.

The Thermosphere Ionosphere Electrodynamics General Circulation Model (TIEGCM) is a self-consistent time-dependent 3-dimensional physics-based model of the Earth's ionosphere-thermosphere system (e.g., [24]). The model uses a pre-known ionosphere-thermosphere state and solves the momentum and continuity equations as it is driven by external forcing from solar and geomagnetic activities (from above) and tidal forcing (from below).

The model includes the most important aspects of the chemistry of the thermosphere—major species O, O_2 and N_2 and minor species $N(^4S)$, $N(^2D)$ and NO—and solves the coupled momentum, continuity and thermodynamics equations to derive the thermospheric dynamics. The largest source of charged particles that make up the ionosphere is from photoionization, the process where solar photons in the soft X-ray ultraviolet, extreme ultraviolet and far ultraviolet energy ionize particles in the Earth's atmosphere. The model uses the chemical reaction rates of the thermospheric particles and a wavelength- and solar activity-dependent solar irradiance proxy model to calculate the resulting ionospheric plasma density across the globe. The loss of plasma density is determined by the height-dependent neutral composition and chemistry, primarily via the process of dissociative recombination; i.e., the capturing of electrons by positively charged ions. The balance between the production and loss of particles, in addition to the coupling between the plasma and the neutral winds, controls the overall spatial and temporal dynamics of the ionosphere (see Fig. 4).

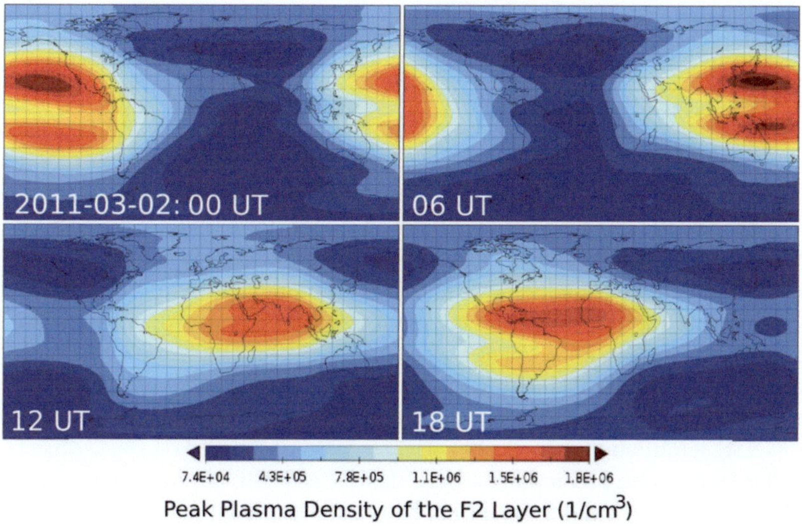

Fig. 4 Four snapshots of the global ionospheric peak plasma density given by the TIEGCM, showing the high degree of spatial and temporal variability in the global plasma distribution

The geomagnetic activity input in the model primarily controls the coupling between the magnetosphere and the high-latitude regions of the ionosphere. This magnetospheric input—among other things, such as particle precipitation—largely consists of spatially varying electric potential (and therefore, electric field) patterns that are imposed on the ionosphere where they drive plasma drift/convection. This plasma convection causes heating in the thermosphere, which alters the neutral wind dynamics in the global system.

The parameterization of the solar and geomagnetic activities by the model induces daily variability upon the model, and can therefore be used to understand the physical factors that drive the ionosphere's short-term variability. The inclusion of thermospheric chemistry and the self-consistency of the model also allows the calculation of the full flux-tube integrated R-T instability growth rate, as demonstrated recently by Carter et al. [9–11].

It has been shown that the TIEGCM, in particular, is useful in not only describing the regional ionospheric conditions leading to EPB growth in various longitude sectors around the globe, but also in forecasting daily EPB occurrence several hours in advance [11]. Physics-based ionosphere-thermosphere modeling has been increasing in complexity over recent decades and has now proved to be a very useful tool in understanding the drivers of the low-latitude ionosphere dynamics that are linked to the occurrence of EPBs.

8 Summary

This research project made significant contributions to space-related applications, e.g. space situational awareness, the tracking and navigation of space objects, the surveillance and orbit prediction of space debris for space safety and to the monitoring of space weather, meteorology and climate change. As a result of this project GNSS RO is now assimilated into the Australian weather forecasting system improving weather forecasting and benefiting all Australians [16]. From the GNSS RO research comparisons with radiosondes the GNSS RO technique has been found to be important for monitoring short term climate trends [17]. The new ray tracing techniques have been used to determine the effects of transverse refractive gradients caused by storms on the GNSS signal paths to the ground and to LEO satellites [23]. The space weather EPB research has led to predictive mathematical algorithms outperforming other GNSS scintillation models [11]. The research performed on space objects, the surveillance and orbit prediction of space debris has led to impressive outcomes and >$20M of funding from the Australian Government to continue this research.

The project also highlights what can be achieved through a wide, inter/intra-disciplinary international collaboration. A variety of research outcomes and substantial results, mathematical models and algorithms were achieved fulfilling the projects aim and objectives.

Acknowledgments The authors would like to gratefully acknowledge the Australian government's research grant support for this project through the Australian Space Research Program.

References

1. Bennett, J.C., Sang, J.: Modelling the evolution of the low-Earth orbit debris population. In: Cairns, I., Short, W. (eds.) 11th Australian Space Science Conference, Canberra, Australia, pp. 165–178 (2011)
2. Bennett, J.C., et al.: Improving low-Earth orbit predictions using two-line element data with bias correction. In: Advanced Maui Optical and Space Surveillance Technologies Conference. Maui, Hawaii (2012)
3. Bennett, J.C., et al.: Accurate orbit predictions for debris orbit manoeuvre using ground-based lasers. Adv. Space Res. **52**(11), 1876–1887 (2013)
4. Bennett, J.C., et al.: A re-analysis of the 2009 Iridium-Cosmos predicted miss distance using two-line element derived orbits. In: Cairns, W.S.I. (ed.) 12th Australian Space Science Conference, pp. 191–200. National Space Society of Australia, Melbourne, Australia (2013)
5. Bennett, J.C., et al.: An analysis of very short-arc orbit determination for low-Earth objects using sparse optical and laser tracking data. Adv. Space Res. **55**(2), 617–629 (2015)
6. Bilitza, D., Reinisch, B.: International reference ionosphere 2007: improvements and new parameters. Ad. Space Res. **42**, 599–609 (2007)
7. Chen, J., Bennett, J.A., Dyson, P.L.: Automatic fitting of quasi-parabolic segments to Ionospheric profiles with application to ground range estimation for single-station location. J. Atmos. Terr. Phys. **52**, 277–288 (1990)
8. Carter, B.A., Zhang, K., Norman, R., Kumar, V.V., Kumar, S.: On the occurrence of equatorial F-region irregularities during solar minimum using radio occultation measurements. J. Geophys. Res. Space Phys. **118**, 892–904 (2013). doi:10.1002/jgra.50089

9. Carter, B.A., Yizengaw, E., Retterer, J.M., Francis, M., Terkildsen, M., Marshall, R., Norman, R., Zhang, K.: An analysis of the quiet time day-to-day variability in the formation of post-sunset equatorial plasma bubbles in the Southeast Asian region. J. Geophys. Res. Space Phys. **119**, 3206–3223 (2014). doi:10.1002/2013JA019570
10. Carter, B.A., Retterer, J.M., Yizengaw, E., Groves, K., Caton, R., McNamara, L., Bridgwood, C., Francis, M., Terkildsen, M., Norman, R., Zhang, K.: Geomagnetic control of equatorial plasma bubble activity modeled by the TIEGCM with Kp. Geophys. Res. Lett. **41**, 5331–5339 (2014). doi:10.1002/2014GL060953
11. Carter, B.A., Retterer, J.M., Yizengaw, E., Wiens, K., Wing, S., Groves, K., Caton, R., Bridgwood, C., Francis, M., Terkildsen, M., Norman, R., Zhang, K.: Using solar wind data to predict daily GPS scintillation occurrence in the African and Asian low-latitude regions. Geophys. Res. Lett. **41**, (2014). doi:10.1002/2014GL062203
12. Caton, R.G., Carrano, C.S., Alcala, C.M., Groves, K.M., Beach, T., Sponseller, D.: Simulating the effects of scintillation on transionospheric signals with a two-way phase screen constructed from ALTAIR phase-derived TEC. Radio Sci. **44**, 9 (article id RS0A12) (2009)
13. Hamilton, W.R.: The mathematical papers of Sir William Rowan Hamilton. In: Conway, A.W., Synge, J.L. (eds.) Geometrical Optics, vol. 1, p. 164. Cambridge University Press, Cambridge (1931)
14. Johnson, N.L., et al.: History of on-orbit fragmentations. In: Orbital Debris Program Office, 14th edn. NASA Johnson Space Center, Houston, TX (2008)
15. Kessler, D.J., Cour-Palais, B.G.: Collision frequency of artificial satellites: the creation of a debris belt. J. Geophys. Res. **83**(A6), 2637–2646 (1978)
16. Le Marshall, J., Xiao, Y., Norman, R., Zhang, K., Rea, A., Cucurull, L., Seecamp, R., Steinle, P., Puri, K., Le, T.: The beneficial impact of radio occultation observations on Australian region forecasts. Aust. Meteorol. Oceanogr. J. **60**, 121–125 (2010)
17. Le Marshall, J., Xiao, Y., Norman, R., Zhang, K., Rea, A., Cucurull, L., Seecamp, R., Steinle, P., Puri, K., Fu, E., Le, T.: The application of radio occultation observations for climate monitoring and numerical weather prediction in the Australian Region. Aust. Meteorol. Oceanogr. J. **62**, 323–334 (2012)
18. Liou, J.-C., Anilkumar, A.K., Bastida, B., Hanada, T., Krag, H., Lewis, H., Raj, M.X.J., Rao, M.M., Rossi, A., Sharma, R.K.: Stability of the Future LEO Environment—An IADC comparison study, 6th European Conference on Space Debris, Darmstadt, Germany (2013). doi:10.13140/2.1.3595.6487
19. Norman, R.J., Dyson, P.L., Bennett, J.A.: Quasicubic-segmented ionospheric model. IEE Proc. Microw. Antennas Propag. **143**, 323–327 (1996)
20. Norman, R.J., Cannon, P.S.: A two-dimensional analytic ray tracing technique accommodating horizontal gradients. Radio Sci. **32**, 387–396 (1997)
21. Norman, R.J., Le Marshall, J., Carter, B.A., Wang, C.-S., Gordon, S., Zhang, K.: A new pseudo three-dimensional segment method analytical ray tracing (3-D SMART) technique. IEEE Trans. Antennas Propag. **60**, 5818–5824 (2012)
22. Norman, R.J., Bennett, J.A., Dyson, P.L., Le Marshall, J., Zhang, K.: A ray-tracing technique for determining ray tubes in anisotropic media. IEEE Trans. Antennas Propag. **61**, 2664–2675 (2013)
23. Norman, R.J., Le Marshall, J., Rohm, W., Carter, B.A., Liu, C., Zhang, K.: The impact of severe weather events on GPS signal propagation. IEEE JSTARS **PP**(99), 1–7 (2014). doi:10.1109/JSTARS.2014.2344091
24. Qian, L., Burns, A.G., Emery, B.A., Foster, B., Lu, G., Maute, A., Richmond, A.D., Roble, R.G., Solomon, S.C., Wang, W.: The NCAR TIE-GCM. In: Huba, J., Schunk, R., Khazanov, G. (eds.) Modeling the Ionosphere-Thermosphere System, pp. 73–83. Wiley, Chichester (2014). doi:10.1002/9781118704417.ch7
25. Rossi, A., et al.: Collisional evolution of the Earth's orbital debris cloud. J. Geophys. Res. **99**(E11), 23195–23210 (1994)
26. Sang, J., Bennett, J.C., Smith, C.H.: Estimation of ballistic coefficients of low altitude debris objects from historical two line elements. Adv. Space Res. **52**(1), 117–124 (2013)

27. Sang, J., Bennett, J.C., Smith, C.: Experimental results of debris orbit predictions using sparse tracking data from Mt. Stromlo. Acta Astronaut. **102**, 258–268 (2014)
28. Sang, J., Bennett, J.C.: Achievable debris orbit prediction accuracy using laser ranging data from a single station. Adv. Space Res. **54**(1), 119–124 (2014)
29. Secan, J.A., Bussey, R.M., Fremouw, E.J., Basu, S.: An improved model of equatorial scintillation. Radio Sci. **30**, 607–617 (1995). doi:10.1029/94RS03172
30. Su, S.Y., Kessler, D.J.: Contribution of explosion and future collision fragments to the orbital debris environment. Adv. Space Res. **5**(2), 25–34 (1985)
31. Sultan, P.J.: Linear theory and modeling of the Rayleigh-Taylor instability leading to the occurrence of equatorial spread F. J. Geophys. Res. **101**, 26875–26891 (1996)
32. Vallado, D.A.: Fundamentals of Astrodynamics and Applications, 3rd edn. Microcosm Press, Hawthorne and Springer, New York (2007)
33. Yeh, K.C., Liu, C.-H.: Radio wave scintillations in the ionosphere. Proc. IEEE **70**, 324–360 (1982). doi:10.1109/PROC.1982.12313

The Effect of Heaped and Sloped Powder Layers on Ejection Times and the Residence-Time Distribution of a Conical Mass-Flow Hopper

Samuel Irvine, Luke Fullard and Clive Davies

Abstract The flow of a hypothetical Coulomb material flowing under gravity from a conical mass-flow hopper is modelled using stress field theory. The assumptions inherent for a Coulomb material can be combined with the assumption of radial flow within the hopper to determine the velocity profile within the hopper. From the velocity profile, ejection times and residence time distributions may be calculated. Since, in a real granular system, the powder layer interface is generally not flat, but sloped at some angle, (nominally the angle of repose), the residence time distribution and ejection times will be dependent on the initial geometry of the powder layers. Residence time distributions and ejection times are calculated for a given granular material in a conical mass-flow hopper firstly for the case of flat layers, secondly for the case where the powder forms a conical heap at the angle of repose, and thirdly for the case when the powder is sloping against a wall. It is found that the shape of the powder layers greatly changes the residence time distribution and ejection times in the system, and needs to be considered when performing residence time measurements in the industrial setting.

Keywords Granular flow · Mixing · Hopper flow · Residence time · Radial solution · Continuum model

Nomenclature

g	Acceleration due to gravity (ms^{-1})
p	Dimensionless quantity given by Eq. (3) (.)
q	Quantity given by Eq. (3) (mskg^{-1})

S. Irvine · L. Fullard (✉)
Institute of Fundamental Sciences, Massey University, Palmerston North, New Zealand
e-mail: L.Fullard@Massey.ac.nz

S. Irvine
e-mail: sammanam1993@gmail.com

C. Davies
School of Engineering and Advanced Technology,
Massey University, Palmerston North, New Zealand
e-mail: C.Davies@massey.ac.nz

R.S. Anderssen et al. (eds.), *Applications + Practical Conceptualization + Mathematics = fruitful Innovation*, Mathematics for Industry 11,
DOI 10.1007/978-4-431-55342-7_16

r	Radial coordinate (m)
t	Time (s)
T	Exit time for a given particle (s)
v	Velocity (ms^{-1})
A_r	Heap angle (.°)
$C(t)$	Concentration of some tracer (m^{-3})
$E(t)$	Residence time distribution (.)
$F(t)$	F-curve (concentration curve) (.)
Q	Volumetric flow rate per unit width (m^2s^{-1})
α	Hopper half angle (.°)
$\dot{\gamma}$	Off diagonal terms in the rate of strain tensor (s^{-1})
$\dot{\varepsilon}$	Diagonal terms in the rate of strain tensor (s^{-1})
θ	Polar angle (zenith angle) (.°)
ρ_b	Bulk density (kgm^{-3})
σ	Normal stress components (Pa)
τ	Shear stress components (Pa)
ϕ	Effective angle of internal friction (.°)
ϕ_W	Angle of wall friction (.°)
χ	Azimuthal angle (.°)
ψ^*	Angle from Mohr's circle (.°)
ω	Angle defined by Eq. (12) (.°)

1 Introduction

The mixing of layers of powder in industrial hopper systems frequently occurs in many granular processing systems, making modelling such a problem of great financial significance for industry. Additionally, these flows can be quantified experimentally on-site by residence-time distribution experiments. Such an experiment is usually performed by measuring the ejection of a tracer which is initially assumed to enter the conical hopper as a thin flat layer. Alternatively, a step-change experiment may be perfomed, where one powder fills the hopper, and another fills the silo above. Again, the concentration of the upper powder layer is measured as a function of time, and the residence time distribution may be obtained from these results [8]. However, in granular systems layers of powder are rarely flat, and are usually inclined at some angle—a distinction from Newtonian fluids. If the hopper-silo system is loaded centrally from above, then the powder layer will form conical heaps, with the peak occurring in the middle of the hopper. If the powder is loaded into the hopper by impact with the side wall, then the powder may form sloped layers, where the maximum height of the powder layer is against the wall at one location, but at a minimum at the wall directly opposite; see Fig. 1. Both the heaped and sloped situation require that the experimental data for the residence time distribution be treated carefully, since a flat layer assumption is no longer valid. In this work, we aim to study the effect of heaped and sloped layers on the residence time distribution, with the goal being to inform industry to treat experimental results appropriately. Some

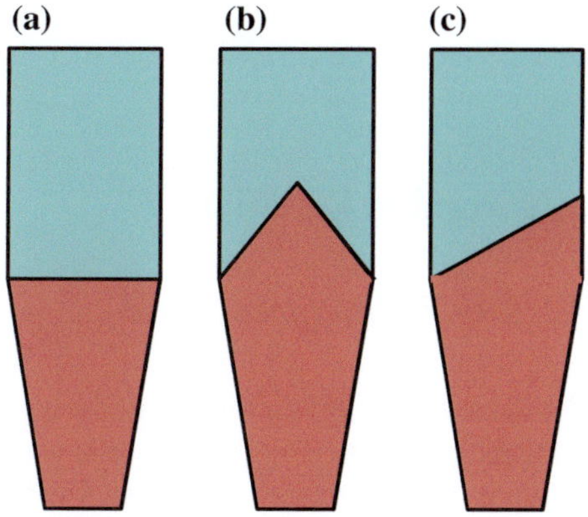

Fig. 1 A 2D sketch of the 3D situation being modelled. **a** Flat powder layers. **b** Heaped layers. **c** Sloped layers

authors have developed models for the velocity profile in a conical hopper using modelling [1, 3–5, 11], or stress field modelling [2, 6, 9, 10, 12]. Here we use the stress field model to calculate the velocity profile in a conical hopper, then evaluate the ejection times and residence time distributions for each of the three layered cases; the heaped, sloped, and horizontal layer case. We note that Johanson [7] has used the residence time distribution to evaluate the efficiency of a cone-in-cone blender.

2 System and Assumptions

The hopper-silo system which is modelled in this paper consists of a conical hopper of a given half angle α, below a cylindrical silo feeding into the hopper. The exit orifice at the bottom of the hopper is assumed to be circular. The granular material contained in the hopper-silo system is modelled as an ideal Coulomb material. Other common assumptions are also made to obtain predictions for stress and velocity profiles, see [10]. These include:

- The flow is in steady-state. The transient development of velocity in the system is not modelled.
- The flow is assumed incompressible and there is no plastic deformation of the individual particles.
- Shear and normal stresses are linearly related, and the systen is everywhere at incipient yield.
- The principle of co-axiality holds.
- The flow within the hopper is radial, while the flow in the silo is modelled as a plug.
- The flow is axially symmetric about the line $\theta = 0$.

While we acknowledge that this model and assumptions will not be valid for every situation which may occur in a hopper system (such as the presence of dilation zones) we are confident that this will suffice for comparison of layer geometry. Such an analysis may be performed for other theoretically derived velocity fields.

3 Governing Equations

3.1 The Radial Stress Field

For determining the stress equations, it is convenient to use spherical coordinates, r, θ, and χ, where $\theta = 0$ defines the vertical centre line of the hopper-silo system, and $r = 0$ is the imaginary apex of the hopper. Jenike [6] developed the stress field theory, which was presented concisely in Nedderman [10]. Assuming that stresses in a hopper system tend to asymptote linearly with distance from the apex, the equations of static equilibrium can be found;

$$\left.\begin{aligned} \frac{\partial \sigma_{rr}}{\partial r} + \frac{2\sigma_{rr} - \sigma_{\theta\theta} - \sigma_{\chi\chi}}{r} + \frac{1}{r}\frac{\partial \tau_{\theta r}}{\partial \theta} + \frac{\tau_{\theta r}}{r}\cot(\theta) + \rho_b g \cos(\theta) = 0 \\ \frac{\partial \tau_{\theta r}}{\partial r} + \frac{1}{r}\frac{\partial \sigma_{\theta\theta}}{\partial \theta} + \frac{3\tau_{\theta r}}{r} + \frac{\sigma_{\theta\theta} - \sigma_{\chi\chi}}{r}\cot(\theta) - \rho_b g \sin(\theta) = 0 \end{aligned}\right\} \tag{1}$$

where σ is the normal stress, τ the shear stress, ρ_b the bulk density, and g the acceleration due to gravity. Using Mohr circle analysis, Nedderman [10], the stress components can be found to be

$$\left.\begin{aligned} \sigma_{rr} &= p(1 + \sin(\phi)\cos(2\psi^*)) \\ \sigma_{\theta\theta} &= p(1 - \sin(\phi)\cos(2\psi^*)) \\ \sigma_{\chi\chi} &= p(1 + \kappa \sin(\phi)) \\ \tau_{\theta r} &= -\tau_{r\theta} = p\sin(\phi)\sin(2\psi^*) \end{aligned}\right\} \tag{2}$$

where $\kappa = +1$ for a conical hopper, ϕ is the angle of internal friction of the powder, p is the distance from the Coulomb line to the σ axis on a stress Mohr circle, and ψ is the angle measured anticlockwise from x-plane to the major principle plane on a Mohr circle, Nedderman [10].

In order to obtain a dimensionless stress field, a parameter q is defined such that

$$p = \rho_b g r q. \tag{3}$$

Since q and ψ^* are functions of θ only, $\frac{\partial q}{\partial r} = \frac{\partial \psi^*}{\partial r} = 0$. Using q, the equations governing the stress field are given by

$$\left.\begin{aligned} A\frac{dq}{d\theta} + B\frac{d\psi^*}{d\theta} + C = 0, \\ D\frac{dq}{d\theta} + E\frac{d\psi^*}{d\theta} + F = 0, \end{aligned}\right\} \tag{4}$$

where the A–F terms are given by

$$\left.\begin{aligned} &A = \sin(\phi)\sin(2\psi^*), \\ &B = 2q\sin(\phi)\cos(2\psi^*), \\ &C = q[1 + \sin(\phi)(4\cos(2\psi^*) - \kappa + \sin(2\psi^*)\cot(\theta))] + \cos(\theta), \\ &D = 1 - \sin(\phi)\cos(2\psi^*), \\ &E = 2q\sin(\phi)\sin(2\psi^*), \\ &F = q\sin(\phi)(4\sin(2\psi^*) - \cos(2\psi^*)\cot(\theta) - \kappa\cot(\theta)) - \sin(\theta). \end{aligned}\right\} \tag{5}$$

By rearranging (4) it is possible to obtain equations for the derivatives of q and ψ^*, which are given by

$$\left.\begin{aligned} \frac{dq}{d\theta} = \frac{CE - BF}{BD - AE}, \\ \frac{d\psi^*}{d\theta} = \frac{AF - CD}{BD - AE}, \end{aligned}\right\} \tag{6}$$

which is a system which can be solved numerically via an ordinary differential equation solver.

3.1.1 Possible Numerical Issues

One issue which can occur is when the denominator,

$$BD - AE = 2q\sin(\phi)(\cos(2\psi^*) - \sin(\phi)), \tag{7}$$

become zero. This occurs when $\psi^* = 135° + \phi/2$, so parameters must be chosen such that this situation is avoided.

A second issue is that on the vertical, $\psi^* = 90°$, which makes the C and F terms defined by the equations in (5) indeterminate. When this occurs, L'Hôpital's rule can be applied to reduce the A–F terms to

$$\left.\begin{aligned} A &= 0, \\ B &= -2q\sin(\phi), \\ C &= 1 + q[1 - \sin(\phi)(5 + 2\frac{d\psi^*}{d\theta})] + \cos(\theta), \\ D &= 1 + \sin(\phi), \\ E &= 0, \\ F &= 0, \end{aligned}\right\} \quad (8)$$

which simplifies the system to

$$\left.\begin{aligned} \frac{dq}{d\theta} &= 0, \\ \frac{d\psi^*}{d\theta} &= \frac{1 + q(1 - 5\sin(\phi))}{4q\sin(\phi)}. \end{aligned}\right\} \quad (9)$$

Hence, the system (9) is solved on the centerline, and system (6) elsewhere.

3.2 Boundary Conditions

In order for the ordinary differential equation solver to compute the problem, boundary conditions are required. The boundary conditions for ψ^* can be found via a Mohr circle.

The first boundary condition for ψ^* is on the vertical line $\theta = 0$. On this line the major principal stress is horizontal, hence $\psi^* = 90°$.

$$\psi^* = 90° \text{ when } \theta = 0° \quad (10)$$

The second condition of ψ^* is at the wall of the hopper. It is given by

$$\psi^* = \psi^*_W = 90° + \frac{1}{2}(\omega + \phi_W) \text{ when } \theta = \alpha° \quad (11)$$

where α is the hopper half angle and ω is an angle defined using Mohr's circles by

$$\sin(\omega) = \frac{\sin(\phi_w)}{\sin(\phi)}, \quad (12)$$

In order to compute the problem, the second ψ^* condition is unnecessary, however an additional condition on q is required at the centre of the hopper. In order to find this value, the shooting method is used. The shooting method is performed by assuming the value of q at the centre line, computing the system, then checking the computed value of ψ^* at the wall against the surplus boundary condition. The assumed value can then be adjusted, and the process repeated until the error in the q boundary condition is within acceptable bounds.

3.3 The Radial Velocity Field

In order to model ejection times, the velocity distribution is required. The velocity distribution can be found be combining the assumptions of incompressible powder, the principle of co-axiality, and the equations relating stress and strain with velocity [10]. In spherical coordinates the incompressible assumption is given by

$$\frac{\partial v_r}{\partial r} + 2\frac{v_r}{r} + \frac{1}{r}\frac{\partial v_\theta}{\partial \theta} + \frac{v_\theta \cot(\theta)}{r} = 0, \tag{13}$$

the principle of co-axiality by

$$\left(\frac{\partial v_r}{\partial r} - \frac{v_r}{r} - \frac{1}{r}\frac{\partial v_\theta}{\partial \theta}\right)\tan(2\psi^*) = r\frac{\partial}{\partial r}\left(\frac{v_\theta}{r}\right) + \frac{1}{r}\frac{\partial v_r}{\partial \theta}, \tag{14}$$

and the equations relating stress and strain with velocity are given by

$$\dot{\varepsilon}_{rr} = -\frac{\partial v_r}{\partial r},\ \dot{\varepsilon}_{\theta\theta} = -\frac{v_r}{r} - \frac{1}{r}\frac{\partial v_\theta}{\partial \theta},\ \dot{\varepsilon}_{\chi\chi} = -\frac{v_r}{r} - \frac{v_r \cot(\theta)}{r},$$

$$\dot{\gamma}_{r\theta} = -\dot{\gamma}_{\theta r} = r\frac{\partial}{\partial r}\left(\frac{v_\theta}{r}\right) + \frac{1}{r}\frac{\partial v_r}{\partial \theta}. \tag{15}$$

These equations can be solved to give the velocity distribution given particular boundary conditions. However, in general it is difficult to determine the correct boundary conditions to use. For this model the velocity field within the hopper is assumed to be radial (for which there is experimental evidence for, at least in the lower half of the hopper [9]). When this assumption is combined with the above equations, the velocity is found to be

$$v_r = -\frac{f(\theta)}{r^2},\ v_\theta = 0,\ v_\chi = 0. \tag{16}$$

where $f(\theta)$ is a function of θ to be found. Using these relations, the stress and strain relationships simplify to

$$\left.\begin{aligned} \dot{\varepsilon}_{rr} &= -\frac{\partial v_r}{\partial r} = -\frac{2f(\theta)}{r^3}, \\ \dot{\varepsilon}_{\theta\theta} &= \dot{\varepsilon}_{\chi\chi} = -\frac{v_r}{r} = -\frac{f(\theta)}{r^3}, \\ \dot{\gamma}_{r\theta} &= \dot{\gamma}_{\theta r} = -\frac{1}{r}\frac{\partial v_r}{\partial \theta} = -\frac{2f'(\theta)}{r^3}, \end{aligned}\right\} \tag{17}$$

where

$$f'(\theta) \equiv \frac{df(\theta)}{d\theta}. \tag{18}$$

The Mohr circle gives the equation

$$\frac{f'(\theta)}{f(\theta)} = -3\tan(2\psi^*), \tag{19}$$

hence

$$f(\theta) = AF(\theta), \tag{20}$$

where A is a constant of integration and

$$F(\theta) = \exp\left(-3\int_0^\theta \tan(2\psi^*)d\theta\right). \tag{21}$$

To eliminate the constant of integration A, the model can be scaled to match the volumetric flow rate. A can be found from

$$Q = -\int_0^\alpha 2\pi r \sin(\theta) v_r r d\theta = 2\pi A \int_0^\alpha F(\theta)\sin(\theta)d\theta. \tag{22}$$

For the velocity in the silo, it is assumed that it is plug flow, meaning that every point moves down at the same constant velocity. This velocity is determined by the volumetric flow rate and area of the hopper-silo boundary.

3.4 Exit Time

Given a particle at any point in the hopper, the exit time of the particle can be found by integrating the velocity over the path the particle will take. Using the velocity model from the previous section, the following is obtained:

$$T = \frac{r^3 - r_0^3}{3f(\theta)}, \tag{23}$$

where T is the time taken for the particle to reach the orifice, r is the radial distance from the imaginary apex of the hopper, and r_0 is the distance from the imaginary apex to the orifice of the hopper.

For any chosen point in the silo, (r, θ, χ), the exit time is determined by combining the time taken for it to vertically drop to the hopper-silo boundary,

$$T = \frac{d_s A_s}{Q} + \frac{r_s^3 - r_0^3}{3f(\theta)}, \tag{24}$$

where d_s is the vertical distance from the point to the hopper-silo boundary, A_s is the area of the horizontal silo cross-section, Q is the volumetric flow rate, and r_s, r_0, and θ are measured from the point on the hopper-silo boundary directly below (r, θ, χ), but otherwise have the same definition as from Eq. (23).

3.5 The Residence Time Distribution

One measure of mixing used in industry is the residence time distribution. One reason for its use is the simplicity in which it may be measured. A simple tracer, or simple concentration step-change experiment [8] may be performed, and the output data measured and processed to find the residence time distribution, $E(t)$. The residence time distribution is easily calculated from the following formula,

$$E(t) = \frac{dF(t)}{dt}, \tag{25}$$

where

$$F(t) = \frac{C(t)}{C_0}, \tag{26}$$

$C(t)$ is the concentration measured at time t, and C_0 is the original tracer concentration such that $0 \leq F(t) \leq 1$. Flows which are plug-like have a narrow $E(t)$, and hence are mixed less than flows with a wide residence time distribution. In the context of this modelling work, the residence time distribution may be found directly from the calculation of exit time, Eq. (24). At any given time t we can calculate what volume of discharged powder originated from above the flat, heaped, or sloped layer, and then normalise by the discharged volume. In this way, the composition of discharged powder may be simply calculated and composed into a residence time plot.

4 Results

Using the theory outlined in earlier sections of this work, we can predict the residence time distributions for flat, heaped and sloped powder layers, and note differences between. The parameters used in this study are outlined in Table 1. We note that any combination of parameters that maintain a mass-flow in the hopper could be used.

Table 1 Parameters used for the simulations

Parameter	Value
Q	0.015 (m^3s^{-1})
H	1 (m)
D_0	0.15 (m)
ϕ	28°
α	25°
ϕ_w	15°

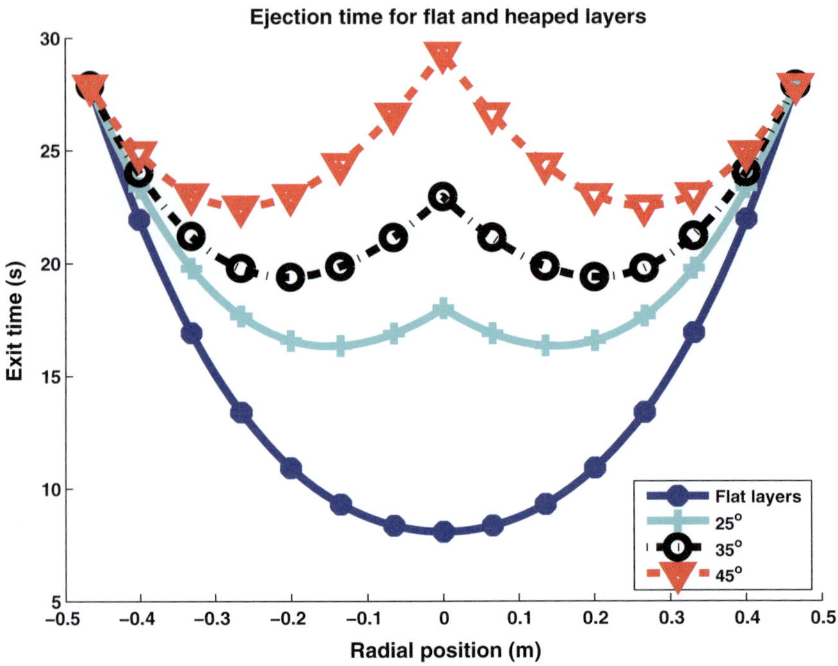

Fig. 2 The ejection times for the heaped layers at 25°, 35°, and 45°, compared with the flat layer case

The parameters chosen are for example. Q, the volumetric flow rate, is chosen from experiential knowledge of realistic flow rates in such systems. The hopper height, H, outlet diameter, D_0, and half angle, α, may be chosen to match any given hopper of interest. The internal angle of friction, ϕ is typically in the range $20° \leq \phi \leq 50°$. The wall friction angle ϕ_W is usually less than 20°, and the angle of repose for a powder is generally $20° \leq A_r \leq 50°$. In this study we consider flat layers, and layers which are inclined in a heap or are sloped at 25°, 35°, and 45° respectively.

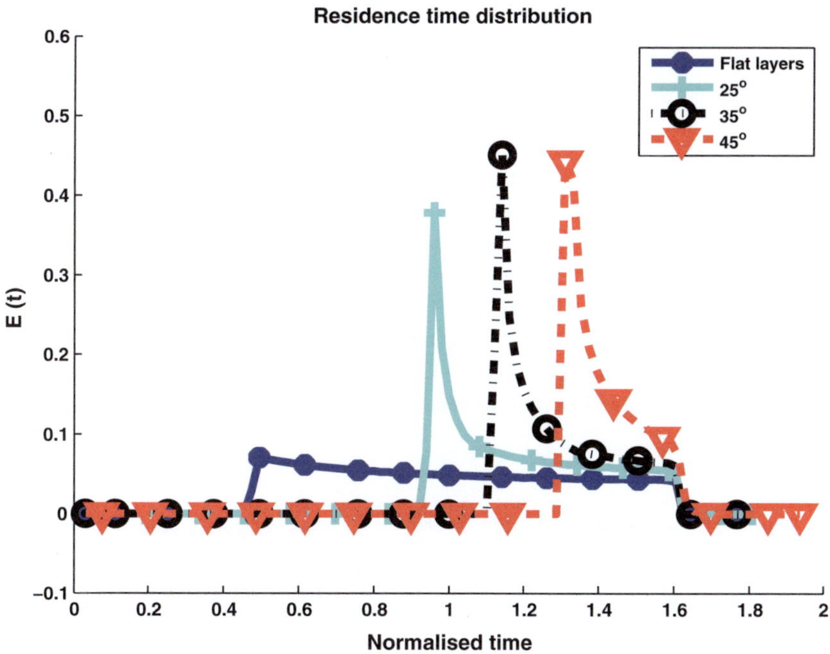

Fig. 3 The residence time distributions for flat layers and layers heaped at 25°, 35°, and 45° respectively

4.1 Heaped Layers

In the case of heaped layers, as in Fig. 1b, the system maintains its axisymmetry about the centerline. Therefore, the analysis of ejection times may be completed in 2D since, for every cross section of the system, the flow is assumed identical. Figure 2 shows the ejection times for the flat layer case compared to the case where the heap is inclined at an angle 25°, 35°, or 45°. In the center of the hopper, where the layers are at maximum height, it is known that the velocity is also greatest. The velocity profile for each of the heaped layer cases is the same, but the layer geometry is changed by the angle of repose. Therefore, we have the simple situation where the ejection time is changed because the particles in a heaped layer which are closer to the center of the hopper have to travel a greater distance to be discharged. It is clear from the figure that there are substantial differences from the flat layer case. Figure 3 is the residence time distribution for each of the cases. The time axis has been normalised by the mean residence time for the flat layer case. This figure is, theoretically, what would be observed during a residence time experiment in an industrial system. Because the velocity is greatest in the center of the hopper, particles in or near this position will be discharged first in the flat layer case. This may not be so in the heaped layer case, since they have further to travel. This is evident in the residence time distribution plot. The heaped layers are delayed in being detected at the orifice. As the angle of repose increases over the realistic range studied, so does the delay increase. This also has the effect of narrowing the spread of the distribution when compared with the flat layer case. Note again that the velocity distribution in each of these cases is equal, but it is the geometry of the powder layers which causes the difference in measured residence time distribution. If such results were obtained during a physical experiment the geometry of powder layers must be accounted for. Neglecting this would greatly change the conclusions drawn from the experiment. Since the velocity distribution is the same in each case, the time particles spend in the hopper is also identical in each case. However, one may conclude from experimental measurement that the flow is quite plug-like if the angle of repose was high and unaccounted for, when, in fact, the flat layer distribution is quite spread. The conclusion is that measured residence time distribution needs to be interpreted based on the initial setup of the experiment.

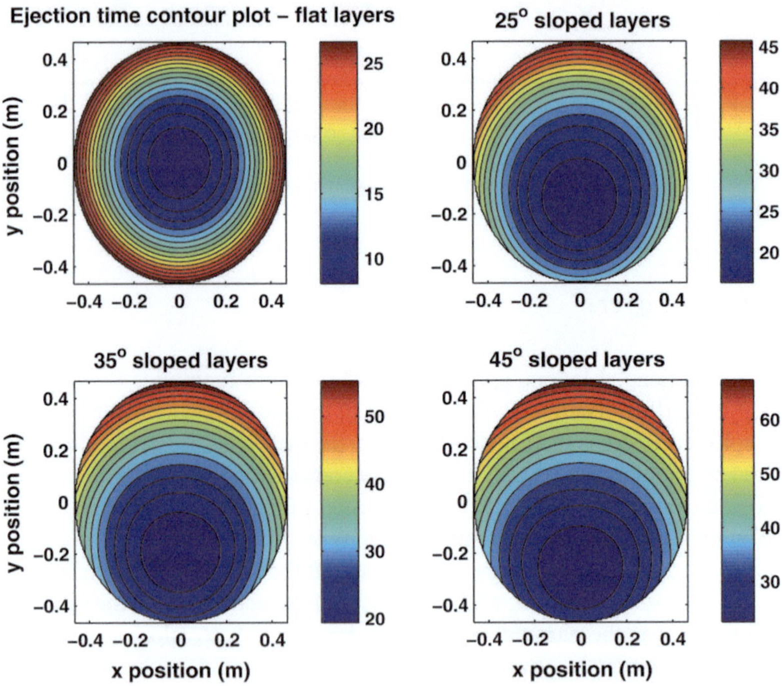

Fig. 4 Exit time contour plots for flat and sloped layers. *Upper left* flat layers. *Upper right* sloped at 25°. *Lower left* sloped at 35°. *Lower right* sloped at 45°

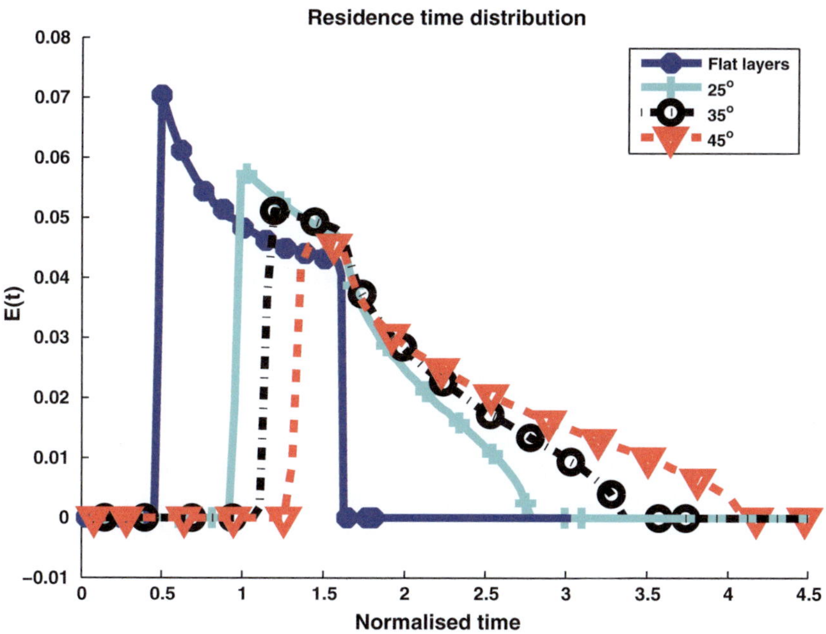

Fig. 5 The residence time distribution for the flat and sloped cases

4.2 Sloped Layers

If a system is loaded by bouncing particles off the hopper walls, a sloped layer situation may arise such as is depicted in Fig. 1c. We note here that the axisymmetry of the problem is destroyed, hence we cannot simply represent the exit time graph using a single curve. In Fig. 4 we have plotted four contour plots of the discharge time of the sloped layer interface for each of the sloped cases. Please note the vastly different scales between each of the plots. Figure 4 indicates great differences between the flat layer case and a sloped layer case. In the flat layer case, the exit time is symmetrical. This symmetry is destroyed when a sloped layer is introduced, as indicated by the figure. For the sloped layer case, the largest ejection time always occurs at the wall (top of each plot) at which the layer is at maximum hight. The minimum ejection time is a function of slope angle. As the slope angle is increased the position of minimum ejection time moves towards the wall. This is due to the increased distance particles must move as the angle of slope is increased. The average ejection time also increases as the angle is increased. Figure 5 is the residence time distribution of the flat layer case compared to the sloped layer cases. Again, the time has been normalised to the average discharge time for the flat layer case. It is obvious that the sloping of layers is changing the observed residence time distribution. The layer is delayed in being ejected and is wider spread compared to the flat layer case. Again, if these results were found experimentally, care must be taken with interpretation, otherwise erroneous conclusions regarding the time in the hopper may be drawn. If the slope of the layers is not taken into account when interpreting results, one might conclude that there is more time spent in the hopper than the flat layer case suggests.

5 Conclusion

In this modelling work, stress field theory and the assumption of radial flow have been used to evaluate ejection times and residence time distributions of flat, heaped, and sloped powder layers in a conical mass-flow hopper-silo system. Heaped and sloped layers commonly occur in industrial practice, however, the experimental analysis of the residence time distribution is designed to give information for flat layer geometries. In this work we have shown numerically that the amplitude and spread of the residence time distribution is greatly affected by the initial geometry of the powder layers. In particular, it was found that for realistic angles of repose, the spread of the residence time distribution was smaller in the case of heaped and larger for sloped layers when compared to flat layers. This implies that during a physical residence time experiment the granular material will appear to pass through the hopper more slowly for non-flat powder layers. This simple result has far-reaching implications for industrial practice. It is suggested that experimental results from residence time experiments be treated with caution—one must be sure of what it is that is actually measured. Corrections must be made to account for the fact that in a granular system the layer interfaces are non-flat, otherwise the predicted residence time profiles will be erroneous and quality control negatively affected. We note that this may be avoided by loading the hopper fully, flattening the powder layers, then loading the tracer on top of the flat layer, ensuring an accurate, flat layered residence time experiment. One must do this with caution, however, as this procedure has the potential to change the stress distribution in the system by rearrangement of the particle packing, and hence, change the velocity profile.

It should be noted that the assumptions made in the modelling work can be questioned. For example, the assumption that the flow in the silo is plug-like may or may not be true depending on the hopper design. Also, the assumption of radial flow in the converging hopper is likely to be true for a section of the hopper only. However, this is a theoretical work, and the results may be treated as instructional, rather than absolute. The analysis may be repeated with more accurate velocity profiles to obtain quantitative results. The key message of the work is that one must take into account the non-flat layer geometry when analysing experimental results.

References

1. Choi, J., Kudrolli, A., Bazant, M.Z.: Velocity profile of granular flows inside silos and hoppers. J. Phys.: Condens. Matter **17**(24), S2533 (2005)

2. Fullard, L.A., Davies, C.E.: A brief investigation into ejection times from a conical mass flow hopper-coulomb and conical model difference. In: Powders and Grains 2013: Proceedings of the 7th International Conference on Micromechanics of Granular Media, vol. 1542, pp. 1254–1257. AIP Publishing (2013)
3. Fullard, L., Davies, C., Wake, G.: A kinematic model for powder mixing in mass flow discharge from a conical hopper. In: Chemeca 2012: Quality of Life Through Chemical Engineering: 23–26 Sept 2012, p. 685. Wellington, New Zealand (2012)
4. Fullard, L.A., Davies, C.E., Wake, G.C.: Modelling powder mixing in mass flow discharge: a kinematic approach. Adv. Powder Technol. **24**(2), 499–506 (2013)
5. Gremaud, P., Matthewsy, J.V., Shearerz, M.: Similarity solutions for granular flows in hoppers. In: Proceedings of the SIAM/AMS Conference on Nonlinear PDEs, Dynamics and Continuum Physics (1998)
6. Jenike, A.W.: A theory of flow of particulate solids in converging and diverging channels based on a conical yield function. Powder Technol. **50**(3), 229–236 (1987)
7. Johanson, K.: Predicting cone-in-cone blender efficiencies from key material properties. Powder Technol. **170**(3), 109–124 (2006)
8. Levenspiel, O.: Chemical Reaction Engineering, 2nd edn. Wiley, New York (1966)
9. Moreea, S.B.M., Nedderman, R.M.: Exact stress and velocity distributions in a cohesionless material discharging from a conical hopper. Chem. Eng. Sci. **51**(16), 3931–3942 (1996)
10. Nedderman, R.M.: Statics and Kinematics of Granular Materials. Cambridge University Press, Cambridge (1992)
11. Nedderman, R.M., Tüzün, U.: A kinematic model for the flow of granular materials. Powder Technol. **22**(2), 243–253 (1979)
12. Pitman, E.B.: Stress and velocity fields in two- and three-dimensional hoppers. Powder Technol. **47**(3), 219–231 (1986)

Uniqueness and Stability for Double Crystals in the Plane

Eriko Shinkawa

Abstract We study a mathematical model of small double crystals, that is, two connected regions in $\mathbf{R}^{n+1}$ with prescribed volumes and with surface tension depending on the direction of the each point of the surface. Each double crystal is a critical point of the anisotropic surface energy which is the integral of the surface tension over the surface. We derive the first and the second variation formulas of the energy functional. For $n = 1$ and a certain special energy density function, we classify the double crystals in terms of symmetry and the given areas. Also, we prove that some of the double crystals are unstable, that is they are not local minimizers of the energy.

Keywords Anisotropic · Uniqueness · Stability · Double crystal

1 Introduction

There was a long-standing conjecture which was called the double bubble conjecture. It says that the standard double bubble provides the least-perimeter way to enclose and separate two given volumes, here the standard double bubble is consisting of three spherical caps meeting along a common circle at 120 degree angles. This conjecture had been believed since about 1870, and was proved in 2002. The existence of the minimizer was proved by Almgren [3] in 1976. (This paper proved, more general case, minimizing surface enclosing k prescribed volumes in $\mathbf{R}^{n+1}$, using geometric measure theory.) In 1993, the double bubble conjecture was proved in the plane by Foisy et al. [2] advised by Frank Morgan. For higher dimensional case, Hutchings [5] proved that any minimizer is axially symmetric and he also obtained a bound of the number of connected components of the two regions of a minimizer. Using these results, finally in 2002, the double bubble conjecture was proved by Hutchings et al. [6] in $\mathbf{R}^3$, and a student of Morgan extended it to higher dimensions [1].

E. Shinkawa (✉)
Graduate School of Mathematics, Kyushu University,
744, Motooka, Nishi-ku, Fukuoka 819-0375, Japan
e-mail: e-shinkawa@math.kyushu-u.ac.jp

R.S. Anderssen et al. (eds.), *Applications + Practical Conceptualization + Mathematics = fruitful Innovation*, Mathematics for Industry 11,
DOI 10.1007/978-4-431-55342-7_17

Double bubbles are a mathematical model of soap bubbles. The energy functional is the total area of the surface. On the other hand, when we think about a mathematical model of anisotropic substance like crystals, we need to consider the energy density function $\gamma : S^n \to \mathbf{R}^+$ depending on the normal direction N of the surface, where $S^n := \{X \in \mathbf{R}^{n+1} \mid \|X\| = 1\}$ is the n-dimensional unit sphere in $\mathbf{R}^{n+1}$. γ is called an anisotropic energy density function, and its integral $\mathscr{F} = \int_\Sigma \gamma(N) d\Sigma$ over the surface Σ is called an anisotropic (surface) energy. The surface is a constant anisotropic mean curvature (CAMC) surface if it is a critical point of the anisotropic energy for all volume preserving variations. CAMC surfaces are a generalization of CMC (constant mean curvature) surfaces.

In this paper, we extend the double bubble problem to a double crystal (DC) problem, that is, we minimize the anisotropic energy instead of the surface area. The solutions are a mathematical model of multiple crystals.

There were some previous researches relating to the DC problem. Gary [4] determined the energy-minimizer for the case where each anisotropic energy density function γ_i $(i = 0, 1, 2)$ is constant (we consider three surfaces, so we need three kinds of anisotropic energy density functions). Hence, his γ_i's are isotropic. His work also means that he gave a new proof of the double bubble conjecture. Besides, for $\gamma := \gamma_1 = \gamma_2 = \gamma_0$ that γ is a norm on $\mathbf{R}^2$, Morgan et al. [11] determined the shapes of the all minimizers for the case of $\gamma_i(\nu_1, \nu_2) = |\nu_1| + |\nu_2|$ $(i = 0, 1, 2)$ $((\nu_1, \nu_2) \in S^1)$.

Recall that there is a unique hypersurface that minimizes $\mathscr{F}$ among all closed hypersurfaces enclosing the same volume (cf. [13]). This surface is known as the Wulff shape. In this paper, we assume that the Wulff shape is smooth. We will derive the first variation formula for the anisotropic energy $\mathscr{F}$ (Theorem 1), and obtain the conditions for a surface Σ to be a double crystal (Theorem 2). Also, we will obtain the second variation formula for $\mathscr{F}$ (Theorem 3) and obtain the condition for a double crystal to be stable.

For $n = 1$, we will consider a special energy density function $\gamma := \gamma_1 = \gamma_2 = \gamma_0$ satisfying

$$\gamma(\nu_1, \nu_2) = \left(\nu_1^{2p} + \nu_2^{2p}\right)^{1-\frac{1}{2p}} / \sqrt{\nu_1^{4p-2} + \nu_2^{4p-2}}.$$

We classify the double crystals in terms of symmetry and the given areas. Also, we prove that some of the double crystals are unstable, that is they are not local minimizers of the energy.

We will explain our problem more precisely in Sect. 2. In Sect. 3, we derive the first and the second variation formulas of the anisotropic surface energy. In Sect. 4, we study the DC problem in the plane.

This paper is essentially a part of the author's doctoral dissertation [12].

2 Preliminaries

In this section, first we introduce some fundamental facts about CAMC surfaces (for details, see [8]). Then, we formulate the DC problem.

Let $\gamma : S^n \to \mathbf{R}^+$ be a positive smooth function on the unit sphere S^n in $\mathbf{R}^{n+1}$. We call this function γ an anisotropic energy density function. Let Σ be an n-dimensional oriented compact C^∞ manifold with or without boundary. And let $X : \Sigma \to \mathbf{R}^{n+1}$ be an immersion with Gauss map (unit normal) $N : \Sigma \to S^n$ be its Gauss map. The anisotropic energy of X is defined as

$$\mathscr{F}(X) = \int_\Sigma \gamma(N) d\Sigma,$$

where $d\Sigma$ is the volume form on Σ induced by X. Any smooth variation $\tilde{X} : \Sigma \times [-\varepsilon_0, \varepsilon_0] \to \mathbf{R}^{n+1}$ $(\varepsilon_0 > 0)$ of X can be represented as $\tilde{X}(*, \varepsilon) = X_\varepsilon = X + \varepsilon(Z + \varphi N) + \mathscr{O}(\varepsilon^2)$, where Z is tangent to X. The first variation of $\mathscr{F}$ for this variation is (cf. Proof of Proposition 3.1 in [8])

$$\begin{aligned} \delta\mathscr{F} &:= \frac{d}{ds}\Big|_{\varepsilon=0} \mathscr{F}(X_\varepsilon) \\ &= \int_\Sigma \varphi(\mathrm{div}_\Sigma D\gamma - nH\gamma) d\Sigma + \oint_{\partial\Sigma} -\varphi\langle D\gamma, \nu\rangle + \gamma\langle Z, \nu\rangle\, ds, \end{aligned} \qquad (1)$$

where D is the gradient on S^n, H is the mean curvature of X, ν is the outward pointing unit conormal of X along $\partial\Sigma$, and ds is the $(n-2)$-dimensional volume form of $\partial\Sigma$. $\Lambda := -\mathrm{div}_\Sigma D\gamma + nH\gamma$ is called the anisotropic mean curvature of X. X is called a Constant Anisotropic Mean Curvature (CAMC) hypersurface when $\Lambda \equiv$ constant. We remark that X is CAMC if and only if $\delta\mathscr{F} = 0$ for all compactly-supported $(n+1)$-dimensional-volume-preserving variations. For $\gamma \equiv 1$, we get $\Lambda = nH$. It means that CAMC surface is a generalization of CMC surface.

It is known that there is a unique (up to translation in $\mathbf{R}^{n+1}$) minimizer of $\mathscr{F}$ among all closed hypersurfaces enclosing the same volume (cf. [13]), and it is a rescaling of the so-called Wulff shape. The Wulff shape (we denote it by W) is a closed convex hypersurface defined by

$$W := \partial \bigcap_{N \in S^n} \{w \in \mathbf{R}^{n+1} | \langle w, N\rangle \le \gamma(N)\}$$

When the W is smooth and strictly convex (that is, all principal curvatures are positive with respect to the inward normal. This condition is equivalent to the condition that $A := D^2\gamma + \gamma \cdot 1$ is positive definite at earh $N \in S^n$, where $D^2\gamma$ is the Hessian of γ on S^n, and 1 is the identity map on $T_N S^n$. This condition is called the convexity condition), W can be parametrized as an embedding $\Phi : S^n \to W \subset \mathbf{R}^{n+1}$:

$$\Phi(N) = D\gamma + \gamma(N)N.$$

The anisotropic mean curvature of W is n with respect to the inward normal.

From now on, we assume that the convexity condition is satisfied.

For later use, we give a new representation of the 1st variation formula:

Lemma 1 *The first variation of $\mathscr{F}$ for the variation $X_\varepsilon = X + \varepsilon Y + \mathscr{O}(\varepsilon^2)$ is*

$$\delta\mathscr{F} = -\int_\Sigma \varphi\Lambda\, d\Sigma + \oint_{\partial\Sigma} \langle \Phi, -\varphi\nu + fN\rangle\, ds,$$

where $\varphi := \langle Y, N\rangle$ and $f := \langle Y, \nu\rangle$.

Proof We compute the integrand of the second term of (1).

$$\begin{aligned} -\varphi\langle D\gamma, \nu\rangle + \gamma\langle Z, \nu\rangle &= \langle -\varphi(D\gamma + \gamma N), \nu\rangle + \gamma f \\ &= \langle \Phi, -\varphi\nu\rangle + \langle D\gamma + \gamma N, N\rangle f \\ &= \langle \Phi, -\varphi\nu + fN\rangle, \end{aligned}$$

which proves the desired result. □

If $n = 1$, curves with constant anisotropic mean curvature are completely determined as follows:

Lemma 2 *Let $n = 1$ and $X : \mathbf{R} \supset I \to \mathbf{R}^2$ be a curve parametrized by arc-length. Then,*

$$\Lambda = \kappa/\kappa_W,$$

where Λ is the anisotropic mean curvature of X, κ is the curvature of X, and κ_W is the curvature of the Wulff shape W.

Proof We denote by θ a point $e^{i\theta}$ in S^1. Then, the Wulff shape W is represented by an embedding $\Phi : S^1 \to \mathbf{R}^2$ defined as

$$\Phi(\theta) = \gamma_\theta(\theta)(-\sin\theta, \cos\theta) + \gamma(\theta)(\cos\theta, \sin\theta).$$

Set $X(s) = (x(s), y(s))$. Then, the Gauss map N of X is

$$N(s) = (-y'(s), x'(s)) =: (\cos\theta(s), \sin\theta(s)).$$

Hence, the anisotropic mean curvature Λ of X is

$$\Lambda(s) = -\gamma_{\theta s} - \kappa\gamma = -\gamma_{\theta\theta}\theta_s - \kappa\gamma = -\kappa(\gamma_{\theta\theta} + \gamma). \tag{2}$$

On the other hand,

$$\frac{d\Phi}{d\theta} = (\gamma_{\theta\theta} + \gamma)(-\sin\theta, \cos\theta),$$

$$\frac{d^2\Phi}{d\theta^2} = (\gamma_{\theta\theta\theta} + \gamma_\theta)(-\sin\theta, \cos\theta) - (\gamma_{\theta\theta} + \gamma)(\cos\theta, \sin\theta).$$

Hence, by elementary calculations, the curvature κ_W of W with respect to the outward pointing unit normal is

$$\kappa_W = \frac{-1}{\gamma_{\theta\theta} + \gamma}. \tag{3}$$

(2) with (3) gives the desired formula. □

Proposition 1 *If the anisotropic mean curvature Λ of a curve X is constant, then either*

1. *X is (a part of) a straight line (when $\Lambda = 0$), or*

2. *X is a part of the Wulff shape up to translation and homothety (when $\Lambda \neq 0$).*

Proof By Lemma 2, the curvature of X is $\kappa = \Lambda\kappa_W$. Hence, by the fundamental theorem for plane curves, we obtain the desired result. □

Remark 1 For $n \geq 2$, we have great many varieties of CAMC hypersurfaces. For example, [10, Sect. 5] gives two parameter family of axisymmetric CAMC surfaces.

Let us explain our problem more precisely. Let Σ_1, Σ_2, Σ_0 be three piecewise smooth oriented connected compact hypersurfaces in $\mathbf{R}^{n+1}$ with common boundary C such that $\Sigma_1 \cup \Sigma_0$ (resp. $\Sigma_2 \cup \Sigma_0$) encloses a region R_1 (resp. R_2) with prescribed volume V_1 (resp. V_2), and let γ_i be energy density functions on Σ_i. We study the following anisotropic energy of the surface $\Sigma := \Sigma_1 \cup \Sigma_2 \cup \Sigma_0$:

$$\mathscr{F}(\Sigma) := \sum_{i=0}^{2} \int_{\Sigma_i} \gamma_i(N_i)\,d\Sigma_i, \tag{4}$$

where $N_i : \Sigma_i \to S^n$ is the unit normal vector field along Σ_i (we refer to Fig. 1 about the directions of N_i) and $d\Sigma_i$ is the n-dimensional volume form on Σ_i. The volumes V_i of the region R_i is given by

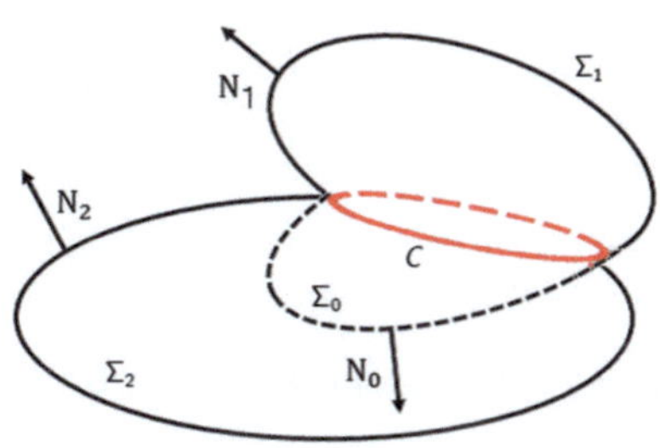

Fig. 1 An admissible surface Σ in $\mathbf{R}^3$. The *red curve* C is the common boundary of Σ_1, Σ_2 and Σ_0. We always assume that Σ_0 is in the *middle*

$$V_1 = \frac{1}{n+1}\left\{\int_{\Sigma_1}\langle x_1, N_1\rangle d\Sigma_1 + \int_{\Sigma_0}\langle x_0, N_0\rangle d\Sigma_0\right\},$$
$$V_2 = \frac{1}{n+1}\left\{\int_{\Sigma_2}\langle x_2, N_2\rangle d\Sigma_2 - \int_{\Sigma_0}\langle x_0, N_0\rangle d\Sigma_0\right\}.$$

Our problem is to study the minimizers of $\mathscr{F}$ among Σ's such that R_1, R_2 have prescribed volumes V_1, V_2, respectively.

3 Variation Formulas

Throughout this section, $\Sigma = \Sigma_1 \cup \Sigma_2 \cup \Sigma_0$ is such the union of smooth hypersurfaces Σ_0, Σ_1, and Σ_2 with common boundary C as in the last part of Sect. 2. We derive the first variation formula for the functional $\mathscr{F}$ defined by (4), and obtain the conditions for critical points.

Let $\tilde{X} : \Sigma \times [-\varepsilon_0, \varepsilon_0] \to \mathbf{R}^{n+1}$ $(\varepsilon_0 > 0)$ be a variation of $X : \Sigma \to \mathbf{R}^{n+1}$. $\tilde{X}$ is called an admissible variation if the two volumes V_1, V_2 are preserved. Such $\tilde{X}$ can be represented as $\tilde{X}(x, \varepsilon) = X_\varepsilon = X + \varepsilon Y + \mathscr{O}(\varepsilon^2)$, and Y is called an admissible variation vector field of X. If Y is admissible, then

$$\delta V_1 := \left.\frac{d}{d\varepsilon}\right|_{\varepsilon=0} V_1(X_\varepsilon) = \int_{\Sigma_1}\langle Y, N_1\rangle d\Sigma_1 + \int_{\Sigma_0}\langle Y, N_0\rangle d\Sigma_0 = 0, \tag{5}$$
$$\delta V_2 := \left.\frac{d}{d\varepsilon}\right|_{\varepsilon=0} V_2(X_\varepsilon) = \int_{\Sigma_2}\langle Y, N_2\rangle d\Sigma_2 - \int_{\Sigma_0}\langle Y, N_0\rangle d\Sigma_0 = 0. \tag{6}$$

hold. By a suitable reparametrization of $\tilde{X}$, we may assume that, at each point on C, Y is orthogonal to the ($(n-1)$-dimensional) tangent space of C. Then, the boundary condition implies the following:

$$Y = \langle Y, N_1\rangle N_1 + \langle Y, \nu_1\rangle \nu_1 = \langle Y, N_2\rangle N_2 + \langle Y, \nu_2\rangle \nu_2 = \langle Y, N_0\rangle N_0 + \langle Y, \nu_0\rangle \nu_0 \tag{7}$$

hold on C, where ν_i is the outward pointing conormal vector for Σ_i along C.

Lemma 3 *Let $\varphi_i, f_i : \Sigma_i \to \mathbf{R}$ be smooth functions on Σ_i satisfying*

(i) $\displaystyle\int_{\Sigma_1}\varphi_1\, d\Sigma_1 + \int_{\Sigma_0}\varphi_0\, d\Sigma_0 = 0, \int_{\Sigma_2}\varphi_2\, d\Sigma_2 - \int_{\Sigma_0}\varphi_0\, d\Sigma_0 = 0,$
(ii) $\varphi_1 N_1 + f_1\nu_1 = \varphi_2 N_2 + f_2\nu_2 = \varphi_0 N_0 + f_0\nu_0$ *on* C.

Then there exists an admissible variation such that the normal (resp. conormal to C) component of the variation vector field Y are $\varphi_i N_i$ (resp. $f_i\nu_i$).

Proof We give functions $h_i : \Sigma_i \to \mathbf{R}$ $(i = 1, 2)$ such that $\int_{\Sigma_i} h_i d\Sigma_i \neq 0$ holds and each h_i has compact support on the interior of Σ_i. And we extend each function h_i to 0 on $\Sigma \setminus \Sigma_i$. On the other hand, set

$$Y := \varphi_i N_i + f_i \nu_i \quad \text{on } \Sigma_i,\ i = 0, 1, 2.$$

Then, Y gives a variation vector field of Σ. Set

$$X(s, t_1, t_2) := X + sY + t_1 h_1 N_1 + t_2 h_2 N_2,$$
$$V_i(s, t_1, t_2) := V_i(X(s, t_1, t_2)), \quad i = 1, 2.$$

Set $V_1^0 := V_1(0, 0, 0)$, $V_2^0 := V_2(0, 0, 0)$. Consider the following simultaneous equations.

$$V_1(s, t_1, t_2) = V_1^0, \quad V_2(s, t_1, t_2) = V_2^0.$$

Differentiate V_1, V_2 at $(s, t_1, t_2) = (0, 0, 0)$ to obtain

$$\frac{\partial V_1}{\partial s}(0,0,0) = \int_{\Sigma_1} \varphi_1 d\Sigma_1 + \int_{\Sigma_0} \varphi_0 d\Sigma_0 = 0, \quad \frac{\partial V_2}{\partial s}(0,0,0) = \int_{\Sigma_2} \varphi_2 d\Sigma_2 - \int_{\Sigma_0} \varphi_0 d\Sigma_0 = 0,$$

$$\frac{\partial V_i}{\partial t_j}(0,0,0) = \delta^i_j \int_{\Sigma_i} h_j \, d\Sigma_i \begin{cases} \neq 0, \ i = j, \\ = 0, \ i \neq j. \end{cases}$$

Therefore, by the implicit function theorem, in there exist a neighborhood I of $s = 0$ and smooth functions $t_1 = t_1(s)$, $t_2 = t_2(s)$ such that $t_1(0) = 0$, $t_2(0) = 0$, $\tilde{V}_1(s) := V_1(s, t_1(s), t_2(s)) = V_1^0$, $\tilde{V}_2(s) := V_1(s, t_1(s), t_2(s)) = V_2^0$ $(s \in I)$. Then,

$$0 = \tilde{V}_i{}'(s) = (V_i)_s + (V_i)_{t_1} t_1'(s) + (V_i)_{t_2} t_2'(s), \quad (i = 1, 2)$$

hold. Hence,

$$t_1'(0) = -\frac{(V_1)_s(0,0,0) + (V_1)_{t_2}(0,0,0) t_2'(0)}{(V_1)_{t_1}(0,0,0)} = 0,$$
$$t_2'(0) = -\frac{(V_2)_s(0,0,0) + (V_2)_{t_1}(0,0,0) t_1'(0)}{(V_2)_{t_2}(0,0,0)} = 0.$$

Consequently,

$$X(s, t_1(s), t_2(s)) = X + sY + t_1(s) h_1 N_1 + t_2(s) h_2 N_2 = X + sY + \mathcal{O}(s^2)$$

is an admissible variation of Σ, and so we obtain the desired result. □

Using Lemma 1, we immediately obtain the following:

Theorem 1 (First variation formula) *For a variation $X_\varepsilon = X + \varepsilon Y + \mathcal{O}(\varepsilon^2)$ of Σ, the first variation of the anisotropic energy $\mathcal{F}$ is*

$$\delta \mathcal{F} := \left.\frac{d}{d\varepsilon}\right|_{\varepsilon=0} \mathcal{F}(X_\varepsilon) = \sum_{i=0}^{2} \left[-\int_{\Sigma_i} \varphi_i \Lambda_i \, d\Sigma_i + (-1)^i \oint_C \langle \Phi_i, -\varphi_i \nu_i + f_i N_i \rangle \, dC \right], \quad (8)$$

where $\Phi_i = D\gamma_i + \gamma_i N_i$, $\varphi_i = \langle Y, N_i\rangle$, $f_i = \langle Y, \nu_i\rangle$ *on* C, *and the orientation of* C *is chosen so that it is the positive orientation for* Σ_1.

Definition 1 Each critical point of $\mathscr{F}$ for all admissible variations is called a *double crystal*.

Theorem 2 *A hypersurface* Σ *is a double crystal if and only if there hold:*

1. *For* $i = 0, 1, 2$, *the anisotropic mean curvature* Λ_i *is constant, and* $-\Lambda_1 + \Lambda_2 + \Lambda_0 = 0$ *holds, and*
2. *at each point* ζ *on* C, $\Phi_0 - \Phi_1 + \Phi_2$ *is in the* $(n-1)$*-dimensional linear subspace determined by the tangent space* $T_\zeta C$ *of* C *at* ζ.

Corollary 1 *Assume* $\gamma_i \equiv 1$, $i = 0, 1, 2$. *Then,* Σ *is a double bubble if and only if*

1. *For* $i = 0, 1, 2$, *the mean curvature* H_i *is constant, and* $-H_1 + H_2 + H_0 = 0$ *holds, and*
2. *at each point on* C, $N_0 - N_1 + N_2 = 0$.

Proof of Theorem 2 Assume that $\Sigma = \Sigma_0 \cup \Sigma_1 \cup \Sigma_2$ is a double crystal. Then, Σ_1 is a critical point of $\mathscr{F}$ for all admissible variations that fix $\Sigma = \Sigma_0 \cup \Sigma_2$. Hence, Λ_1 is constant. Similarly, Λ_2 is constant. Now consider any variation $\Sigma_0(\varepsilon)$ of Σ_0 that fixes $\partial\Sigma_0$. Then, the variation vector field of $\Sigma_0(\varepsilon)$ can be extended to an admissible variation vector field of Σ. In fact, $\Sigma_0(\varepsilon)$ can be represented as

$$X_\varepsilon = X + \varepsilon\varphi_0 N_0 + \mathscr{O}(\varepsilon^2),$$

where $\varphi_0 = 0$ on C. It is obvious that we can find functions φ_1, φ_2, $f_1 = 0$, and $f_2 = 0$ satisfying (i) and (ii) in Lemma 3. So, by Lemma 3, there exists an admissible variation of Σ whose variation vector field is an extension of $Y_0 := \varphi_0 N_0$. We obtain, using Theorem 1, (5), and (6),

$$0 = \delta\mathscr{F} = \delta\mathscr{F} + \Lambda_1\delta V_1 + \Lambda_2\delta V_2 = \int_{\Sigma_0} (\Lambda_1 - \Lambda_2 - \Lambda_0)\langle Y_0, N_0\rangle\, d\Sigma_0.$$

Hence, $\Lambda_1 - \Lambda_2 - \Lambda_0 = 0$ holds, which proves the condition 1. Now, assume that the condition 2 does not hold. Then, there exists a non-empty open set U of C such that $(\Phi_0 - \Phi_1 + \Phi_2)(\zeta) \notin T_\zeta C$ for any $\zeta \in U$. Then, we can define a non-zero vector field $\tilde{Y}$ on C with support in U such that $\tilde{Y}$ is orthogonal to C at any $\zeta \in U$ and

$$\oint_C \langle \Sigma_{i=0}^2 (-1)^i \Phi_i, \tilde{Y}\rangle\, dC \neq 0$$

holds. Clearly, $\tilde{Y}$ can be represented as

$$\tilde{Y} = -\varphi_i \nu_i + f_i N_i, \quad i = 0, 1, 2,$$

and $Y := \varphi_i N_i + f_i \nu_i$ can be extended to an admissible variation vector field along Σ. Here we used Lemma 3 again as above. We obtain

$$0 = \delta \mathscr{F} = \delta \mathscr{F} + \Lambda_1 \delta V_1 + \Lambda_2 \delta V_2 = \oint_C \langle \Sigma_{i=0}^2 (-1)^i \Phi_i, \tilde{Y} \rangle \, dC \neq 0,$$

which is a contradiction.

Conversly, assume that the conditions 1 and 2 hold. Then, again by using Theorem 1, (5), and (6), for any admissible variation, we have

$$\delta \mathscr{F} = \delta \mathscr{F} + \Lambda_1 \delta V_1 + \Lambda_2 \delta V_2 = 0.$$

Hence, the hypersurface is a double crystal. □

Definition 2 A double crystal Σ is said to be stable if the second variation $\delta^2 \mathscr{F}$ is nonnegative for all admissible variations, and otherwise it is said to be unstable.

Theorem 3 (Second variation formula) *Let $\Sigma = \Sigma_1 \cup \Sigma_2 \cup \Sigma_0$ be a double crystal. Then for any admissible variational vector field Y, the second variation of the anisotropic energy $\mathscr{F}$ is given by*

$$\delta^2 \mathscr{F} = \sum_{i=0}^{2} \left[- \int_{\Sigma_i} \varphi_i L[\varphi_i] d\Sigma_i + (-1)^i \oint_C \varphi_i \langle A_i \nabla \varphi_i - f_i A_i dN_i(\nu_i), \nu_i \rangle \, dC \right], \quad (9)$$

where L is the self-adjoint Jacobi operator

$$L[\varphi_i] := \operatorname{div}(A_i \nabla \varphi_i) + \langle A_i dN_i, dN_i \rangle \varphi_i,$$

$A_i := D^2 \gamma_i + \gamma_i \cdot 1$, $\varphi_i = \langle Y, N_i \rangle$, and $f_i = \langle Y, \nu_i \rangle$ on C.

Proof The first variation formula (Theorem 1) gives

$$\delta \mathscr{F} := \left. \frac{d}{d\varepsilon} \right|_{\varepsilon=0} \mathscr{F}(X_\varepsilon) = \sum_{i=0}^{2} \left[- \int_{\Sigma_i} \varphi_i \Lambda_i \, d\Sigma_i + (-1)^i \oint_C \langle \Phi_i, \tilde{Y} \rangle \, dC \right],$$

where $\tilde{Y} = -\varphi_i \nu_i + f_i N_i$. Hence, any volume-preserving variation, at a double crystal Σ, we obtain

$$\delta^2 \mathscr{F} = \delta(\delta \mathscr{F} + \Lambda_1 \delta V_1 + \Lambda_2 \delta V_2) = \sum_{i=0}^{2} \left[- \int_{\Sigma_i} \varphi_i \delta \Lambda_i \, d\Sigma_i + (-1)^i \oint_C \langle \delta \Phi_i, \tilde{Y} \rangle \, dC \right].$$

Note that $\delta \Lambda_i = L[\varphi_i]$ holds (cf. [8]). Also, we compute, on C,

$$\begin{aligned} \langle \delta \Phi_i, -\varphi_i \nu_i + f_i N_i \rangle &= -\varphi_i \langle \delta \Phi_i, \nu_i \rangle = -\varphi_i \langle A_i(-\nabla \varphi_i + dN_i(f_i \nu_i)), \nu_i \rangle \\ &= -\varphi_i \langle -A_i \nabla \varphi_i + f_i A_i dN_i(\nu_i), \nu_i \rangle. \end{aligned}$$

□

4 Double Crystals in the Plane

In this section, we assume $n = 1$ and apply the above discussion to a certain special energy density function on S^1. The Wulff shape corresponding to this energy density function is a smooth square (see Fig. 2). We will discuss the critical points (i.e. double crystals) and their stability.

From Proposition 1 and Theorem 2, we immediately obtain the following:

Theorem 4 *For $n = 1$, $\Sigma = \Sigma_0 \cup \Sigma_1 \cup \Sigma_2$ is a double crystal if and only if there hold:*

(i) Each Σ_i is, up to translation, a part of a rescaling of the Wulff shape corresponding to γ_i.
(ii) $\Phi_0 - \Phi_1 + \Phi_2 = 0$ on the common boundary C (C is a set of two points).

From now on, if we do not say anything special, we assume that the energy density functions $\gamma_i : S^1(\subset \mathbf{R}) \to \mathbf{R}$ are the following special ones.

$$\gamma(\nu_1, \nu_2) := \gamma_{(p)}(\nu_1, \nu_2) := \gamma_i(\nu_1, \nu_2) = \left(\nu_1^{2p} + \nu_2^{2p}\right)^{1-\frac{1}{2p}} / \sqrt{\nu_1^{4p-2} + \nu_2^{4p-2}}, \quad i = 0, 1, 2, \tag{10}$$

where p is any fixed positive integer. Then the Wulff shape is given by

$$\Phi(\theta) := (\cos^{2p}\theta + \sin^{2p}\theta)^{-\frac{1}{2p}}(\cos\theta, \sin\theta).$$

4.1 Classifications of DC for a Special Energy Density Function

From now on, without loss of generality, we assume $V_1 \leq V_2$. If $\Lambda_i \neq 0$, from (i) in Theorem 4, Σ_i is represented by

Fig. 2 The Wulff shapes $W_{(p)}$ for the energy density $\gamma_{(p)}$ in (10). $W_{(1)}$ is a *circle*. When p approaches infinity, $W_{(p)}$ converges to a *cube*

$$X_1(\theta) = -\frac{1}{\Lambda_1}(\cos^{2p}\theta + \sin^{2p}\theta)^{-\frac{1}{2p}}(\cos\theta, \sin\theta) + (a_1, b_1), \quad \alpha_1 \le \theta \le \beta_1, \tag{11}$$

$$X_2(\theta) = -\frac{1}{\Lambda_2}(\cos^{2p}\theta + \sin^{2p}\theta)^{-\frac{1}{2p}}(\cos\theta, \sin\theta) + (a_2, b_2), \quad \beta_2 \le \theta \le \alpha_2, \tag{12}$$

$$X_0(\theta) = -\frac{1}{\Lambda_0}(\cos^{2p}\theta + \sin^{2p}\theta)^{-\frac{1}{2p}}(\cos\theta, \sin\theta) + (a_0, b_0), \quad \beta_0 \le \theta \le \alpha_0, \tag{13}$$

where $\alpha_0, \alpha_1, \alpha_2$ correspond to one of the two points in the common boundary C, and $\beta_0, \beta_1, \beta_2$ correspond to the other point in C. By the second condition in Theorem 4, we have

$$\begin{cases} f(\theta_0)\cos\theta_0 - f(\theta_1)\cos\theta_1 + f(\theta_2)\cos\theta_2 = 0, \\ f(\theta_0)\sin\theta_0 - f(\theta_1)\sin\theta_1 + f(\theta_2)\sin\theta_2 = 0, \end{cases} \quad (\theta_i = \alpha_i, \beta_i), \tag{14}$$

where $f(\theta) = (\cos^{2p}\theta + \sin^{2p}\theta)^{-\frac{1}{2p}}$.

We can prove the following results about geometry of the double crystals.

Lemma 4 [7] *There are uniquely determined functions φ, $\psi : \mathbf{S}^1 \to \mathbf{R}$ such that $\theta_2 = \varphi(\theta_1)$ and $\theta_0 = \psi(\theta_1)$ satisfy (14).*

Lemma 5 [7] *For double crystals, we have the following results about the relationship between α_i and β_i.*

(I) If $\alpha_1 + \beta_1 = 2n_1\pi$ ($n_1 \in \mathbf{Z}$), then $\alpha_i + \beta_i = 2n_i\pi$, ($n_i \in \mathbf{Z}$, $i = 0, 2$).
(II) If $\alpha_1 + \beta_1 = (2n_1 + 1/2)\pi$ ($n_1 \in \mathbf{Z}$), then $\alpha_i + \beta_i = (2n_i + 1/2)\pi$, ($n_i \in \mathbf{Z}$, $i = 0, 2$).
(III) If $\alpha_1 + \beta_1 = (2n_1 + 1)\pi$ ($n_1 \in \mathbf{Z}$), then $\alpha_i + \beta_i = (2n_i + 1)\pi$, ($n_i \in \mathbf{Z}$, $i = 0, 2$).
(IV) If $\alpha_1 + \beta_1 = (2n_1 + 3/2)\pi$ ($n_1 \in \mathbf{Z}$), then $\alpha_i + \beta_i = (2n_i + 3/2)\pi$, ($n_i \in \mathbf{Z}$, $i = 0, 2$).
(V) If $\beta_1 = \alpha_1 + \pi$, then $\beta_0 = \alpha_0 - \pi$ and $\beta_2 = \alpha_2 - \pi$.

Lemma 5 gives the following result about symmetry of double crystals:

Theorem 5 [7] *About the five types of the double crystals in Lemma 5, we have the following three types of symmetry (up to translation and homothety) (see Fig. 3).*

Type 1 Symmetry with respect to either a horizontal line or a vertical line.
Type 2 Symmetry with respect to the $\pm\pi/4$ rotation of the horizontal line.
Type 3 Rotational symmetry with respect to the center point of the smallest cube. In this case, the two bigger Wulff shapes are double size of the smallest one.

Actually, double crystals of Type (I) and (III) have Type 1 symmetry, double crystals of Type (II) and (IV) have Type 2 symmetry, and double crystals of Type (V) have Type 3 symmetry.

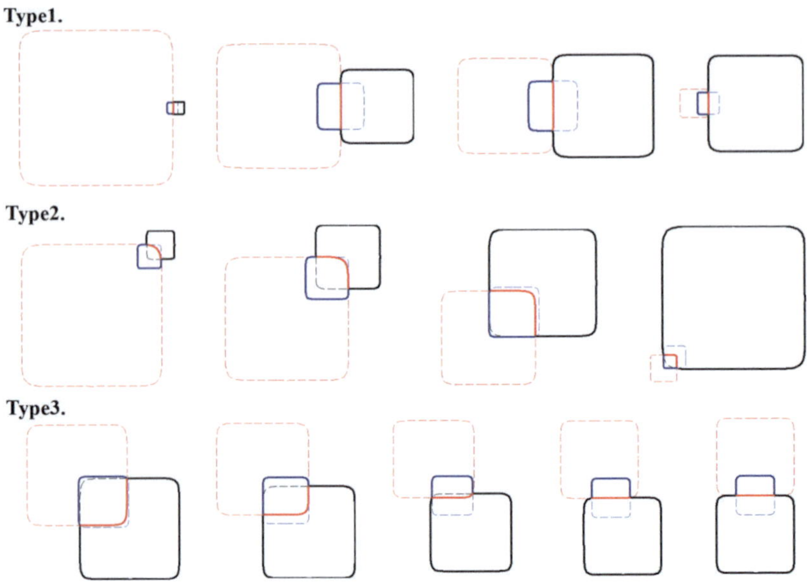

Fig. 3 These figures show the three types in Theorem 5 according to $\rho = V_2/V_1$

Remark 2 In Type 1 and 2, ρ can be any number bigger than or equal to 1. On the other hand, in Type 3, ρ can take numbers in the interval [3, 8]. In fact, in Type 3, two bigger Wulff shapes (black and red shape in Fig. 3) are double size of the smallest one (blue shape in Fig. 3).

4.2 Stability for Special Energy Density Function

In this section we discuss the stability of the three types of double crystals appeared in Theorem 5.

First we give a result about instability of some double crystals which was essentialy proved in [11].

Lemma 6 *Set* $\gamma_\infty(\nu_1, \nu_2) = |\nu_1| + |\nu_2|$ $(\nu_1, \nu_2) \in S^1$, *and consider an anisotropic surface energy* $\mathscr{F}(X) = \int_\Sigma \gamma_\infty(N)\, d\Sigma$. *Consider the three types of shapes in Fig. 4. Then we can decrease the anisotropic energy of these shapes without changing the enclosed areas (Fig. 5).*

Proof Note that the anisotropic energy of a horizontal or vertical edge is equal to its length, and the anisotropic energy of a diagonal edge is equal to $\sqrt{2}$ times its length. Figures 6, 7 and 8 show how the anisotropic energy is decreased without changing the enclosed area. □

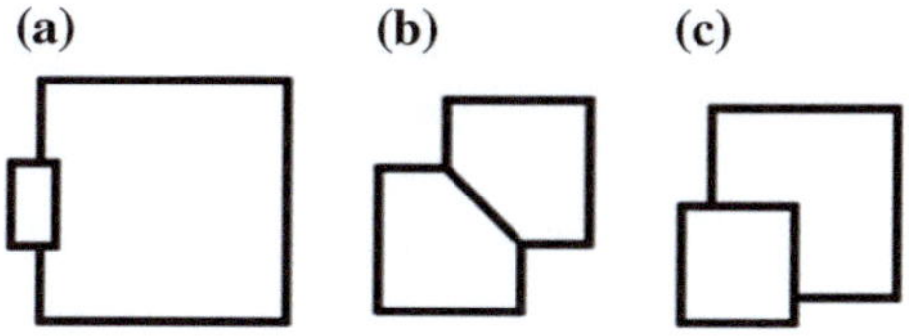

Fig. 4 The three types of shapes of which we can decrease the anisotropic energy without changing the enclosed areas

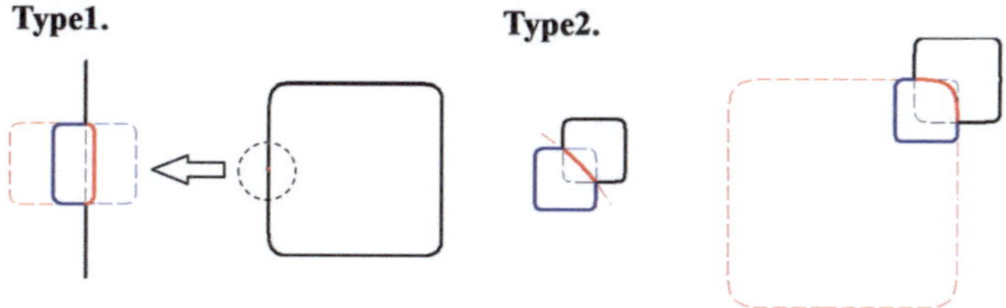

Fig. 5 Unstable examples corresponding to Fig. 4

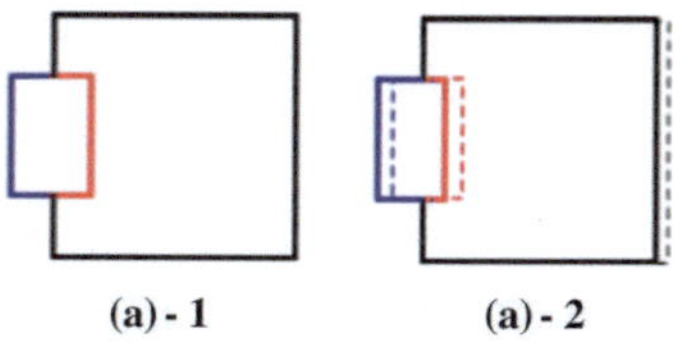

Fig. 6 The anisotropic energy of *(a)-1* is decreased without changing the enclosed area when it is changed like *(a)-2*

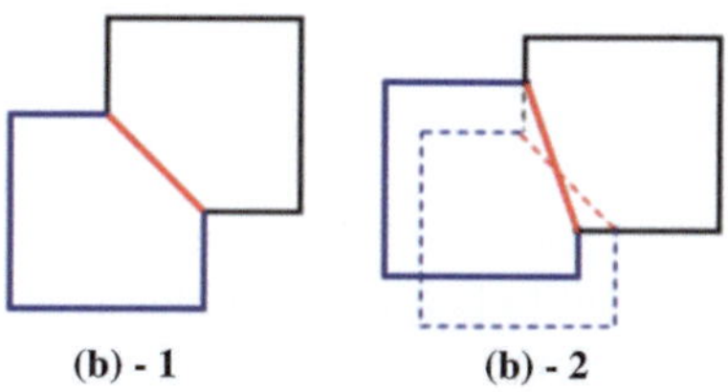

Fig. 7 The anisotropic energy of *(b)-1* is decreased without changing the enclosed area when it is changed like *(b)-2*

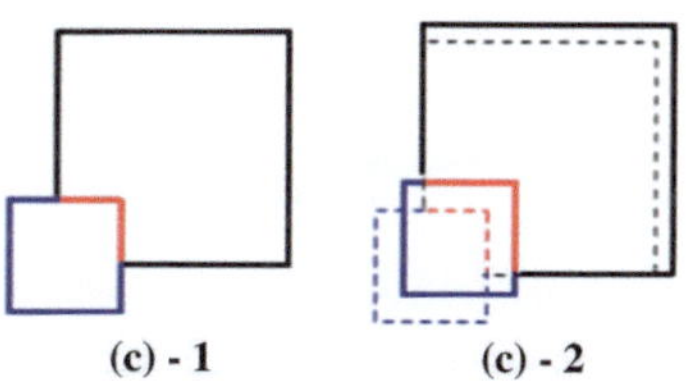

Fig. 8 The anisotropic energy of *(c)-1* is decreased without changing the enclosed area when it is changed like *(c)-2*

Let us think about the stability of the double crystals for our energy $\gamma = \gamma_{(p)}$ defined in (10). Define two angles ζ and η so that

$$f(\zeta)\cos(\zeta) = \frac{1}{2}, \quad f(\eta)\sin\left(\eta + \frac{\pi}{4}\right) = 2^{\frac{-1-p}{2p}}$$

holds.

Recall that the Wulff shape $W_{(p)}$ for the energy density $\gamma_{(p)}$ in (10) converges to the Wulff shape for γ_∞ when p approaches infinity (Fig. 2). By Lemma 6 and an approximation procedure (Fig. 5), we can show the following:

Proposition 2 *For sufficiently large p, we have the following result about the instability of double crystals. Double crystals of type 1 in Theorem 5 are unstable if* $\frac{\pi}{4} < \alpha_0 < \zeta$. *Double crystals of type 2 are unstable if* $\frac{\pi}{4} \leq \alpha_0 < \frac{\pi}{2} - \eta$.

We apply Theorem 3 (second variation formula) to the 2-dimensional case. For admissible variation vector field Y, We obtain

$$\delta^2\mathscr{F} = -\sum_{i=0}^{2}\int_{\Sigma_i} q_i L[q_i]\, d\Sigma_i + [\Lambda_i p_i q_i - A_i(q_i)_t q_i]_a^b, \tag{15}$$

where

$p_i = \langle Y, \nu_i\rangle$ and $q_i = \langle Y, N_i\rangle$. We expect that we will be able to prove the following conjecture by using (15).

Conjecture 1 Except the cases in Proposition 2, double crystals of types 1–3 are stable.

Let the Wulff shape be a square. Then the energy minimizing shape is one of the three types in Fig. 9 according to $\rho = V_2/V_1$. We expect that, by using the variational method, we will be able to obtain not only the absolute minimum but also local minimums. It is important to get local minimums because the physical state sometimes takes a local minimum.

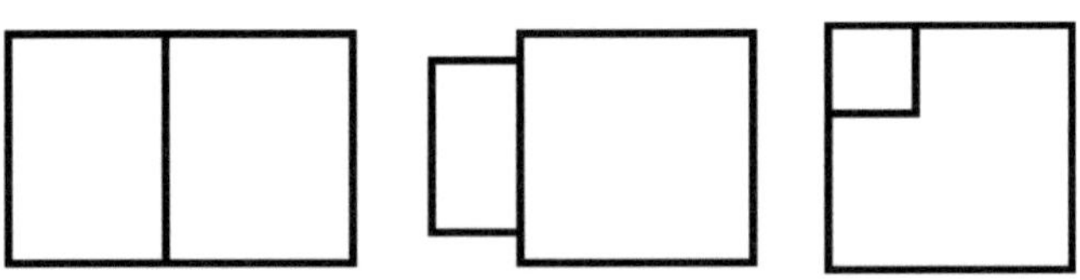

Fig. 9 The *right side* figure is $\rho \leq 2$ and both R_1 and R_2 are rectangular. The ratio of *middle* figure is $2 \leq \rho \leq \rho_0 := \frac{43+30\sqrt{2}}{16}$ and R_1 is square and R_2 is rectangular (this ratio of edge length is 1 : 2). The ratio of *left side* figure $\rho \leq \rho_0$, and both R_1 and R_2 are squares

References

1. Ben, W.R.: Proof of the double bubble conjecture in $\mathbf{R}^n$. J. Geom. Anal. **18**(1), 172–191 (2008)
2. Foisy, J., Alfaro, M., Brock, J., Hodges, N., Zimba, J.: The standard double soap bubble in $\mathbf{R}^2$ uniquely minimizes perimeter. Pac. J. Math. **159**(1), 47–59 (1993)
3. Frederick, J., Almgren, Jr.: Existence and regularity almost everywhere of solutions to elliptic variational problems with constraints. Mem. Amer. Math. Soc. **4**(165), viii+199 (1976)
4. Gray, R.L.: Double bubbles for immiscible fluids in $\mathbf{R}^n$. J. Geom. Anal. **24**(1), 190–204 (2014)
5. Hutchings, M.: The structure of area-minimizing double bubbles. J. Geom. Anal. **7**(2), 285–304 (1997)
6. Hutchings, M., Morgan, F., Ritoré, M., Ros, A.: Proof of the double bubble conjecture. Ann. Math. **155**, 459–489 (2002)
7. Koiso, M., Shinkawa, E.: Geometry of anisotropic double crystals, in preparation
8. Koiso, M., Palmer, B.: Geometry and stability of surfaces with constant anisotropic mean curvature. Indiana Univ. Math. J. **54**, 1817–1852 (2005)
9. Koiso, M., Palmer, B.: Stability of anisotropic capillary surfaces between two parallel planes. Calc. Var. Partial. Differ. Equ. **25**(3), 275–298 (2006)
10. Koiso, M., Palmer, B.: Anisotropic umbilic points and Hopf's theorem for surfaces with constant anisotropic mean curvature. Indiana Univ. Math. J. **59**(1), 79–90 (2010)
11. Morgan, F., French, C., Greenleaf, S.: Wulff cluster in $\mathbf{R}^2$. J. Geom. Anal. **8**(1), 97–115 (1998)
12. Shinkawa, E.: Uniqueness and stability for double crystals. Ph. D. thesis, Kyushu University, Japan (2015)
13. Taylor, J.E.: Crystalline variational problems. Bull. Am. Math. Soc. **84**, 568–588 (1978)

Modeling and Control of Fiber Orientation in Papermaking Machines

Takashi Sasaki

Abstract An algorithm for online control of the fiber orientation angle obtained in paper machines was developed. The control algorithm utilizes local linear models reflecting the change in fiber orientation angle as a function of the edge flow rates and slice bolt adjustments of the head box. These local linear models were derived from the results of fluid simulations based on the nonlinear Navier-Stokes equation. By the steepest descent method, we obtained the control algorithm that optimizes the estimate function calculated from these local linear models. The control algorithm was incorporated in the computer control system of an actual paper machine, and significant improvement in productivity and paper quality over extended operation has been realized.

Keywords Modeling · Papermaking · Fluid simulation · Fiber orientation

1 Introduction and Process Description

Various algorithms have been developed to control paper quality in papermaking machines. Owing to the complex paper quality properties and the papermaking process, these control algorithms are intrinsically based on process modeling techniques that use thermal dynamics, fluid dynamics, control theory, mathematics, etc. In this study, we introduce a methodology of fiber orientation (FO) control that was developed based on a process modeling technique that uses fluid dynamics and optimizing control theory.

Papermaking machines are essentially very large dryers, which, like other steam engine driven machines such as the steam locomotive and steamboat, were invented during the Industrial Revolution. Paper machines continuously convert pulp materials with a water content of 99.7 % into paper products with a water content of 3–5 %, which are then used in our daily life.

T. Sasaki (✉)
IA PF PBC P&W Solution Department, Yokogawa Electric Corporation, Tokyo, Japan
e-mail: T.Sasaki@jp.yokogawa.com

R.S. Anderssen et al. (eds.), *Applications + Practical Conceptualization + Mathematics = fruitful Innovation*, Mathematics for Industry 11,
DOI 10.1007/978-4-431-55342-7_18

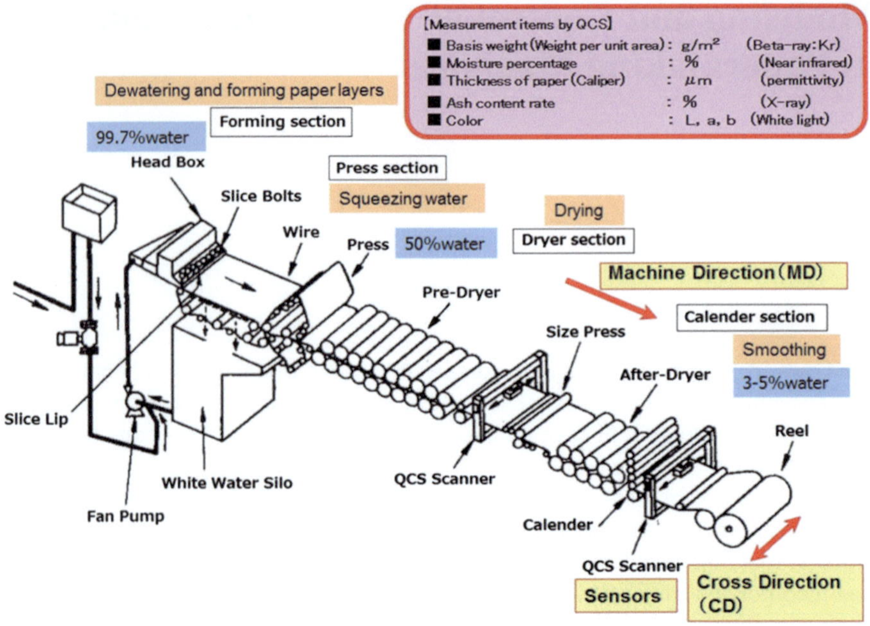

Fig. 1 Overview of a paper machine

As shown in Fig. 1, paper machines consist of four main sections: a forming section, pressing section, dryer section, and calender section. Firstly, various pulp materials, such as needle leaf pulp, broad leaf pulp, and waste paper pulp, are blended in a stock preparation process according to predefined proportions. After processing, the pulp material is sent to the head box, which sits at the fore of the forming section. At this stage, the concentration of pulp in the slurry is approximately 0.3 % (that is, its water content is 99.7 %). The head box is several meters wide, and has a 10 mm vertical gap through which the pulp material is jetted at high speed (around 60 km/h, depending on the paper machine) onto a moving polyester fiber fabric loop denoted as the wire. The pulp material is dewatered while being conveyed on the wire. Important paper qualities, such as basis weight, uniformity, and orientation of the fiber dispersion, are established along the wire.

The remaining processing steps are as follows. The press section squeezes out much of the water, reducing the water content to 50 % at the outlet of the press section.

The dryer section uses saturated steam to reduce the water content to only 3–5 %.

In the calender section, the paper surface is smoothed and the paper thickness (caliper) is adjusted to ensure uniformity as the paper passes through several metal rolls under pressure and friction.

Finally, the paper product is formed into a roll by the reel at the end of the paper machine. The roll is quite large, with a diameter typically around 2 m.

The control performance and paper quality of paper machines are specified according to two directions: the machine direction (MD), in which the paper moves, and the cross direction (CD), which is perpendicular to the MD.

Various paper qualities such as basis weight (weight per unit area), moisture percent, thickness (caliper), ash content, and color are measured by quality control system (QCS) sensors installed in the middle of the dryer section and just before the reel. The sensors are mounted on cross-directional scanners, and continuously collect measurements as they move back-and-forth across the paper.

The data measured in one scan by QCS sensors are denoted as the "profile." Specifically, the profile is defined as a set of measured data obtained in one scan and usually the number of the data is 50–200. The profile is regarded as a section of a paper sheet, and it reflects important paper qualities, such as printability and shape stability. Therefore, profiles have to be controlled to remain as flat as possible by control functions in QCS to realize high productivity and quality in paper production.

2 Development of On-Line Fiber Orientation Measurement and Control

FO angles are measured on both sides of the paper by our QCS. The upper side of the paper is denoted as the felt side and the lower side is denoted as the wire side. The difference in FO angles between the felt and wire sides (Figs. 2 and 3) is said to be the crucial factor generating the twist-curl of sheets, which negatively affects the dimensional stability [1–4]. Previously, we developed the world's first online FO sensor system, which made it possible to measure the FO of both the felt and wire side surfaces of the paper [5]. This sensor was incorporated in the QCS system, and, through the measurements and calculations, FO angle profiles of both felt and wire

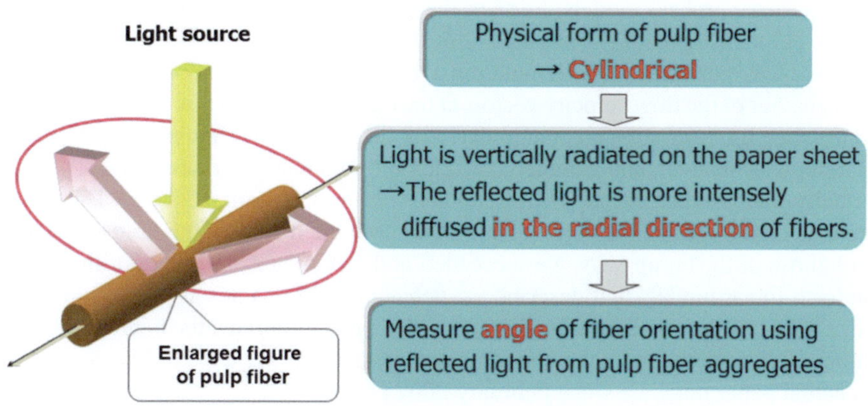

Fig. 2 Principle of the fiber orientation measurement

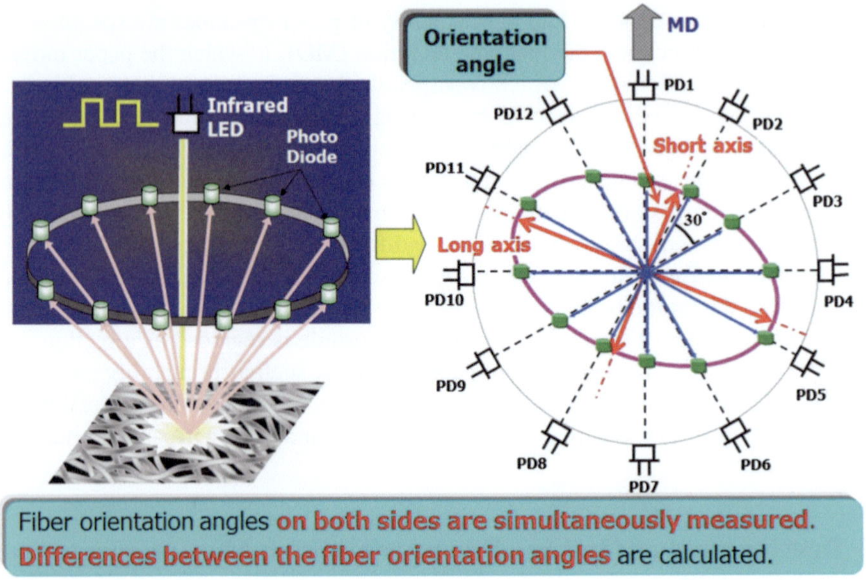

Fig. 3 Configuration of the fiber orientation sensor

sides are produced. The sensor was launched in 1999, and has been valued highly by our customers.

Several earlier studies related to FO control have been published [6–8]. In these studies, the change in the FO angle caused by the adjustment of process parameters are discussed based on the data gathered in paper machines through inductive approaches, but precise investigations based on physical models have not been discussed. The present paper describes the process model of the change in the FO angle profile based on fluid analysis and constructs an optimizing control algorithm on the basis of these process models. Further, the effectiveness of the control utilizing the online FO measurement is presented.

Firstly, we performed fluid simulation within a head box, and thoroughly analyzed the behavior of the flow velocity vectors at the outlet of the head box which is denoted as the slice lip. Secondly, we manipulated the edge flow rates and slice lip actuators in a stepwise manner, and gathered considerable data that describes how FO profiles actually change when the edge flow rates and slice lip actuators are varied. These tests are called bump tests. Thirdly, by combining both the results of the fluid simulation with those of the bump tests, we succeeded in developing a control algorithm based on local linear models that describe the behavior of FO angles as a function of the edge flow rate and slice lip adjustment. Lastly, we incorporated the control software in our QCS system, and have been implementing the continuous online control for an actual paper machine since February 2007 [9, 10].

3 Fiber Orientation Measurement

3.1 Correlation Between Fiber Orientation and Twist-curl

Paper possesses the characteristic of two-sidedness, which reflects differences between the properties of the felt and wire sides of the paper. A paper sheet with strong two-sidedness can be easily curled with a change in moisture content. Consequently, copy paper is required to have reduced two-sidedness to ensure dimensional stability for uniform operation in copy machines. Differences between the FO angles of the felt and wire sides of the paper is one of the principal factors generating two-sidedness that has been experimentally determined to have a strong correlation with the twist-curl of the paper [9, 10]. Therefore, adequate control of the dimensional stability of paper products first requires the simultaneous measurement of the FO angles of both the felt and wire sides of the paper, followed by a suitable control method based on how FO profiles actually change with changing process parameters.

3.2 Measurement Principle of the Fiber Orientation Sensor

The measurement principle of the FO sensor is based on the fact that light reflected by the cellulose fibers of paper surfaces has anisotropic properties. More specifically, as illustrated in Fig. 2, the reflected light is more intensely diffused in the radial direction of a fiber than in the axial direction. As is illustrated in Fig. 3, the FO sensor is equipped with an infrared light source and a circular array of light-detecting elements. The intensities of signals detected in these elements can be approximated by an ellipse as in Fig. 3. This ellipse is considered to represent the anisotropy of dispersion of fibers on the sheet surface. Here, the fiber index is defined as the ratio of the long axis to the short axis, whereas the FO angle is defined as the angle between the short axis and the machine direction. Mounting FO sensors on both the felt and wire sides of the paper provides for measurement of the difference of the FO angles between the two sides. This is quite an important quality factor for maintaining dimensional stability, which is especially crucial for copy paper.

4 Development of Fiber Orientation Control

To minimize the difference in FO angles between the felt and wire sides, we developed an online control of FO angles by applying a process model based on fluid simulation techniques.

4.1 Fluid Simulation Within the Head Box

The pulp material flows in the slice channel through the turbulence generator, as shown in Fig. 4. Consequently, our control algorithm is based on process simulations within the head box. The velocity distribution within the slice channel is simulated by computational fluid dynamics (CFD) software. Here because the pulp concentration is only 0.3 %, the flow can be approximated by that of water. The behavior of the flow is numerically calculated based on the Navier-Stokes equation, and it is essentially nonlinear. The flow rate at the edges of the slice channel is controlled by two edge flow valves: one installed at the driving side (DS) and the other on the tender side (TS). The flow rates provide the initial conditions of the fluid simulation. The simulation calculates the distribution of the jet velocity of the pulp material at the outlet of the slice channel. After being ejected from the slice lip, the jet lands on the wire, and the FO is determined by the way the sheet layers form on the wire. However, between the outlet of the slice lip and the sheet forming process on the wire, we can assume that the CD velocity component of the jet is preserved. Under this assumption, the FO angle can be approximately calculated by the jet velocity vector at the outlet of the slice channel. Thus, by varying the edge flow rates, we can adjust the jet velocity vectors at the outlet of the slice channel, which, in turn, adjusts the FO angle.

When viewed from the side, the slice channel is tapered, as shown in the lower figure of Fig. 5. This geometry accelerates the flow speed as well as makes the flow more uniform. The top lip at the outlet is a long metal sheet, where the slice bolts are vertically mounted at 100–140 mm intervals as shown in the figure at right. By adjusting these slice bolts, the shape of the top lip can be adjusted; thus, the shape of the outlet of the slice lip can be varied with micrometer precision (Fig. 1). Because the shape of the outlet of the slice lip corresponds to the boundary conditions of the fluid simulation, the simulations can be applied to the control of the velocity vectors at the outlet of the slice lip by adjusting these slice bolts.

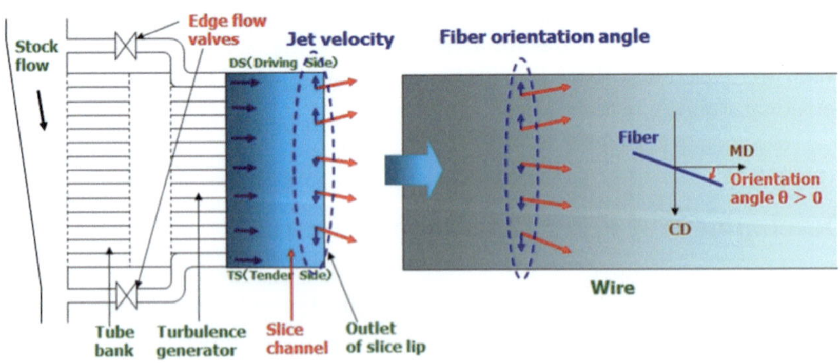

Fig. 4 Description of the model for the fluid simulation within the head box

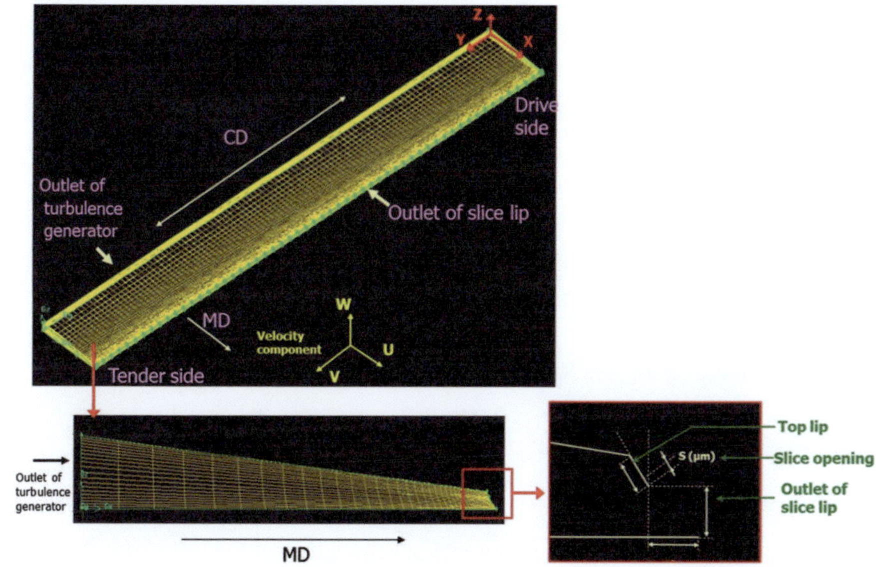

Fig. 5 Outlet shape and *side view* of the slice channel

4.2 Local Velocity Models Affecting the Fiber Orientation Angle Profile

Based on the simulation results for various edge flow rates and slice lip openings, we can obtain equations that describe local variations in the MD and CD velocities as a function of the edge flow rate and slice bolt adjustments.

Firstly, we describe how the FO angle changes locally with respect to edge flow rates. For example, as illustrated in the top figure of Fig. 6, when the DS edge flow is increased, the MD velocity is increased at the inlet of the slice channel. However, due to momentum diffusion, a component of the MD velocity is diffused to that of a CD velocity directed inwardly, resulting in inwardly diverted jet velocities. As a result, the overall change in the MD velocity at the slice lip outlet is negligible compared to the change in the CD velocity.

In contrast, when the TS edge flow is decreased, the jet velocities are directed outward. This is further illustrated by the graphs of the MD and CD velocities as a function of the DS and TS edge flow rates shown in bottom figure of Fig. 6. Here, the horizontal axis indicates the number of slice bolts and N denotes the total number of the slice bolts. The vertical axis indicates the change in each velocity component. While the MD velocity is nearly constant, the CD velocity is proportional to the change in the edge flow rate, whereas the magnitude of the change decreases linearly towards the center. Beyond the center, the CD velocity is regarded as constant. From these simulation results, we constructed local linear models reflecting the change

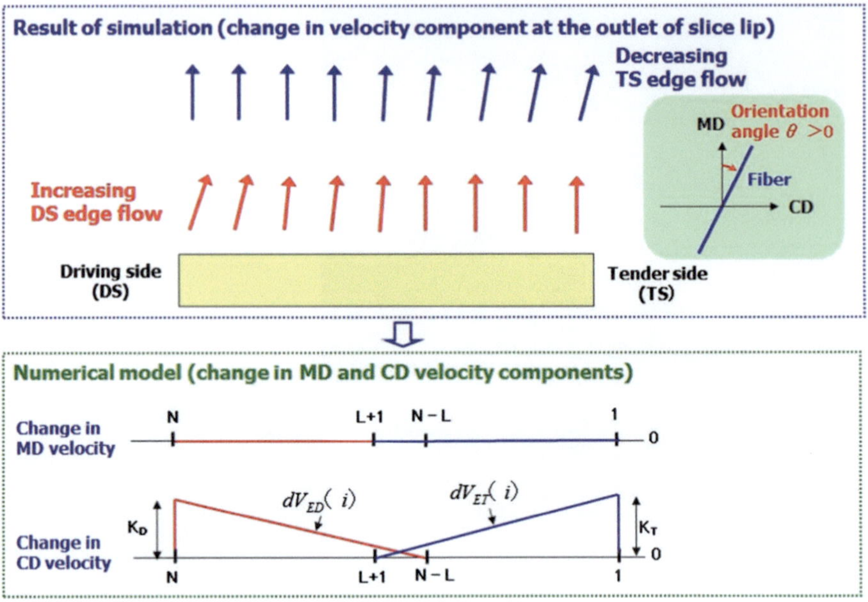

Fig. 6 Changes in MD/CD velocity as a function of edge flow rates

in the CD velocity as a function of the edge flow rates. The velocity changes are expressed by the following equations.

$$dV_{ET}(i) = \begin{cases} -\frac{L+1-i}{L} \cdot K_T \cdot dE_T & (\text{when } i \leq L) \\ \mathbf{0} & (\text{when } L < i) \end{cases} \tag{1}$$

$$dV_{ED}(i) = \begin{cases} \frac{i-(N-L)}{L} \cdot K_D \cdot dE_D & (\text{when } N-L \leq i) \\ \mathbf{0} & (\text{when } i < N-L) \end{cases} \tag{2}$$

dE_T: Change in opening of TS edge flow valve (%)
dE_D: Change in opening of DS edge flow valve (%)
$dV_{ET}(i)$: Change in CD velocity at the ith slice bolt by manipulation of TS edge flow valve (m/s)
$dV_{ED}(i)$: Change in CD velocity at the ith slice bolt by manipulation of DS edge flow valve (m/s)
N: Number of slice bolts, L: Response width

We now evaluate changes in the local velocity vectors with respect to changes in the slice lip opening.

Assuming that the slice lip is that shown in the top figure of Fig. 7, the jet velocities change as if the jets were directed into the widest part of the slice lip opening.

The changes in the resulting MD and CD velocities are plotted in the lower figure of Fig. 7. The MD velocity linearly decreases towards the center of the slice

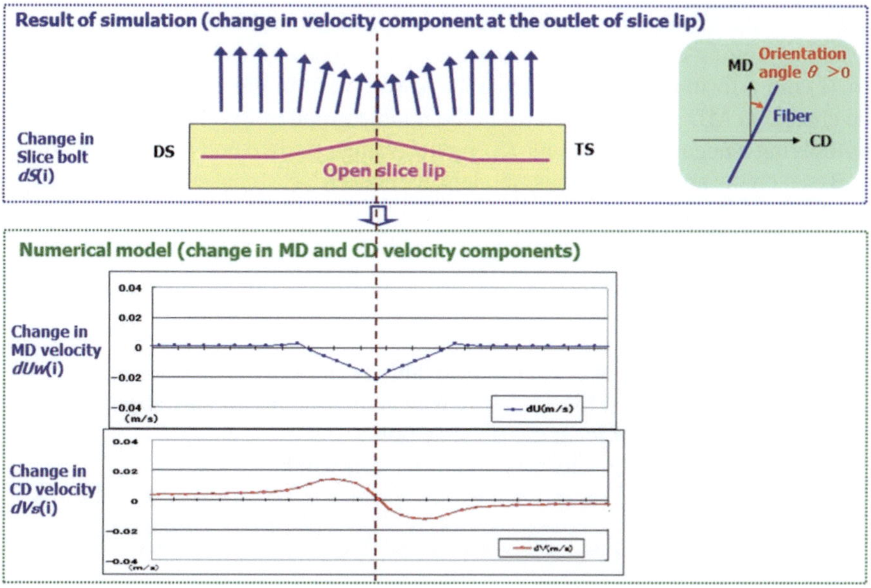

Fig. 7 Changes in MD/CD velocity as a function of the slice lip opening

lip opening where the opening is widest. Therefore, the MD velocity decreases in proportion to the width of the slice lip opening. Inversely, the MD velocities increase in proportion to the decreasing slice lip width. On the other hand, the changing CD velocity follows an S curve. Of paramount importance here is that the extents of increase and decrease are equal, and cancel each other; therefore, the summation of the CD velocity changes over the full width is zero.

From these simulation results, we constructed local linear models expressing the changes in CD and MD velocities as a function of the slice bolt settings.

$$dT(i) = dS(i-1) - dS(i)(i = 2, \ldots, N)$$

$$dT_m(i) = \frac{1}{7}\sum_{k=-3}^{+3} dT(i+k)$$

$$dT_{mm}(i) = \frac{1}{7}\sum_{k=-3}^{+3} dT_m(i+k)$$

$$dV_S(i) = K_V \cdot dT_{mm}(i) \quad (3)$$

$$dU_W(i) = K_U \cdot dS(i) \quad (4)$$

$dS(i)$: Change in setting of the ith slice bolt (μm)
$dV_S(i)$: Change in CD velocity by manipulation of the ith slice bolt (m/s)
$dU_W(i)$: Change in MD velocity by manipulation of the ith slice bolt (m/s)

Here, if the change in slice bolt $dS(i)$ has the form in the upper graph in Fig. 7, by taking moving average twice as in Eq. (3), the shape of the change in CD velocity $dV_s(i)$ has a form of an S curve as in the lower graph in Fig. 7. On the other hand, the change in MD velocity $dU_W(i)$ in the middle graph in Fig. 7 is obtained just by multiplying a negative coefficient K_U by the change in slice bolt $dS(i)$ as in Eq. (4).

Based on the above local linear models, we can derive a local model that describes the change in the FO angle on the wire side as follows.

$$FO(i) = \frac{180}{\pi} \times \arctan\left(\frac{V(i)}{U_W(i)}\right) \tag{5}$$

$V(i)$: CD Velocity of jet at the ith slice bolt
$U_W(i) = U_Y(i) - \overline{U_Y(i)}/R$: Relative MD velocity of jet to wire speed
$U_Y(i)$: MD velocity of jet at the ith slice bolt
$\overline{U_Y(i)}$: Average of MD velocity of jet in the full width
R: Jet-to-wire ratio (=1.01)

FO angle profile F is defined as a vector which has $FO(i)$s as entries in Eq. (6).

$$F = \begin{bmatrix} FO(1) \\ FO(2) \\ \vdots \\ FO(N) \end{bmatrix} \tag{6}$$

We can linearize the expression (5) by differentiation. Thus, we compute the local changes in the FO angles caused by changes in MD and CD velocities according to Eq. (7) based upon the edge flow rate and the slice bolt adjustments, as given in Eqs. (1)–(4).

$$\begin{aligned} dFO(i) &= \frac{180}{\pi} \times \left\{ \frac{\partial}{\partial U_W(i)}\left(\arctan\left(\frac{V(i)}{U_W(i)}\right)\right) \cdot dU_W(i) + \frac{\partial}{\partial V(i)}\left(\arctan\left(\frac{V(i)}{U_W(i)}\right)\right) \cdot dV(i) \right\} \\ &= \frac{180}{\pi} \times \left\{ \frac{-V(i)}{U_W(i)^2 + V(i)^2} \cdot dU_W(i) + \frac{U_W(i)}{U_W(i)^2 + V(i)^2} \cdot (dV_S(i) + dV_{ET}(i) + dV_{ED}(i)) \right\} \end{aligned} \tag{7}$$

$dU_W(i)$: Change in MD velocity due to slice lip (m/s)
$dV_S(i)$: Change in CD velocity due to slice lip (m/s)
$dV_{ET}(i)$: Change in CD velocity due to TS edge flow (m/s)
$dV_{ED}(i)$: Change in CD velocity due to DS edge flow (m/s)

The general linearized local model given by Eq. (8) has been obtained from Eqs. (1)–(7) for describing the local change in the FO angle profile given by the vector on the left-hand side of Eq. (8) as a function of edge flow rates and slice bolt settings given by the vector on the right-hand side of Eq. (8) in conjunction with the coefficient matrix K. This is a simple first-order expression.

$$\begin{bmatrix} dFO(1) \\ dFO(2) \\ \vdots \\ dFO(N) \end{bmatrix} = K \begin{bmatrix} dS(1) \\ dS(2) \\ \vdots \\ dS(N) \\ dE_T \\ dE_D \end{bmatrix} \tag{8}$$

$dFO(i)$: Change in FO angle on the wire side at the ith slice bolt
$dS(i)$: Change in setting of the ith slice bolt (μm)
dE_T: Change in opening of TS edge flow valve (%)
dE_D: Change in opening of DS edge flow valve (%)

4.3 Results of Bump Tests

Through the fluid simulations, we obtained the local linear models reflecting the change in the FO angle profile as a function of edge flow rate and slice bolt adjustments in Eq. (8). In addition, we performed bump tests of these actuators by gathering actual FO angle profile data as a function of variations in actuator settings. The bump tests confirmed that the obtained local linear models satisfactorily approximate the actual changes in the FO angle measurement profiles obtained in the bump tests.

4.4 Vertical Distribution of Velocity and Control Strategies

In Eq. (5), $FO(i)$ represents the FO angle on the wire side at the ith slice bolt. We have to consider more thoroughly how the difference between FO angles of the felt and wire sides is derived from Eq. (5). During the CFD analyses, the vertical distributions of the MD velocities at the outlet of the slice lip were calculated, as shown in Fig. 8. Seeing that the fibers accumulate and form the sheet layers as they land on the wire, it is presumed that the relative MD velocity (U_F) of the fibers on the felt side surface to the fibers in the lower layer is negative. In contrast, on the wire side surface, the relative MD velocity (U_W) is presumed to be positive under a jet to wire ratio of 1.01. Here, a jet to wire ratio means the ratio of the averaged MD velocity of the fiber to the rotating speed of the wire. Under these conditions, if the FO angle on the wire side (θ_W in Fig. 8) is positive, then the FO angle on the felt side (θ_F) is negative, and consequently, the difference of the FO angle between the felt and wire sides ($\theta_F - \theta_W$) is always negative. Conversely, if the FO angle on the wire side (θ_W) is negative, the difference of the FO angle between the felt and wire sides ($\theta_F - \theta_W$) is always positive. From these analyses, we can conclude that the difference of FO angles between the felt and wire sides becomes close to zero if

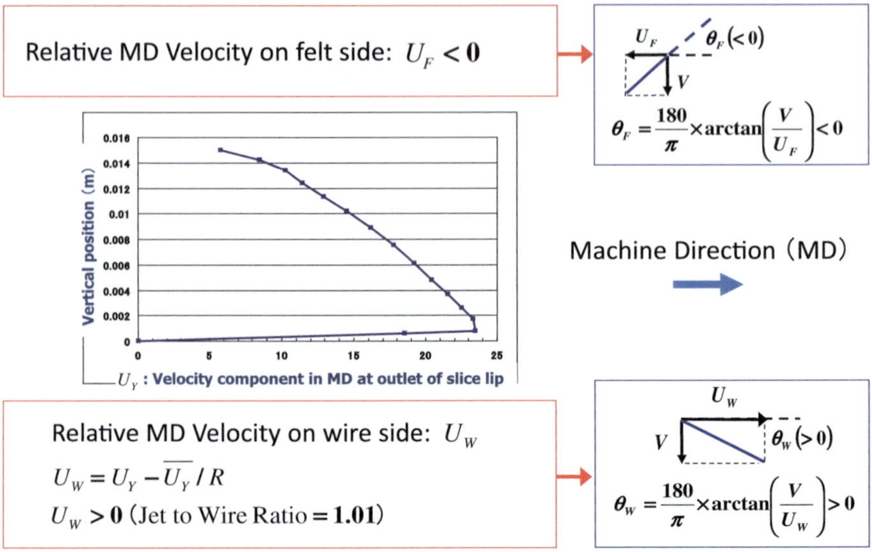

Fig. 8 Vertical distribution of MD velocities at the outlet of the slice lip

and only if the FO angle profile on the wire side defined by Eq. (5) becomes close to zero. Therefore, our control strategy is as follows: by making the FO angle profile in Eq. (5) close to zero through the control calculation, we can minimize the difference between the FO angles of the felt and wire sides, and consequently, produce a paper product with the least two-sidedness and a minimum twist-curl.

4.5 Fiber Orientation Control Calculation

From Eq. (8), we can easily obtain the setting values to minimize the absolute value of the entries in the FO angle profile given in Eq. (6) by applying optimization control theory. The control target is that the FO angle profile becomes zero. Consequently, minimization of the squared sum J of all entries in the FO angle profile would cause all entries in the FO angle profile to reside close to zero. Therefore, we adopt J as an estimate function of the optimization control.

The settings vector on the right-hand side of Eq. (8) that reduces J most rapidly is calculated by the steepest descent method, which is given by taking the gradient of J with respect to the manipulated variables. Therefore, in our framework, the settings that most rapidly reduce the entries in FO angle profile to values close to zero are simply given by the transpose of FO angle profile multiplied by the transpose of the coefficient matrix K as follows.

$$J = \sum_{i=1}^{N} FO(i)^2 \tag{9}$$

$$\begin{aligned}[dS(1), dS(2), \ldots, dS(N), dE_T, dE_D]^t &= -\frac{\varepsilon}{2}\nabla J \\ &= -\frac{\varepsilon}{2}\left[\frac{\partial J}{\partial S(1)}, \cdots, \frac{\partial J}{\partial S(N)}, \frac{\partial J}{\partial E_T}, \frac{\partial J}{\partial E_D}\right]^t \\ &= -\varepsilon \times K^t \times [FO(1), FO(2), \ldots, FO(N)]^t \end{aligned} \tag{10}$$

$\varepsilon > 0$: Control gain (a small positive number)
$d\underline{S} = [dS(1), dS(2), \ldots, dS(N), dE_T, dE_D]^t$: Change in manipulated value
$\underline{F} = [FO(1), FO(2), \ldots, FO(N)]^t$: FO angle profile

Because the linear model in Eq. (8) describes the local behavior of the FO angle profile, we must apply the control calculation given in Eq. (10) iteratively; consequently, the absolute value of each entry in the FO angle profile converges to a minimal value.

4.6 Control Results and Benefits

This control algorithm was incorporated in our QCS system software, and the described online FO control has been executed in an actual paper machine since the beginning of 2007.

Immediately after machine starts to run, the coarse control is executed to minimize the difference profile of FO angles between the felt and wire sides (the θ_{F-W} profile) as rapidly as possible. Since profile is a set of measured data, the quality of profile is usually represented by the 2-sigma of the data set. Through the coarse control, the convergence time of the 2-sigma of the θ_{F-W} profile is reduced by 70 % compared with manual operations. During steady state production, the fine control is executed to maintain paper quality as stable as possible. By means of the fine control over extended operation, the 2-sigma of the θ_{F-W} profile is improved by 30 % on average compared with manual operations.

Moreover, it has been confirmed that these improvements of the θ_{F-W} profile have resulted in the improvement of twist-curl by 40 % on average compared with manual operations.

5 Conclusions

In this paper, we introduced a methodology for fiber orientation (FO) control that was developed based on the process modeling technique, and the resulting online fiber orientation control algorithm was applied to an actual paper machine. The control algorithm was based on local linear models reflecting the change in the FO angle as a function of the edge flow rate and slice bolt adjustments, which were derived from the results of fluid simulations based on the nonlinear Navier-Stokes equation. The control algorithm realized significant improvement in productivity and quality over extended operation since February 2007 in an actual paper machine.

References

1. Nonomura, F., Abe, Y., Takeuchi, N.: A Study on the Curling Behavior of Paper Resulting from Heatroller Heating: International Paper Physics Conference, San Diego, California (1999)
2. Abe, Y., Todoroki, H., Takeuchi, N., Sakamoto, A.: Method for measuring fiber orientation on paper surface. Appita J. **3C41**, 625–632 (1996)
3. Hellstrom, A.: The two sides of fiber orientation. In: TAPPI Practical Papermaking Conference, Milwaukee, Wisconsin (2005)
4. Amirthalingam, R.: Curl and twist modeling for monitoring and control. In: Control Systems 2002, Stockholm, Sweden (2002)
5. Fukuoka, K., Hatano, A., Abe, Y., Takeuchi, N.: Development of online system for measuring fiber orientation on paper surface. In: TAPPI Engineering Conference, Atlanta, Georgia (2000)
6. Subbarayan, R., Tran, P., Chen, S.: Control of fiber orientation of a paper sheet using on-line measurements. In: Control Systems 2002, Stockholm, Sweden (2002)
7. Kniivila, J., Nuyan, S.: Focusing on the edges of profile control. In: Control Systems 2004, Quebec City, Canada (2004)
8. Chen, S., Hellstrom, A.: Online measurement and control of sheet dimensional stability twist and curl. In: Control Systems 2006, Tampere, Finland (2006)
9. Yamamoto, J., Todoroki, T., Ono, K., Ochi, T., Sasaki, T., Sano, H.: New automatic control system for fiber orientation and improvement of the quality of copy paper. In: PaperCon '08, Dallas, Texas (2008)
10. Sasaki, T., Sano, H., Yamamoto, J., Todoroki, T., Ono, K., Ochi, T.: Online fiber orientation measurement and control. In: Proceedings of Control Systems Pan-Pacific Conference, pp. 117–121, 16–18 June 2008, Vancouver, Canada (2008)

Recent Methods for Reconstructing Missing Data in Multispectral Satellite Imagery

Farid Melgani, Grégoire Mercier, Luca Lorenzi and Edoardo Pasolli

Abstract One of the major limitations of passive sensors is their high sensitivity to weather conditions during the image acquisition process. The resulting images are frequently subject to the presence of clouds, which makes the image partly useless for assessing landscape properties. The common approach to cope with this problem attempts to remove the clouds by substituting them with cloud-free estimations. The cloud removal problem can be viewed as an image reconstruction/restoration issue, in which it is aimed at recovering an original scene from degraded or missing observations. Two cloud removal approaches are detailed and discussed in this chapter. The first one is a single-channel method for the reconstruction in a sequence of temporal optical images. Given a contaminated image of the sequence, each area of missing measurements is recovered by means of a contextual prediction process that reproduces the local spectro-temporal relationships. The second approach exploits the Compressive Sensing (CS) theory, which offers the capability to recover an unknown sparse signal with a linear combination of a small number of elementary samples. The two reconstruction approaches are evaluated experimentally on a real multitemporal multispectral remote sensing image.

Keywords Cloud removal · Compressive sensing · Genetic algorithms · Image reconstruction · Optical imagery · Satellite image time series · Sparse representation

---

F. Melgani (✉) · L. Lorenzi
Department of Information Engineering and Computer Science,
University of Trento, 38123 Trento, Italy
e-mail: melgani@disi.unitn.it

G. Mercier
Télécom Bretagne, Technopôle Brest-Iroise, 29238 Brest, France
e-mail: gregoire.mercier@telecom-bretagne.eu

E. Pasolli
Centre for Integrative Biology, University of Trento, 38123 Trento, Italy
e-mail: edoardo.pasolli@gmail.com

R.S. Anderssen et al. (eds.), *Applications + Practical Conceptualization + Mathematics = fruitful Innovation*, Mathematics for Industry 11,
DOI 10.1007/978-4-431-55342-7_19

1 Introduction

One of the major limitations of passive sensors is their high sensitivity to weather conditions during the image acquisition process. The resulting images are frequently subject to the presence of clouds, whose extent depends on the season and the geographic position of the study region. For instance, in Canada, from 50 to 80 % of the Earth's surface can be obscured by clouds in mid-morning [1]. Depending on the application and the end-user requirements, clouds can be viewed: as a source of information for measuring important parameters such as cloud liquid water useful in meteorological forecasting and hydrological studies [2]; or as a source of contamination that makes the image partly useless for assessing landscape properties. In the latter case, which represents the subject of this paper, clouds distort the spectral response of land covers, thereby resulting in missing data for high-frequency passive sensors including multispectral optical sensors and microwave radiometers. The common approach to cope with this problem first detects the contaminated regions and, in a second instance, attempts to remove the clouds by substituting them with cloud-free estimations.

The cloud removal problem, which is the focus of this contribution, can be viewed as an image reconstruction/restoration issue, in which it is aimed at recovering an original scene from degraded or missing observations. Image reconstruction/restoration has been intensively and extensively studied in various application fields, such as radio astronomy, biomedical engineering, and machine vision, because of its practical importance as well as theoretical interest [3]. In the remote sensing field, significant attention has been devoted to the reconstruction/restoration of images subject to various problems, such as acquisition blur and geometric distortions, phase distortions, resampling problems, or problems related to applications like buried object detection. By contrast, less attention has been paid to the specific problem of cloud removal. Among the relatively few works available in the literature, one can find for instance [4, 5]. The main drawbacks that can be identified from these cloud removal algorithms are: sensor-dependence, very high temporal resolution, ground cover type-dependence, cloud-type dependence, or/and high methodological complexity. Two alternative recent approaches, based on linear contextual prediction and compressive sensing [6, 7], respectively, that aim at circumventing most of these drawbacks, are detailed and discussed in this paper.

2 Problem Formulation

We consider a set of multitemporal multispectral images $I^{(i)}$ acquired over the same geographical area by an optical sensor at times t_i (with $i \in S = \{1, 2, \ldots, T\}$). Let us suppose that the images have been registered. We assume that (1) the images of the sequence may convey changes in the spectral appearance of objects on the ground and (2) they are characterized by an almost similar spatial structure. The last assumption can be considered realistic if the acquisition dates are close to each other (i.e., high-temporal resolution) or if the spatial dynamics of the geographical area under analysis is slow compared to the total time interval of the sequence (e.g., forest, mountainous, and urban areas). Moreover, we assume that the images have first been processed to generate a sequence of cloud/noncloud classification maps $M^{(i)}$ $(i \in S)$ by using an automatic cloud-masking method or simply by photointerpretation. Given $M^{(i)}$, cloudy and non cloudy areas are represented by $\Omega^{(i)}$ and $\Phi^{(i)}$, respectively, subject to $I^{(i)} = \Omega^{(i)} \cup \Phi^{(i)}$. The specific problem of the detection of clouds (and their shadow) is not dealt with in this paper. The objective of the investigated methods is to reconstruct any area contaminated by clouds (or by cloud shadows) for each image of the sequence. Therefore each classification map $M^{(i)}$ will be used to guide the cloud removal process.

For the first investigated approach based on linear contextual prediction, image channels are processed separately. Let us denote by $X^{(i)}$ $(i \in S)$ one of the available T single-channel temporal images. We denote by C a cloudy area of the image $X^{(i)}$. The reconstruction problem of C in $X^{(i)}$ can be expressed as a problem of generating an image $Y^{(i)}$ such that:

$$Y^{(i)}(u, v) = \begin{cases} X^{(i)}(u, v) \text{ if } (u, v) \notin C \\ f\left[X^{(k)}(u, v),\ k \in S_c\right], \text{ otherwise} \end{cases} \tag{1}$$

where $f[\cdot]$ represents a contextual prediction function and S_c stands for the subset of indices corresponding to images $X^{(k)}$ $(k \neq i)$ that are cloud-free in the spatial area $N(C)$ including C and its neighborhood. In other words, any image of the sequence obscured by the presence of a cloud in $N(C)$ will not contribute to the reconstruction process of C. In each image $X^{(k)}$ $(k \in S_c)$, the spatial area $N(C)$ can be subdivided into two cloud-free areas C and $\overline{C}$ such that $N(C) = C \cup \overline{C}$ and $C \cap \overline{C} = \emptyset$ where C represents the spatial area that corresponds to the cloudy area in $X^{(i)}$ and $\overline{C}$ stands for the neighboring spatial area.

For the second approach based on compressive sensing, we simplify the problem by supposing to have just two images $(T = 2)$ and we relax the above channel-based processing constraint (all channels are processed together). At this level, we make the hypothesis that the image $I^{(2)}$ has clouds, while the image $I^{(1)}$ is cloud-free. We assume that any pixel $\mathbf{x}^{(1)} \in \Omega^{(1)}$ can be expressed as a linear combination of pixels in region $\Omega^{(1)}$ of $I^{(1)}$. In other words, in $I^{(1)}$ we have:

$$\mathbf{x}^{(1)} = \Phi^{(1)} \cdot \boldsymbol{\alpha},\ \forall \mathbf{x}^{(1)} \in \Omega^{(1)} \tag{2}$$

where $\boldsymbol{\alpha}$ is an unknown weight vector associated with the considered pixel $\mathbf{x}^{(1)}$ and having the same dimension as the number of pixels belonging to $\Phi^{(1)}$. The problem at this point is to infer $\boldsymbol{\alpha} = f\left(\Phi^{(1)}, \mathbf{x}^{(1)}\right)$. Once $\boldsymbol{\alpha}$ is computed, if we assume that $I^{(1)}$ and $I^{(2)}$ are temporally close, it will be possible to reuse the $\boldsymbol{\alpha}$ coefficients to reconstruct the spatially corresponding pixel in the missing area $\Omega^{(2)}$, adopting the previous formulation for $I^{(2)}$, i.e., $\hat{\mathbf{x}}^{(2)} = \Omega^{(2)} \cdot \boldsymbol{\alpha}$. In other words, for each pixel

$\mathbf{x}^{(1)} \in \Omega^{(1)}$, we evaluate $\boldsymbol{\alpha}$, and in a second moment we reuse this weight vector to return an estimation of $\mathbf{x}^{(2)} \in \Omega^{(2)}$:

$$\text{From } I^{(1)} : \boldsymbol{\alpha} = f\left(\Phi^{(1)}, \mathbf{x}^{(1)}\right)$$

$$\text{To } I^{(2)} : \hat{\mathbf{x}}^{(2)} = \Omega^{(2)} \cdot \boldsymbol{\alpha} \tag{3}$$

where $f(\cdot)$ represents an estimation function. We recall that, differently from the first approach, all image channels are processed simultaneously.

3 Linear Contextual Prediction Approach

The first approach is a single-channel method for the reconstruction of cloud-contaminated areas in a sequence of temporal optical images. It is based on the assumptions that spectral non-stationarity is allowed while the spatial structure of the image should be identical over the image sequence. Given a contaminated image of the sequence, each area of missing measurements is recovered by means of a contextual prediction process that reproduces the local spectro-temporal relationships. These latter are deduced from the cloud-free areas in the spatial neighborhood of the contaminated area over the available series of temporal images. The contextual prediction process is implemented in two steps. First, a prediction system is trained to learn over $\overline{C}$ the temporal relationship between the set of available images $X^{(k)}$ ($k \in S_c$) that are cloud-free in $N(C)$ on the one hand and the image $X^{(i)}$ on the other. This is done by means of an ensemble of linear predictors, each trained in an unsupervised way over a local temporal region that is spectrally homogeneous in each temporal image of the sequence. In order to obtain such regions, each temporal image is locally classified in an unsupervised way by means of the Expectation-Maximization (EM) algorithm [8] assuming the data classes are Gaussian. The number of data classes is estimated automatically by minimizing the Minimum Descriptive Length (MDL) criterion [9]. Once the training is completed, the prediction system is used to provide an estimate of each contaminated pixel of image $X^{(i)}$ in C, based on the spatially corresponding pixel values in images $X^{(k)}$ ($k \in S_c$).

3.1 Contextual Prediction Process

The complexity of the relationship between images $X^{(k)}$ ($k \in S_c$) and image $X^{(i)}$ in $N(C)$ will depend mainly on the complexity of their statistical distribution, which is conditioned by the quantity and quality of ground-cover classes in $N(C)$ at each date t_k ($k \in S_c \cup \{i\}$). In multispectral imagery, the assumption that the distribution of images can be approximated as a mixture of normally distributed samples is generally well-accepted. Accordingly, the probability distribution function (pdf) of each image $X^{(k)}$ ($k \in S_c$) in $N(C)$ can be written as:

$$p_k(x) = \sum_{m=1}^{M_k} P(\omega_m^k) \cdot p(x \mid \omega_m^k) \tag{4}$$

where $P(\omega_m^k) = P_m^k$ and $p(x|\omega_m^k) = N(\mu_m^k, \sigma_m^k)$ are the prior probability and the conditional pdf associated with the mth gaussian mode in the $N(C)$ region of the kth image, respectively. Constant M_k stands for the number of modes characterizing the related pdf $p_k(x)$, while μ_m^k and σ_m^k are mean and standard deviation parameters, respectively.

Given a multitemporal pixel vector $\boldsymbol{x} = [x_1, x_2, \ldots, x_K]$ (K is the cardinality of S_c), such that x_j represents the pixel value in the jth image of the temporal sequence $X^{(k)}[k \in S_c; k = p(j)$, where $p(\cdot)$ is a mapping of the integers $\{1, 2, \ldots, K\}$ into $S_c]$ and $x_j \in \omega_{n_j}^j$ $(n_j \in \{1, 2, \ldots, M_j\})$, the contextual prediction function $f[\cdot]$ can be expressed as follows:

$$y = f[\mathbf{x}] = \tilde{f}\left[x_1, x_2, \ldots, x_K \mid x_1 \in \omega_{n_1}^1, x_2 \in \omega_{n_2}^2, \ldots, x_K \in \omega_{n_K}^K\right] \tag{5}$$

where $\tilde{f}[\cdot]$ is a multitemporal mapping associated with the combination of modes $\left(\omega_{n_1}^1, \omega_{n_2}^2, \ldots, \omega_{n_K}^K\right)$. Accordingly, for each possible multitemporal combination of modes, a prediction function $\tilde{f}[\cdot]$ needs to be defined.

Under a linear prediction model, the function $\tilde{f}[\cdot]$ associated with the multitemporal combination of modes $\left(\omega_{n_1}^1, \omega_{n_2}^2, \ldots, \omega_{n_K}^K\right)$ can be written as:

$$y = \tilde{f}\left[x_1, x_2, \ldots, x_K \mid x_1 \in \omega_{n_1}^1, x_2 \in \omega_{n_2}^2, \ldots, x_K \in \omega_{n_K}^K\right] = \sum_{j=1}^{K} \beta_j \cdot x_j \tag{6}$$

where β_j stands for the weight assigned to the jth mode of the combination. The determination of the weight values can be addressed by the minimum square error pseudo-inverse technique. This consists of solving the following system of R linear equations with K unknown variables $(R > K)$:

$$\begin{bmatrix} x_1^1 \; x_2^1 \ldots x_K^1 \\ x_1^2 \; x_2^2 \ldots x_K^2 \\ \ldots\ldots\ldots\ldots \\ x_1^R \; x_2^R \ldots x_K^R \end{bmatrix} \cdot \begin{bmatrix} \beta_1 \\ \beta_2 \\ \ldots \\ \beta_K \end{bmatrix} = \begin{bmatrix} y_1 \\ y_2 \\ \ldots \\ y_R \end{bmatrix} \quad \Leftrightarrow \quad \underline{P} \cdot \beta = \underline{Y} \tag{7}$$

where R represents the number of multitemporal vectors $\boldsymbol{x}^r = [x_1^r, x_2^r, \ldots, x_K^r]$ $(r \in \{1, 2, \ldots, R\})$ observed in $\overline{C}$ and collected in $\underline{P}$, such that $x_1^r \in \omega_{n_1}^1, x_2^r \in \omega_{n_2}^2, \ldots,$ and $x_K^r \in \omega_{n_K}^K$. The corresponding observations y_r in the cloudy image $X^{(i)}$ (to be reconstructed) are gathered in the target vector $\underline{Y}$. The estimate of the optimal weight vector $\underline{\beta}^*$ is given by the following equation based on the pseudoinverse $\underline{P}^\#$ of the matrix $\underline{P}$:

$$\beta^* = (\underline{P}^t \cdot \underline{P})^{-1} \cdot \underline{P}^t \cdot \underline{Y} = \underline{P}^\# \cdot \underline{Y} \tag{8}$$

The prediction system involved in the contextual multiple linear prediction (CMLP) method is thus made up of an ensemble of linear predictors, each trained to learn the relationship between images $X^{(k)}$ $(k \in S_c)$ and image $X^{(i)}$ over a possible multitemporal combination of classes $\left(\omega_{n_1}^1, \omega_{n_2}^2, \ldots, \omega_{n_K}^K\right)$ in $N(C)$. In addition, we integrate the ensemble with an additional linear predictor, termed global predictor, which is trained over all samples of $\overline{C}$ independently of their class membership. The motivation behind such an integration is that the global predictor is useful to deal with one of the following two possible situations: (1) a combination of classes in C does not exist in the set of feasible combinations of classes identified in $\overline{C}$; or (2) the number of samples collected in $\underline{P}$ and $\underline{Y}$ for a given multitemporal combination of classes available in $\overline{C}$ is not enough to apply Eq. (8).

3.2 Unsupervised Classification with the EM Algorithm

The first step of the contextual prediction process is that of classifying the region $N(C)$ of each image $X^{(k)}$ $(k \in S_c)$ into a set of M_k data classes. This can be done in an unsupervised way (i.e., without the need of training samples) by means of the EM algorithm [8]. The EM algorithm is an iterative procedure that converges to local but usually good Maximum Likelihood (ML) estimates of mixture parameters. It is based on the interpretation of $\widetilde{X}^{(k)} = \left\{x(u,v) \,:\, x(u,v) \in X^{(k)}, k \in S_c, (u,v) \in N(C)\right\}$ as incomplete data where the missing part is $Z^{(k)}$, i.e., its classification map. Assuming that L is the number of pixels in $\widetilde{X}^{(k)}$, the missing part can be evaluated as a set of L labels $Z^{(k)} = \left\{\mathbf{z}_k^{(1)}, \mathbf{z}_k^{(2)}, \ldots, \mathbf{z}_k^{(L)}\right\}$ associated with the L pixels, indicating which class is at the origin of each pixel realization. Each label is a binary vector $\mathbf{z}_k^{(i)} = [z_{k,1}^{(i)}, z_{k,2}^{(i)}, \ldots, z_{k,M_k}^{(i)}]$, such that $z_{k,r}^{(i)} = 1$ $(r \in \{1, 2, \ldots, M_k\})$ if the ith pixel x_k^i of $\widetilde{X}^{(k)}$ belongs to the rth data class ω_r^k, and $z_{k,r}^{(i)} = 0$ otherwise. The complete log-likelihood function, from which it would be possible to estimate the vector of parameters $\Theta^k = [P_1^k, P_2^k, \ldots, P_{M_k}^k, \mu_1^k, \mu_2^k, \ldots, \mu_{M_k}^k, \sigma_1^k, \sigma_2^k, \ldots, \sigma_{M_k}^k]$ if the complete data $\Psi^{(k)} = \{\widetilde{X}^{(k)}, Z^{(k)}\}$ were observed, is given by:

$$\log p(\Psi^{(k)} \mid \Theta^k) = \ell(\Psi^{(k)} \mid \Theta^k) = \sum_{i=1}^{L} \sum_{r=1}^{M_k} z_{k,r}^{(i)} \ln\left[P_r^k \, p(x_k^i \mid \theta_r^k)\right] \tag{9}$$

where $\theta_r^k = [\mu_r^k, \sigma_r^k]$.

The quantity $z_{k,r}^{(i)}$ can be estimated as the conditional expectation of $z_{k,r}^{(i)}$ given the observation x_k^i and the set of parameters Θ^k. The EM algorithm consists of expectation and maximization steps, which are iterated up to convergence. The expectation step is represented by the computations of $z_{k,r}^{(i)}$ $(i = 1, 2, \ldots, L$ and $r = 1, 2, \ldots, M_k)$ using the current estimates of the set of parameters Θ^k. The maximization step allows updating such parameter estimates.

At convergence of the EM algorithm, the final parameter estimates will define completely the Gaussian data classes (modes) available in $\widetilde{X}^{(k)}$. The latter is then transformed into a classification map with minimum error by adopting the well-known maximum a posteriori probability (MAP) decision rule. Since the final estimates of $z_{k,r}^{(i)}$ represent the estimates of the posterior probabilities $P(\omega_r^k \mid x_k^i)$ ($i = 1, 2, \ldots, L$ and $r = 1, 2, \ldots, M_k$), one can assign to each pixel x_k^i of $\widetilde{X}^{(k)}$ the optimal class label $\hat{\omega} \in \Omega = \left\{\omega_r^k : r = 1, 2, \ldots, M_k\right\}$, such that:

$$\hat{\omega} = \arg \max_{\omega_r^k \in \Omega} \{P(\omega_r^k \mid x_k^i)\} \tag{10}$$

3.3 *Estimation of the Number of Classes*

The number of data classes M_k, which is not known a priori, is estimated using the Minimum Description Length (MDL) criterion [9]:

$$MDL(M_k) = -\widetilde{\ell}(\Psi^{(k)} \mid \Theta^k) + \gamma \cdot \kappa \cdot \log(L) \tag{11}$$

where $\widetilde{\ell}(\Psi^{(k)} \mid \Theta^k)$ represents the log-likelihood function value found at convergence of the EM algorithm, κ is the number of parameters in Θ^k, and γ is a constant.

The optimal number of data classes $\hat{M}_k$ in $\widetilde{X}^{(k)}$ is estimated by minimizing the MDL criterion, i.e.,

$$\hat{M}_k = \arg \min_{M_k=1,\ldots,M_{\max}} \{MDL(M_k)\} \tag{12}$$

where $M_{\max}$ is a predefined maximal number of data classes.

4 Compressive Sensing Reconstruction Approach

The second approach exploits the Compressive Sensing (CS) theory recently developed by Donoho [10], which offers the capability to recover an unknown sparse signal with a linear combination of a small number of elementary samples called atoms. The fundamental of the CS theory is the ability to recover with relatively few measurements $\mathbf{x} = D \cdot \boldsymbol{\alpha}$ by solving the following L_0-minimization problem:

$$\min \|\boldsymbol{\alpha}\|_0 \text{ subject to } \mathbf{x} = D \cdot \boldsymbol{\alpha} \tag{13}$$

where D is a dictionary with a certain number of atoms, $\mathbf{x}$ is the original signal which can be represented as a sparse linear combination of these atoms, and the minimization of $\|\cdot\|_0$, the L_0-norm, corresponds to the maximization of the number zeros in $\boldsymbol{\alpha}$, following this formulation: $\|\boldsymbol{\alpha}\|_0 = \#\{i : \alpha_i \neq 0\}$. Equation (13) represents a NP-hard problem, but can be simplified in a relatively easily linear programming solution. Under some reasonable assumptions, minimizing L_1-norm is equivalent to minimizing L_0-norm, which is defined as $\|\boldsymbol{\alpha}\|_1 = \sum_i |\alpha_i|$. Accordingly, it is possible to rewrite (13) as:

$$\min \|\boldsymbol{\alpha}\|_1 \text{ subject to } \mathbf{x} = D \cdot \boldsymbol{\alpha} \tag{14}$$

The CS formulation can be adapted to the investigated reconstruction problem as described in Sect. 2 of this Chapter. More details are provided in [7].

Several algorithms have been proposed in the literature to solve the optimization problem expressed in (13). In the following, first, we show how two common CS solutions, the orthogonal matching pursuit (OMP) and the basis pursuit (BP) algorithms, can be formulated for a cloud-contaminated image reconstruction problem. Then, we illustrate an alternative solution for solving the CS problem under a L_0-norm perspective, by exploiting the capabilities of genetic algorithms.

4.1 CS Solutions

The basis pursuit (BP) principle represents a traditional solution to solve the CS problem [11]. It suggests a convexification of the problem by using the L_1-norm instead of L_0. This means that the best approximation of the problem becomes equal to a support minimization problem. BP finds signal representations in overcomplete dictionaries by convex, nonquadratic optimization technique. It can be reformulated as a linear programming (LP) problem, and solved using modern interior-point methods, simplex methods, or other techniques, such as homotopy techniques. Note that, if the original signal $\mathbf{x}$ is sufficiently sparse, the recovery via BP is provably exact.

One of the easiest and fastest alternative technique is the orthogonal matching pursuit (OMP) [12], an improved version of the matching pursuit (MP) method. MP finds the atom that has the highest correlation with the signal. It subtracts off the correlated part from the signal and then iterates the procedure on the resulting residual signal. Unfortunately the convergence speed of this algorithm is not fast. To overcome this drawback, an improved solution called orthogonal MP (OMP) was developed. Differently from MP, OMP updates the coefficients of the selected atoms at each iteration so that the resulting residual vectors are orthogonal to the subspace spanned by the selected atoms. When stopped after only few iterations, it generally yields a satisfactory approximation, using only few atoms.

4.2 Genetic Algorithm-Based CS Solution

A third CS strategy that we investigate in this part of the chapter is based on genetic algorithms (GAs) [13]. A genetic optimization algorithm performs a search by regenerating a population of candidate solutions (or individuals) represented by chromosomes. From one generation to the next, the population is improved following biological rules, adopting deterministic and nondeterministic genetic operators. In general, GA involves the following steps. First, an initial population of chromosomes is randomly generated. Then, the goodness of each chromosome is evaluated according to a predefined fitness function representing the aim of the optimization. Once the selection of the best chromosomes is done, the next step is devoted to the reproduction of a new population. This is done by genetic operators such as crossover and mutation operators. All these steps are iterated until some predefined condition is satisfied.

Several multiobjective GA-based approaches have been proposed in the literature. We adopt the nondominated sorting solution (NSGA-II) [14] for its low computational requirements, its aptitude to distribute uniformly the optimal solutions along the Pareto front and its successful application to different remote sensing problems. The design of a multiobjective GA optimization relies upon two components, the chromosome structure and the fitness functions, which encode the considered optimization problem and show the direction to obtain the best solution, respectively.

Concerning the first component, we consider a population of M chromosomes $\boldsymbol{\alpha}_m \in \Re$, $m \in \{1, 2, \ldots, M\}$, where each chromosome is a real vector composed of genes corresponding to the weight vector $\boldsymbol{\alpha}$ defined above in the previous sections. The length w of the chromosome is thus equal to the one of the dictionary D.

Regarding the fitness function, we investigate separately and jointly two fitness functions, i.e., those defining the optimization problem in (13). The first one aims at maximizing the sparsity level by minimizing the L_0-norm of the weight vector $\boldsymbol{\alpha}$, which corresponds to minimize the number of almost nonzero coefficients in $\boldsymbol{\alpha}$:

$$f_1 = \min \|\boldsymbol{\alpha}\|_0. \tag{15}$$

An almost nonzero coefficient is a coefficient exhibiting a value less than a predefined small threshold value. The second fitness function is derived from the constraint in [13] and points to a perfect reconstruction of the considered pixel:

$$f_2 = \min \|D\boldsymbol{\alpha} - \mathbf{x}\|_0. \tag{16}$$

5 Experimental Results

A. Dataset description and experimental design

The two investigated reconstruction approaches have been evaluated experimentally on a real multitemporal multispectral remote sensing image. The dataset was acquired by the Taiwanese optical high resolution FORMOSAT-2 satellite and represent part of the Arcachon basin in the south region of Aquitaine, in France. The images were acquired on the 24th of June and the 16th of July 2009, respectively, and are composed of 400×400 pixels, 4 spectral bands (blue, green, red and near-infrared) with a spatial resolution of 8 m (see Fig. 1).

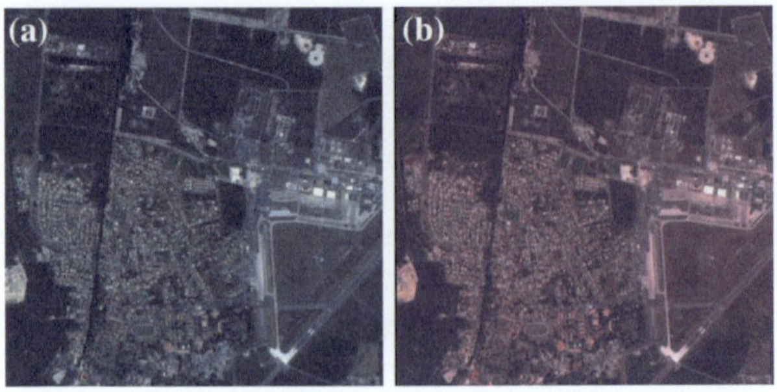

Fig. 1 FORMOSAT-2 images acquired in the Arcachon area on **a** the 24th of June and **b** the 16th of July, 2009

In order to quantify the reconstruction accuracy, the experiments were done in this way: (1) a cloud-free image $I^{(1)}$ is considered; (2) the presence of clouds is simulated by partly obscuring the other image $I^{(2)}$; and (3) the reconstructed image is compared with the original cloud-free image. The simulation study aims at understanding the sensitivity of the four investigated methods (i.e., the CMLP, the OMP, the BP, and the GA) to two aspects: (1) the kind of ground covers obscured and (2) the size of the contaminated area. In terms of comparisons, we considered also a recent work based on multiresolution inpainting (MRI). In order to obtain a detailed assessment of the reconstruction quality, we adopt the well-known peak signal-to-noise ratio (PSNR) measure. Other quantitative criteria are the computation time (in seconds) and the model complexity, namely the number of coefficients required for the reconstruction model. Regarding the dictionaries, we collected directly training samples from the image, by sampling pixels in the source region Φ.

B. Simulation experimental results

Figure 2a shows different masks whose positions were selected in a way to simulate the obscuration of different kinds of ground cover. In particular, mask A covers a region that includes mainly a urban area, mask B obscures an industrial zone, and

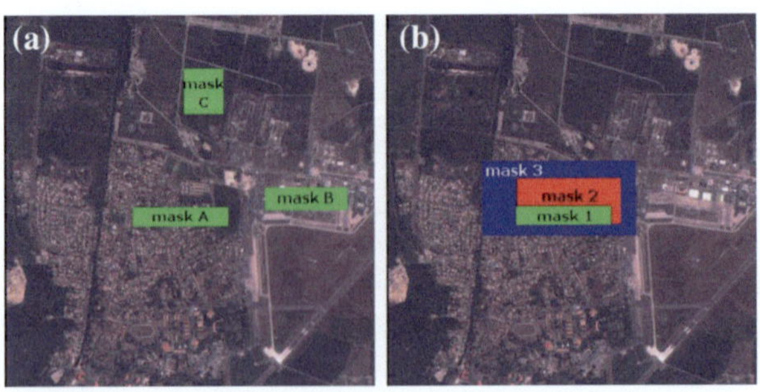

Fig. 2 Masks adopted to simulate **a** the contamination of different ground covers and **b** the different sizes of contamination

Table 1 Quantitative results obtained in the first simulation experiments

	Method	MRI	CMLP	OMP	BP	GA
Mask A	PSNR l_1	–	–	39.41	80.59	42.09
	PSNR l_2	22.54	20.99	23.96	22.22	23.78
	Complexity	–	1	3	294	148
	Time [s]	2856	1	4	66	68621
Mask B	PSNR l_1	–	–	36.33	77.10	43.38
	PSNR l_2	16.05	20.11	20.60	24.74	23.15
	Complexity	–	1	3	168	95
	Time [s]	2517	1	4	59	26312
Mask C	PSNR l_1	–	–	44.28	98.53	45.62
	PSNR l_2	33.77	24.05	31.97	30.67	32.01
	Complexity	–	1	3	301	138
	Time [s]	2898	1	4	60	43193

mask C covers a vegetation area. The experiments were carried out by considering each mask at a time, where each mask is composed by around 2000 pixels and the dictionary by 300 pixels.

Table 1 reports the results of the different reconstruction techniques over different obscured land covers. In greater detail, MRI generally reconstructs the missing data with a good PSNR level, but the corresponding reconstructed images appear visually of poor quality since it does not capture satisfactorily the textural properties of the missing areas. In general, MRI can return visually satisfactory results only when the missing area refers to a uniform region such as vegetation region. CMLP method provides generally satisfactory results in terms of reconstruction error and computation time. To obtain better results, it would need more than two temporal images. Coming now to the CS-based implementations, OMP algorithm produces very sparse reconstruction solutions (around 3 nonzero coefficients). On the one hand, this may be an advantage in terms of computation time. On the other hand, OMP is potentially subject to underfitting problems. On the contrary, BP algorithm may be subject to overfitting problems due to the fact that most of the time it selects a large number of weight coefficients (in general around 300 coefficients). Comparing OMP and BP in terms of computation time, the latter is far less efficient, whereas in terms of PSNR both methods return similar reconstruction values, outperforming CMLP and MRI. Lastly, GA can be viewed as a compromise between the two previous methods. Despite the very long time needed to estimate the reconstruction model, it results sparser than BP, but less parsimonious to OMP (see model complexity columns in Table 1). Its reconstruction error is almost all the time the best or the second best.

Another important test for the five methods consists of assessing their performances by varying the amount of missing data. Figure 2b illustrates the three different masks adopted to simulate different increasing cloud cover sizes. In particular, masks 1 is the same as the mask A adopted in the previous experiments, i.e., it covers about 2000 pixels. To build masks 2 and 3, we multiplied the previous size by 3 and

Table 2 Quantitative results obtained in the second simulation experiments

	Method	MRI	CMLP	OMP	BP	GA
Mask 1	PSNR l_1	–	–	39.41	80.59	42.09
	PSNR l_2	22.54	20.99	23.96	22.22	23.78
	Complexity	–	1	3	294	148
	Time [s]	2856	1	4	66	68621
Mask 2	PSNR l_1	–	–	42.45	80.18	45.46
	PSNR l_2	21.35	21.13	23.21	22.89	23.85
	Complexity	–	1	7	277	140
	Time [s]	6938	1	6	145	99072
Mask 3	PSNR l_1	–	–	42.00	79.53	45.13
	PSNR l_2	19.63	20.83	25.01	21.47	23.03
	Complexity	–	1	3	265	149
	Time [s]	14774	2	19	865	275394

by 6, respectively. Also in these experiments, the adopted dictionaries are composed of 300 pixels belonging to the Φ region. Table 2 reports the results achieved by the different reconstruction techniques and by varying the amount of missing data.

From a quantitative viewpoint, in terms of PSNR, we have similar results as in the previous experiments. MRI still presents problems in reconstructing satisfactorily complex textures. CMLP competes seriously with MRI in terms of computation time and PSNR. However to get higher PSNR values, one needs to resort to CS techniques. Indeed, the CS-based implementations return better results in term of PSNR in all the simulations and present the advantage for not depending on the size of the missing area. The best solution in these experiments in terms of PSNR comes from GA. About the computation time, as expected, it increases as the amount of missing data increases. Results from this viewpoint underline the main weakness of the GA solution, i.e., its expensive computational needs. Regarding the model complexity, we got in these experiments results in line with those of the previous series of experiments.

Figure 3 shows qualitative reconstruction results in RGB composites obtained in the most critical reconstruction scenario (i.e., mask 3). As mentioned before, MRI reconstruction exhibits the worst reconstruction case. CMLP method is capable to obtain a good reconstruction compared with MRI. Regarding the CS reconstruction techniques (OMP, BP, and GA), good reconstructions are obtained.

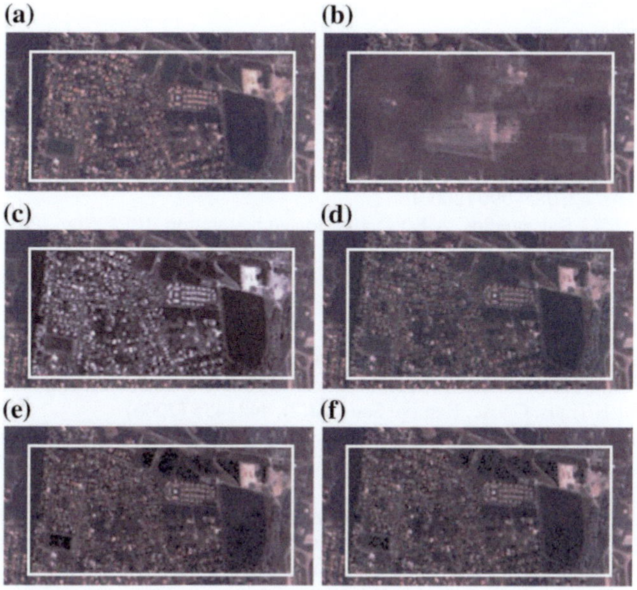

Fig. 3 Color composite (**a**) of the original image and the same image reconstructed after the contamination with the largest simulated mask 3 by **b** MRI, **c** CMLP, **d** OMP, **e** BP, and **f** GA methods

6 Conclusion

In this chapter we have dealt with the problem of the removal of clouds from sequences of multitemporal multispectral optical images. Given a contaminated image of a sequence, each area of missing measurements is reconstructed. Two main approaches have been investigated, the first one based on a contextual prediction system (CMLP) and the second one based on compressive sensing (CS), for which three different solutions based on basis pursuit (BP), orthogonal matching pursuit (OMP) and genetic algorithm (GA) have been considered.

The experimental results stress that the kind of ground cover obscured and the size of the contaminated area only marginally affect the performance of the first reconstruction approach. The latter depends more directly on the representativeness of the samples extracted from the spatial neighborhood of the contaminated area and used to train the predictors. The results point out also the superiority of the CS approach. Comparing the CS solutions, OMP has the advantage to be sparser and significantly faster than BP and GA, but it is the less robust method. GA can be seen as a good compromise between OMP and BP methods, mainly because it is more robust than OMP and more sparse than BP.

References

1. Cihlar, J.: Remote sensing of global change: an opportunity for Canada. In: Proceedings of 11th Canadian Symposium on Remote Sensing, pp. 39–48. Waterloo, Canada (1987)
2. Vasudevan, B.G., Gohil, B.S., Agarwal, V.K.: Backpropagation neural-network-based retrieval of atmospheric water vapor and cloud liquid water from IRS-P4 MSMR. IEEE Trans. Geosc. Remote Sens. **42**, 985–990 (2004)
3. Banham, M.R., Katsaggelos, A.K.: Digital image restoration. IEEE Sign. Process. Mag. **14**, 24–41 (1997)
4. Tseng, D.-C., Tseng, H.-T., Chien, C.-L.: Automatic cloud removal from multi-temporal SPOT images. Appl. Math. Comput. **205**(2), 584–600 (2008)
5. Lin, C.-H., Tsai, P.-H., Lai, K.-H., Chen, J.-Y.: Cloud removal from multitemporal satellite images using information cloning. IEEE Trans. Geosc. Remote Sens. **51**(1), 232–241 (2013)
6. Melgani, F.: Contextual reconstruction of cloud-contaminated multitemporal multispectral images. IEEE Trans. Geosc. Remote Sens. **44**(2), 442–455 (2006)
7. Lorenzi, L., Melgani, F., Mercier, G.: Missing area reconstruction in multispectral images under a compressive sensing perspective. IEEE Trans. Geosc. Remote Sens. **51**(7), 3998–4008 (2013)
8. Moon, T.K.: The expectation-maximization algorithm. Sig. Process. Mag. **13**, 47–60 (1996)
9. Rissanen, J.: Stochastic Complexity in Statistical Enquiry. World Scientific, Singapore (1989)
10. Donoho, D.L.: Compressed sensing. IEEE Trans. Inf. Theor. **52**(4), 1289–1306 (2006)
11. Chen, S.S., Donoho, D.L., Saunders, M.A.: Atomic decomposition by basis pursuit. SIAM J. Sci. Comp. **20**, 33–61 (1999)
12. Pati, Y.C., Rezaiifar, R., Krishnaprasad, P.S.: Orthogonal matching pursuit: recursive function approximation with applications to wavelet decompositions. In: Proceedings of 27th Asilomar Conferences on Signal System and Computers (1993)
13. Deb, K.: Multi-Objective Optimization Using Evolutionary Algorithms. Wiley, Chichester (2001)
14. Srinivas, N., Deb, K.: Multiobjective function optimization using nondominated sorting genetic algorithms. Evol. Comput. **2**(3), 221–248 (1995)

Sensitivity Analysis and Its Numerical Methods for Derivatives of Quadratic Eigenvalue Problems

Delin Chu, Jiang Qian and Roger C.E. Tan

Abstract Derivatives of eigenvalues and eigenvectors of parameter-dependent matrix eigenproblems play a key role in the optimum design of structures in engineering, and in the solution of inverse problems, such as the problem of model updating, which arises, for example, when information on the normal modes of vibration of a structure is used to detect structural damage. Both these applications often involve quadratic eigenvalue problems. Most existing methods for the computation of derivatives of quadratic eigenvalue problems are based on the assumption that repeated eigenvalues have well separated first order derivatives. In this paper we propose new algorithms for computing derivatives of eigenvalues and eigenvectors for quadratic eigenvalue problems under much more general conditions than existing methods, whose effectiveness for repeated or tightly clustered eigenvalues are confirmed by some numerical examples.

Keywords Sensitivity analysis · Eigenvalues · Eigenvectors · Quadratic eigenvalue problems

1 Introduction

Sensitivity analysis of eigenvalues and eigenvectors is important in many engineering problems, such as structural design optimization, model updating, damage detection, quantum mechanics, diffraction grating theory, medical imaging, social network

D. Chu · R.C.E. Tan (✉) · J. Qian
Department of Mathematics, National University of Singapore,
Singapore 119076, Singapore
e-mail: scitance@nus.edu.sg

D. Chu
e-mail: matchudl@nus.edu.sg

J. Qian
Beijing University of Posts and Telecommunications, Beijing 100876, China
e-mail: jqian104@gmail.com

R.S. Anderssen et al. (eds.), *Applications + Practical Conceptualization + Mathematics = fruitful Innovation*, Mathematics for Industry 11,
DOI 10.1007/978-4-431-55342-7_20

theory and many other applications (see [3, 11–13, 15, 21–23] and the references therein). For example, in the analysis of derivatives of spin dynamics simulations with respect to the pulse sequence or spin system parameters, we need to calculate the derivative of an observable operator. This can be easily evaluated once the derivative of a matrix exponential has been found, which can be computed from derivatives of eigenvalues and eigenvectors of the Hamiltonian in the matrix exponential (see [13] for more details). Theoretically, the sensitivity analysis focuses on the continuity and differentiability analysis of eigenvalues and eigenvectors with respect to the parameter, and computationally, it focuses on numerical methods for computing these derivatives.

In this paper, we consider the quadratic eigenvalue problem (QEP) depending on a parameter $\rho \in \mathbf{R}$:

$$(\lambda^2(\rho)M(\rho) + \lambda(\rho)C(\rho) + K(\rho))x(\rho) = 0, \tag{1}$$

where $M(\rho), C(\rho), K(\rho) \in \mathbf{R}^{n\times n}$ are analytic functions of ρ throughout some open interval D_0 containing ρ_0, and $\lambda(\rho) \in \mathbf{C}$ and $x(\rho) \in \mathbf{C}^n$ are eigenvalues and eigenvectors of the QEP (1), respectively. The quadratic eigenvalue problems arise in a wide variety of applications, such as the dynamical analysis of structural mechanical, acoustic systems, fluid mechanics, electrical circuit simulation, etc. A good survey on QEPs can be found in [20]. We consider the case when the QEP (1) has r distinct analytic eigenvalue functions $\lambda_1(\rho), \ldots, \lambda_r(\rho)$ and r analytic eigenvector functions $x_1(\rho), \ldots, x_r(\rho)$ throughout some open interval containing ρ_0, where the r eigenvalue functions coincide at $\rho = \rho_0$, that is, $\lambda_1(\rho_0) = \cdots = \lambda_r(\rho_0) = \lambda_1$ is a repeated eigenvalue of multiplicity r of the QEP (1) at $\rho = \rho_0$. Repeated eigenvalues often occur in problems of design optimizing [11] and mode tracking [6]. We are interested in computing derivatives of these analytic eigenvalue and eigenvector functions at $\rho = \rho_0$.

There exist many numerical methods for computing these derivatives in the literature. Computing derivatives of eigenvalues is generally much easier than the computation of derivatives of eigenvectors. There are different approaches for computing derivatives of eigenvectors, which can generally be classified into modal methods, direct methods and iterative methods. Modal methods expand the derivatives of eigenvectors as linear combinations of modes (eigenvectors). Exact expressions of first order derivatives of eigenvectors for generalized eigenvalue problems are given in [8] by using the modal expansion technique, which is extended to quadratic eigenvalue problems in [25, 26]. These modal methods require the full set of eigenvectors. However, in engineering applications, the information regarding all the eigenvectors may not be readily available. Generally, only the low-frequency modes can be obtained. Some methods then use this incomplete part of modes as the basis vectors of eigenvector derivatives. But the accuracy would then quite depend on the modes being used.

Instead of using all eigenvectors in modal methods, direct methods only require knowledge of eigenvectors being considered. In direct methods, derivatives of eigenvectors are computed by solving some linear systems, whose coefficient matrices are singular. There are two types of methods for solving these singular systems. One is the so called Nelson-type methods, which transform the singular systems into nonsingular systems by removing some redundant rows and columns of the singular coefficient matrices. They originate from the method for generalized eigenvalue problems proposed by Nelson [16]. It is then extended to compute first and second order derivatives of eigenvectors corresponding to distinct eigenvalues of quadratic eigenvalue problems in [7, 9]. Repeated eigenvalues with distinct first order derivatives are discussed in [19]. In [17], the authors consider more general cases when the repeated eigenvalues may have repeated high order derivatives. The other is the bordered matrix methods, or algebraic methods, which transform the singular systems into nonsingular systems by adding some rows and columns to the singular coefficient matrices. Some algebraic methods for distinct eigenvalues of quadratic eigenvalues problems can be found in [4, 5, 10, 14]. They are extended to repeated eigenvalues in some literature, however, there are some flaws in these extensions.

When the dimension of matrices are large, the computational cost of direct methods may be unacceptable. In this case, iterative methods are more preferable, especially when only moderate accuracy is needed. Some simultaneous iteration methods for computing derivatives of several eigenpairs of a single matrices can be found in [2, 18]. These methods are convergent for the eigenpairs whose eigenvalues, after a suitable original shift, are largest in magnitude. An iterative method for quadratic eigenvalue problems is proposed in [24]. However, its convergence is based on some conditions which are not guaranteed to hold.

In this paper, we will extend ideas in [1] to quadratic eigenvalue problems. Our algorithms can compute derivatives, of arbitrary order, of repeated semisimple eigenvalues and corresponding eigenvectors, under the more general cases when repeated eigenvalues may have repeated derivatives. Specifically, we assume that

$$\lambda_j^{(i)}(\rho_0) = \lambda_k^{(i)}(\rho_0), \qquad 1 \le j, k \le r,\ i = 0, 1, \ldots, m-1, \tag{2}$$

$$\lambda_j^{(m)}(\rho_0) \ne \lambda_k^{(m)}(\rho_0), \qquad \text{for all } j \ne k \tag{3}$$

hold for some integer m. Under this assumption, Algorithm 1 will compute derivatives of eigenvalues of order 1 to $m+1$ and the first order derivative of eigenvectors, and then Algorithm 2 can successively compute derivatives, of arbitrary order, of eigenvalues and eigenvectors. Numerical examples show that these algorithms are feasible and reliable. Indeed, the assumptions (2) and (3) can be relaxed. We refer readers to our another paper [17] for more details.

2 Our Algorithms

Let

$$\Lambda(\rho) = \mathrm{diag}(\lambda_1(\rho), \ldots, \lambda_r(\rho)), \quad X(\rho) = [x_1(\rho), \ldots, x_r(\rho)],$$

where $\lambda_1(\rho), \ldots, \lambda_r(\rho)$ and $x_1(\rho), \ldots, x_r(\rho)$ are respectively analytic eigenvalues and eigenvectors functions as in the previous section. Then (1) can be also written as

$$M(\rho)X(\rho)\Lambda(\rho)^2 + C(\rho)X(\rho)\Lambda(\rho) + K(\rho)X(\rho) = 0. \tag{4}$$

Since $M(\rho)$, $C(\rho)$, $K(\rho)$ are not necessarily symmetric, we similarly let

$$Y(\rho) = [y_1(\rho), \ldots, y_r(\rho)]$$

be the matrix of corresponding left eigenvectors, that is,

$$\Lambda(\rho)^2 Y(\rho)^\top M(\rho) + \Lambda(\rho)Y(\rho)^\top C(\rho) + Y(\rho)^\top K(\rho) = 0.$$

To ensure the uniqueness of eigenvectors, we take the following normalization condition on $x_i(\rho)$:

$$y_i^\top(\rho_0)(2\lambda_1(\rho_0)M(\rho_0) + C(\rho_0))x_i(\rho) = 1. \tag{5}$$

At $\rho = \rho_0$, we further take the normalization condition

$$Y^\top(\rho_0)(2\lambda_1(\rho_0)M(\rho_0) + C(\rho_0))X(\rho_0) = I, \quad x_i^*(\rho_0)x_i(\rho_0) = 1. \tag{6}$$

The computed eigenvalues of the quadratic eigenvalue problem (1) at $\rho = \rho_0$ are $\Lambda(\rho_0) = \lambda_1(\rho_0)I$, while the corresponding right and left eigenvectors may not be $X(\rho_0)$ and $Y(\rho_0)$ described as above. Assume that the computed linearly independent right and left eigenvectors corresponding to $\lambda_1(\rho_0)$ are columns of $n \times r$ matrices X_0 and Y_0, respectively. Then there exist nonsingular $r \times r$ matrices C_0 and D_0 such that

$$X(\rho_0) = X_0 C_0, \qquad Y(\rho_0) = Y_0 D_0. \tag{7}$$

Without loss of generality, we assume that the computed X_0 and Y_0 satisfy

$$Y_0^\top(2\lambda_1(\rho_0)M(\rho_0) + C(\rho_0))X_0 = I. \tag{8}$$

Then (6), (7) and (8) show that $D_0^\top C_0 = I$, or equivalently, $D_0^\top = C_0^{-1}$.

Throughout the rest of the paper, all functions are assumed to be evaluated at $\rho = \rho_0$ unless stated otherwise. Differentiating (4) $i(\le m)$ times with respect to ρ and setting $\rho = \rho_0$, we have

$$Q(\lambda)X^{(i)} + \sum_{p=1}^{i-1} \binom{i}{p} W_p X^{(i-p)} + Z_i X + Q'(\lambda)X\Lambda^{(i)} = 0, \tag{9}$$

where

$$\lambda = \lambda_1, \quad Q(\lambda) = \lambda^2 M + \lambda C + K, \quad Q'(\lambda) = 2\lambda M + C,$$

$$W_p = \sum_{q=0}^{p} \sum_{s=0}^{q} \binom{p}{s, q-s, p-q} \Lambda^{(q-s)} \Lambda^{(s)} M^{(p-q)} + \sum_{q=0}^{p} \binom{p}{q} \Lambda^{(q)} C^{(p-q)} + K^{(p)}, \quad p = 0, \ldots, m-1, \tag{10}$$

$$Z_i = \sum_{q=0}^{i-1} \sum_{s=0}^{q} \binom{i}{s, q-s, i-q} \Lambda^{(q-s)} \Lambda^{(s)} M^{(i-q)} + \sum_{s=1}^{i-1} \binom{i}{s} \Lambda^{(i-s)} \Lambda^{(s)} M + \sum_{q=0}^{i-1} \binom{i}{q} \Lambda^{(q)} C^{(i-q)} + K^{(i)}, \quad i = 1, \ldots, m, \tag{11}$$

$(\cdot)^{(i)}$ denotes the ith derivative of $(\cdot)$. Under the assumptions (2) and (3), the following Algorithm 1 will compute $\Lambda^{(i)} (i = 1, \ldots, m+1)$ and X'.

Algorithm 1:

Input: The $n \times n$ coefficient matrices M, C, K and their derivatives, the eigenvalue λ and the corresponding right and left eigenvector matrices $X_0, Y_0 \in \mathbf{C}^{n \times r}$ satisfying (8), and a small positive tolerance τ_1.

Output: $\Lambda^{(i)} (i = 1, \ldots, m+1)$ and X'.

1. Set $i = 1$, $V_0 = X_0$, compute $Z_1 = \lambda^2 M' + \lambda C' + K'$ and $M_1 = -Y_0^T Z_1 X_0$. Compute $\lambda^{(1)} = \text{trace}(M_1)/n$ and set $\Lambda^{(1)} = \lambda^{(1)} I$.
2. **While** $\|M_i - \Lambda^{(i)}\| < \tau_i$ **do**

   Compute $W_i = Z_i + \lambda^{(i)} Q'(\lambda)$.

   Find a solution V_i to the equation $Q(\lambda) V_i = -\sum_{p=1}^{i} \binom{i}{p} W_p V_{i-p}$.

   Compute Z_{i+1} by (11) and compute

   $$M_{i+1} = -Y_0^T \left(\sum_{p=1}^{i} \binom{i+1}{p} W_p V_{i+1-p} + Z_{i+1} X_0 \right).$$

   Compute $\lambda^{(i+1)} = \text{trace}(M_{i+1})/n$ and set $\Lambda^{(i+1)} = \lambda^{(i+1)} I$.
   Set $i = i+1$ and select τ_i.

   End
3. Set $m = i$. Find all eigenvalues and eigenvectors of M_m. Set $\Lambda^{(m)} = \text{diag}(l_1, \ldots, l_r)$ where $l_1, \ldots, l_r$ are eigenvalues of M_m, and set $C_0 = [c_1, \ldots, c_r]$, where c_i are

eigenvectors of M_m, normalized so that $\|X_0 c_i\|_2 = 1$. Compute $X = X_0 C_0$ and $Y = Y_0 (C_0^T)^{-1}$.

4. Find a solution V_m to the equation

$$Q(\lambda)V_m = -\sum_{p=1}^{m-1} \binom{m}{p} W_p V_{m-p} - Z_m X_0 - Q'(\lambda) X \Lambda^{(m)} C_0^{-1},$$

and compute

$$\begin{aligned} M_{m+1} &= -Y^\top \Big(\sum_{p=1}^{m-1} \binom{m+1}{p} W_p V_{m+1-p} C_0 + \ (m+1) Q'(\lambda) V_1 C_0 \Lambda^{(m)} \\ &+ (m+1) Z_m V_1 C_0 + \Big(\sum_{p=0}^{m-1} \binom{m+1}{p} \sum_{q=0}^{p} \binom{p}{q} \Lambda^{(p-q)} \Lambda^{(q)} M^{(m+1-p)} \\ &+ (m+1) \sum_{p=1}^{m-1} \binom{m}{p} \Lambda^{(m-p)} \Lambda^{(p)} M' + \sum_{p=2}^{m-1} \binom{m+1}{p} \Lambda^{(m+1-p)} \Lambda^{(p)} M \\ &+ \sum_{p=0}^{m-1} \binom{m+1}{p} \Lambda^{(p)} C^{(m+1-p)} + K^{(m+1)} \Big) X \\ &+ (m+1)(2\lambda M' + C') X \Lambda^{(m)} + (m+1) M X \Lambda' \Lambda^{(m)} \Big). \end{aligned}$$

5. Set $\Lambda^{(m+1)} = \mathrm{diag}(M_{m+1})$, that is, $\Lambda^{(m+1)}$ is a diagonal matrix whose diagonal elements are corresponding diagonal elements of M_{m+1}.
6. Compute the off-diagonal elements c_{ij} of C_1 by $c_{ij} = \frac{m_{ij}}{(m+1)(l_j - l_i)}$, $i \neq j$, where m_{ij} is the (i, j) element of M_{m+1} and l_i is the i-th diagonal element of $\Lambda^{(m)}$, and compute the diagonal elements of C_1 as the corresponding diagonal elements of $-Y^T Q'(\lambda) V_1 C_0$.
7. Compute $X' = V_1 C_0 + X C_1$.

Remark 1 Although Algorithm 1 is based on (2) and (3), the value of m is not known a priori. The successive small values of the tolerance τ_i in Step 2 is then used to determine m.

Remark 2 Under assumptions (2) and (3), $M_i = \Lambda^{(i)} = \lambda^{(i)} I$ for $i < m$ in exact arithmetic. However, the diagonal elements of computed M_i may not be exactly the same. Hence we set the average of the diagonal elements of M_i as $\lambda^{(i)}$ to minimize the effect of roundoff.

Remark 3 Note that in those equations for V_i and V_m, the coefficient matrix $Q(\lambda)$ is singular with rank $n - r$. The solutions may not be unique, but the uniqueness of eigenvalues and eigenvectors ensures that the computed $\Lambda^{(i)}$ and X' are independent of the choice of V_i.

Algorithm 1 computes $\Lambda^{(i)} (i = 1, \ldots, m+1)$ and X'. Then the following Algorithm 2 will successively compute higher orders derivatives of eigenvalues and eigenvectors. Specifically, for any positive integer $k \geq 2$, assume that we have already obtained $\Lambda, \Lambda', \ldots, \Lambda^{(m+k-1)}$ and $X, X', \ldots, X^{(k-1)}$, Algorithm 2 will compute $\Lambda^{(m+k)}$ and $X^{(k)}$. Thus Algorithms 1 and 2 can successively compute derivatives of eigenvalues and eigenvectors of any order.

Algorithm 2:

Input: The $n \times n$ coefficient matrices M, C, K and their derivatives, the eigenvalues Λ and their derivatives $\Lambda', \ldots, \Lambda^{(m+k-1)}$, the right eigenvectors matrix X and its derivatives $X', \ldots, X^{(k-1)}$ and the left eigenvectors matrix Y.

Output: $\Lambda^{(m+k)}$ and $X^{(k)}$.

1. For $i = k, \ldots, m+k-1$, find a solution V_{ik} to the equation

$$\begin{aligned} Q(\lambda)V_{ik} = & -\sum_{p=0}^{k-1}\binom{i}{p}\sum_{q=0}^{i-p}\binom{i-p}{q}C^{(i-p-q)}X^{(p)}\Lambda^{(q)} - \sum_{p=0}^{k-1}\binom{i}{p}K^{(i-p)}X^{(p)} \\ & -\sum_{p=k}^{i-1}\binom{i}{p}W_{i-p}V_{pk} - \sum_{p=0}^{k-1}\binom{i}{p}\sum_{q=0}^{i-p}\binom{i-p}{q}M^{(i-p-q)}X^{(p)}\sum_{s=0}^{q}\binom{q}{s}\Lambda^{(q-s)}\Lambda^{(s)}. \end{aligned}$$

2. Compute

$$\begin{aligned} M_{m+k} = & -Y^{\top}\left(MX\sum_{s=1}^{m+k-1}\binom{m+k}{s}\Lambda^{(m+k-s)}\Lambda^{(s)} + \sum_{p=1}^{k-1}\binom{m+k}{p}K^{(m+k-p)}X^{(p)}\right. \\ & + \sum_{q=0}^{m+k-1}\binom{m+k}{q}M^{(m+k-q)}X\sum_{s=0}^{q}\binom{q}{s}\Lambda^{(q-s)}\Lambda^{(s)} + \sum_{p=k+1}^{m+k-1}\binom{m+k}{p}W_{m+k-p}V_{pk} \\ & + \sum_{q=0}^{m+k-1}\binom{m+k}{q}C^{(m+k-q)}X\Lambda^{(q)} + \binom{m+k}{k}\left(Z_m V_{kk} + Q'(\lambda)V_{kk}\Lambda^{(m)}\right) \\ & + \sum_{p=1}^{k-1}\binom{m+k}{p}\sum_{q=0}^{m+k-p}\binom{m+k-p}{q}M^{(m+k-p-q)}X^{(p)}\sum_{s=0}^{q}\binom{q}{s}\Lambda^{(q-s)}\Lambda^{(s)} \\ & \left. + \sum_{p=1}^{k-1}\binom{m+k}{p}\sum_{q=0}^{m+k-p}\binom{m+k-p}{q}C^{(m+k-p-q)}X^{(p)}\Lambda^{(q)} + K^{(m+k)}X\right). \end{aligned}$$

3. Set $\Lambda^{(m+k)} = \mathrm{diag}(M_{m+k})$, that is, Λ^{m+k} is a diagonal matrix whose diagonal elements are corresponding diagonal elements of M_{m+k}.
4. Compute the off-diagonal elements c_{ij} of C_{kk} by

$$c_{ij} = \frac{m_{ij}}{\binom{m+k}{k}(l_j - l_i)}, \qquad i \neq j,$$

where m_{ij} is the (i, j) element of M_{m+k} and l_i is the i-th diagonal element of $\Lambda^{(m)}$, and set the diagonal elements of C_{kk} as the corresponding diagonal elements of the matrix $-Y^\top Q'(\lambda)V_{kk}$.

5. Compute $X^{(k)} = V_{kk} + XC_{kk}$.

Starting with Algorithm 1, Algorithm 2 then enables us to successively compute derivatives, of any arbitrary order, of the repeated eigenvalues and corresponding eigenvectors.

Note that both Algorithms 1 and 2 are based on the assumptions (2) and (3). Indeed, these assumptions can be relaxed. We refer readers to [17] for details on more general cases when (2) and (3) do not hold.

3 Numerical Examples

In this section, we present some examples to illustrate the performance of our algorithms. All calculations used MATLAB 7.5.0, with machine epsilon $\varepsilon \approx 2.2 \times 10^{-16}$.

Example 1 Consider the system (1), with $\rho \in \mathbf{R}$, where

$$\begin{aligned} M &= S\mathrm{diag}(\rho+1, \rho-1, 2\rho^2-3\rho-2, \rho^2+4)S, \\ C &= S\mathrm{diag}(-\rho^3-2\rho^2-3\rho-3-\delta(\rho+1)^2, 2\rho^2+\rho-1, \\ &\qquad -5\rho^2-\rho-3, -5\rho^3-\frac{1}{2}\rho^2-20\rho-3)S, \qquad (12) \\ K &= S\mathrm{diag}(\rho^2+\rho+2+\delta(\rho+1), \rho^3+2\rho^2-\rho+6, 2\rho^2-\rho-1, 5\rho+\frac{1}{2})S, \end{aligned}$$

and S is the orthogonal matrix

$$S = \frac{1}{\sqrt{3}}\begin{bmatrix} \cos\rho & 1 & \sin\rho & -1 \\ -\sin\rho & -1 & \cos\rho & -1 \\ 1 & -\sin\rho & 1 & \cos\rho \\ -1 & \cos\rho & 1 & \sin\rho \end{bmatrix}.$$

If $\delta = 0$, the two eigenvalues $\lambda_1(\rho) = \rho^2+\rho+2+\delta(\rho+1)$ and $\lambda_2(\rho) = -\rho-2(\rho-1)^{-1}$ coincide at $\rho_0 = 0$, and the remaining eigenvalues are well separated. Specifically, $\lambda_1(0) = \lambda_2(0) = 2$, $\lambda_1'(0) = \lambda_2'(0) = 1$, and $\lambda_1''(0) = 2 \neq \lambda_2''(0) = 4$. So algorithms proposed in [19, 25] requiring eigenvalues to be well separated fail, while Algorithm 1, with $r = 2$ and $m = 2$, gave estimates of Λ', Λ'' and Λ''' at $\rho_0 = 0$ with relative accuracy approximately 10^{-15} and the relative error $\|\widetilde{X}' - X'\|_F/\|X'\|_F$ in the computed value $\widetilde{X}'$ of X', was approximately 10^{-14}. Algorithm 2, again with $m = 2$, estimated $\Lambda^{(4)}$ and X'' with similar accuracy. We also added some small random perturbations on $M(0)$, $C(0)$ and $K(0)$ to test the robustness of our algorithms. For each Δ, the maximum value of the norms of the

Table 1 Relative errors of computed derivatives of eigenvectors

δ	Nelson's method	Algorithm 1
10^{-14}	4.205×10^{-1}	6.452×10^{-14}
10^{-12}	3.370×10^{-3}	9.148×10^{-12}
10^{-10}	8.038×10^{-5}	9.140×10^{-10}
10^{-8}	8.039×10^{-7}	9.140×10^{-8}

perturbations, the resulting relative error in the computed value of Λ' was less than Δ, while that for X' was less than 20Δ.

When δ is small but nonzero, the two eigenvalues $\lambda_1(0)$ and $\lambda_2(0)$ become close rather than identical. We then apply Algorithm 1 on (12) with several small nonzero δ, and compare the results with those obtained by the modified Nelson's method in [7] (but using the normalization conditions in this paper), which regards these eigenvalues as distinct. The relative errors in the computed X' by these two methods are listed in Table 1. The error with our algorithm is approximately δ, while that with Nelson's method is close to ε/δ. This suggest that when $\delta < \varepsilon^{1/2}$, our algorithm is more preferable than the Nelson's method.

Example 2 Consider the system (1), with S is as in Example 1 and,

$$\begin{aligned}
M &= S\text{diag}(\rho + 1, \rho - 1, 2\rho^2 - 3\rho - 2, \rho^2 + 4)S, \\
C &= S\text{diag}(-2\rho^3 - 3\rho^2 - 3\rho - 3 - (\frac{1}{2}\delta\rho^2 + \delta\rho + \delta^2)(\rho + 1), \\
&\qquad 2\rho^2 + \rho - 1, -5\rho^2 - \rho - 3, -5\rho^3 - \frac{1}{2}\rho^2 - 20\rho - 3)S, \\
K &= S\text{diag}(2\rho^2 + \rho + 2 + \frac{1}{2}\delta\rho^2 + \delta\rho + \delta^2, \rho^3 + 2\rho^2 - \rho + 6, \\
&\qquad 2\rho^2 - \rho - 1, 5\rho + \frac{1}{2})S.
\end{aligned}$$

At $\rho = 0$, the eigenvalues $\lambda_1(\rho) = 2\rho^2 + \rho + 2 + \frac{1}{2}\delta\rho^2 + \delta\rho + \delta^2$ and $\lambda_2(\rho) = -\rho - 2(\rho - 1)^{-1}$ differ by δ^2, and the remaining eigenvalues are well separated. Specifically $\lambda_1'(0) = 1 + \delta, \lambda_2'(0) = 1, \lambda_1''(0) = 4 + \delta, \lambda_2''(0) = 4$, $\lambda_1'''(0) = 0, \lambda_2'''(0) = 12$.

We apply Algorithm 1 with three different choices of τ_i. Choosing $\tau_1 = \tau_2 = \delta/100$ in Algorithm 1 gives $m = 1$ (treating $\lambda_1'(0)$ and $\lambda_2'(0)$ as distinct), choosing $\tau_1 = \tau_2 = 100\delta$ gives $m = 3$ (treating both λ' and λ'' as multiple), while choosing $\tau_1 = 100\delta$ and $\tau_2 = \delta/100$ gives $m = 2$ (treating λ' but not λ'' as multiple). The relative errors of the computed derivatives of eigenvectors for various δ, for $m = 1, 2$ and 3, are listed in Table 2.

Table 2 Relative errors of computed derivatives of eigenvectors

δ	$m = 1$	$m = 2$	$m = 3$
10^{-12}	2.95×10^{9}	6.91×10^{11}	4.39×10^{-12}
10^{-10}	3.81×10^{5}	6.94×10^{9}	4.39×10^{-10}
10^{-8}	3.35×10^{1}	6.94×10^{7}	4.39×10^{-8}
10^{-6}	3.36×10^{-1}	6.94×10^{5}	4.39×10^{-6}
10^{-4}	3.3×10^{-1}	6.95×10^{3}	4.39×10^{-4}

From Table 2 we can see that for small δ, the relative errors for $m = 1$ are $O((\max(\varepsilon, \delta^2))/\delta^2)$, those for $m = 2$ are $O(1/\delta)$, which is even worse, while those for $m = 3$ are just $O(\delta)$. These show that for small δ, Algorithm 1 with $m = 3$ gives much better results than Algorithm 1 with $m = 1$ and $m = 2$. Choosing an appropriate m is vital in this example.

References

1. Andrew, A.L., Tan, R.C.E.: Computation of derivatives of repeated eigenvalues and the corresponding eigenvectors of symmetric matrix pencils. SIAM. J. Matrix Anal. Appl. **20**, 78–100 (1998)
2. Andrew, A.L., Tan, R.C.E.: Iterative computation of derivatives of repeated eigenvalues and the corresponding eigenvectors. Numer. Linear Algebra Appl. **7**, 151–167 (2000)
3. Backstrom L., Lescovec J.,: Supervised random walks: predicting and recommending links in social networks. In: WSDM'11: Proceedings of Fourth ACM Conferences Web Search Data Mining (2011). doi:10.1145/1935826.1935914
4. Choi, K.M., Jo, H.K., Kim, W.H., Lee, I.W.: Sensitivity analysis of non-conservative eigensystems. J. Sound Vib. **274**, 997–1011 (2004)
5. Chouchane, M., Guedria, N., Smaoui, H.: Eigensensitivity computation of asymmetric damped systems using an algebraic approach. Mech. Syst. Sig. Process. **21**, 2761–2776 (2007)
6. Eldred, M.S., Venkayya, V.B., Anderson, W.J.: Mode tracking issues in structural optimization. AIAA J. **33**, 1926–1933 (1995)
7. Friswell, M.I., Adhikari, S.: Derivatives of complex eigenvectors using Nelson's method. AIAA J. **38**, 2355–2357 (2000)
8. Fox, R.L., Kapoor, M.P.: Rates of change of eigenvalues and eigenvectors. AIAA J. **6**, 2426–2429 (1968)
9. Guedria, N., Chouchane, M., Smaoui, H.: Second-order eigensensitivity analysis of asymmetric damped systems using Nelson's method. J. Sound Vib. **300**, 974–992 (2007)
10. Guedria, N., Smaoui, H., Chouchane, M.: A direct algebraic method for eigensolution sensitivity computation of damped asymmetric systems. Int. J. Comput. Methods Eng. **68**, 674–689 (2006)
11. Haug, E.J., Choi, K.K., Komkov, V.: Design Sensitivity Analysis of Structural Systems. Academic Press, New York (1986)
12. Karlsson, J., Ericsson, A., Åström, K.: Shape modelling by optimising description length using gradients and parameterisation invariance. Springer Proc. Math. **6**, 51–90 (2012)
13. Kuprov, I., Rodgers, C.T.: Derivatives of spin dynamics simulations. J. Chem. Phys. (2009). doi:10.1063/1.3267086

14. Lee, I.W., Kim, D.O., Jung, G.H.: Natural frequency and mode shape sensitivities of damped systems: part I, distinct natural frequencies. J. Sound Vib. **223**, 399–412 (1999)
15. Mottershead, J.E., Friswell, M.I.: Model updating in structural dynamics: A survey. J. Sound Vib. **167**, 347–375 (1993)
16. Nelson, R.B.: Simplified calculation of eigenvector derivatives. AIAA J. **14**, 1201–1205 (1976)
17. Qian, J., Andrew, A.L., Chu, D.L., Tan, R.C.E.: Computing derivatives of repeated eigenvalues and corresponding eigenvectors of quadratic eigenvalue problems. SIAM J. Matrix Anal. Appl. **34**, 1089–1111 (2013)
18. Tan, R.C.E., Andrew, A.L.: Computing derivatives of eigenvalues and eigenvectors by simultaneous iteration. IMA J. Numer. Anal. **9**, 111–122 (1989)
19. Tang, J., Ni, W.M., Wang, W.L.: Eigensolutions sensitivity for quadratic eigenproblems. J. Sound Vib. **196**, 179–188 (1996)
20. Tisseur, F., Meerbergen, K.: The quadratic eigenvalue problem. SIAM Rev. **43**, 235–286 (2001)
21. Van Der Aa, N.P.: Sensitivity analysis for grating reconstruction. Eindhoven University of Technology, Thesis (Proefschrift) (2007)
22. Weber, B., Paultre, P., Proulx, J.: Consistent regularization of nonlinear model updating for damage identification. Mech. Syst. Sig. Proc. **23**, 1965–1985 (2009)
23. Weber, B., Paultre, P.: Damage identification in a truss tower by regularized model updating. ASCE J. Struct. Eng. **136**, 307–316 (2010)
24. Xie, H.Q.: An iterative method for partial derivatives of eigenvectors of quadratric eigenvalue problems. BIT **52**, 525–536 (2012)
25. Xie, H.Q., Dai, H.: Derivatives of repeated eigenvalues and corresponding eigenvectors of damped systems. Appl. Math. Mech. (English Edition) **28**, 837–845 (2007)
26. Xie, H.Q., Dai, H.: Calculation of derivatives of multiple eigenpairs of unsymmetrical quadratic eigenvalue problems. Int. J. Comput. Math. **85**, 1815–1831 (2008)

Risk Aggregation with Copula for Banking Industry

Toshinao Yoshiba

Abstract This paper surveys several applications of parametric copulas to market portfolios, credit portfolios, and enterprise risk management in the banking industry, focusing on how to capture stressed conditions. First, we show two simple applications for market portfolios: correlation structures for returns on three stock indices and a risk aggregation for a stock and bond portfolio. Second, we show two simple applications for credit portfolios: credit portfolio risk measurement in the banking industry and the application of copulas to CDO valuation, emphasizing the similarity to their application to market portfolios. In this way, we demonstrate the importance of capturing stressed conditions. Finally, we introduce practical applications to enterprise risk management for advanced banks and certain problems that remain open at this time.

Keywords Copula · Multivariate distribution · Tail dependence · Risk aggregation · Economic capital

1 Introduction

Risk managers in the banking industry evaluate the risks associated with financial asset and credit portfolios based on the assumption that underlying risk factors follow multivariate probability distributions that consist of marginal distributions and correlation structures. Most financial asset returns that determine marginal distributions have heavier tails than Gaussian distributions. In addition, each pair of returns is mutually dependent, particularly in the tail part of the distribution. For example, stock prices will plunge simultaneously in the event of financial turmoil like the Lehman shock of September 2008, while prices correlate moderately under more typical conditions. This suggests that the dependence between stock prices in the lower tail is stronger than in other parts of the multivariate distribution.

T. Yoshiba (✉)
Institute for Monetary and Economic Studies, Bank of Japan, Tokyo, Japan
e-mail: toshinao.yoshiba@boj.or.jp

R.S. Anderssen et al. (eds.), *Applications + Practical Conceptualization + Mathematics = fruitful Innovation*, Mathematics for Industry 11,
DOI 10.1007/978-4-431-55342-7_21

Parametric copulas are widely used in financial risk management to capture the various correlation structures that hold between risk factors. A copula is a function that joins, or couples, univariate distribution functions to a multivariate distribution function, as denoted by C in the equation

$$\Pr(X_1 \leq x_1, \ldots, X_d \leq x_d) = C(\Pr(X_1 \leq x_1), \ldots, \Pr(X_d \leq x_d)), \qquad (1)$$

for the multivariate random vector $(X_1, \ldots, X_d)$ (for examples, see [8] or [16]). A copula extracts the dependence structure from the joint distribution, independent of marginal distributions. We can construct the joint distribution of financial assets returns by specifying the marginal distributions and the parametric copula. Due to their tractability, copulas are used in evaluations of certain portfolio risks involving several assets and in evaluations of collateralized debt obligations (CDOs) comprising several bonds associated with default risks.

A Gaussian copula is the most popular copula for representing correlation structures for asset returns. Although the copula is easy to estimate, it cannot capture lower tail dependencies. One solution is to assume copulas capable of representing correlation structures under stressed conditions.

This paper overviews several applications of copulas in the banking industry to market portfolios, credit portfolios, and enterprise risk management, referring to work by [19, 21, 22]. Using parametric copulas to capture stressed conditions is a particular focus.

This paper is organized as follows. Section 2 overviews risk aggregation for market portfolios for which copulas are used. Section 3 discusses the use of copulas to assess credit portfolio risks. Section 4 describes how advanced banks apply copulas to enterprise risk management. Section 5 gives concluding remarks.

2 Risk Aggregation for Market Portfolios

With regard to risk aggregation for market portfolios, risk managers determine correlation structures for certain risk factors non-parametrically or parametrically. Popular non-parametric methods include historical simulation methods, which capture only patterns of risk factors for both correlation structures and marginal distributions. Popular parametric methods assume a Gaussian copula for the correlation structure of risk factors and a Gaussian distribution for marginal distributions. (See, for example, [9]).

Most banks adopt an unconditional approach to measure relevant market risks for one day, ten days, and other short-term periods. Unconditional approaches assume that the relevant risk factors for the period in question follow the same distribution and do not incorporate time series characteristics, such as serial correlations. Following

Table 1 Bivariate parametric copulas and their rank correlation

Copula	Parameter	Expression $C(u_1, u_2)$	Kendall's tau τ_K
Gaussian	ρ	$\boldsymbol{\Phi}_\rho(\Phi^{-1}(u_1), \Phi^{-1}(u_2))$	$(2/\pi)\arcsin\rho$
t	$\rho,\ \nu$	$\boldsymbol{T}_{\nu,\rho}(T_\nu^{-1}(u_1), T_\nu^{-1}(u_2))$	$(2/\pi)\arcsin\rho$
Clayton	α	$(u_1^{-\alpha} + u_2^{-\alpha} - 1)^{-1/\alpha}$	$\alpha/(\alpha+2)$
Gumbel	γ	$\exp\left\{-((-\ln u_1)^\gamma + (-\ln u_2)^\gamma)^{1/\gamma}\right\}$	$1 - 1/\gamma$

the standard practice of risk measurement in banks, we measure market portfolio risk unconditionally for a one-day period based on daily log returns for risk factors.[1]

2.1 Correlation Structure Between Three Major Stock Indices

Shintani et al. [19] analyzed correlation structures for three major stock indices daily returns in the US (the S&P500), the Eurozone (Euro Stoxx 50), and for Japan (the Nikkei 225). The observation period was from January 2001 to September 2009.

They began by analyzing pairwise lower tail dependence with 5 and 1 % thresholds: $\lambda_L(0.05)$ and $\lambda_L(0.01)$. The lower tail dependence with a certain threshold u is defined as

$$\lambda_L(u) = \Pr[F_2(X_2) \le u | F_1(X_1) \le u] = \frac{C(u, u)}{u}, \tag{2}$$

where $F_1(X_1)$ and $F_2(X_2)$ are marginal distribution functions of the first and the second variable. We adopt an empirical distribution for the marginal distributions $F_1(\cdot)$ and $F_2(\cdot)$. We estimate each parametric copula parameter by equating the sample correlation and theoretic correlation after calculating Kendall's sample rank correlation $\hat{\tau}_K$.

Table 1 summarizes the expressions and the rank correlations for some of the bivariate copulas used in this subsection. Although the Gumbel copula is upper-tail dependent, it can be applied to lower-tail dependent data with the rotated-Gumbel copula. The rotated copula $C(u_1, u_2)$ for a copula $\hat{C}(u_1, u_2)$ is obtained by $(1 - u_1, 1 - u_2) \sim \hat{C}$ and is defined as $C(u_1, u_2) = u_1 + u_2 - 1 + \hat{C}(1 - u_1, 1 - u_2)$. The rank correlation of the rotated copula is equal to that of the original copula. In Table 1, $\boldsymbol{\Phi}_\rho(\cdot, \cdot)$ denotes a bivariate standard Gaussian distribution function with correlation ρ. $\boldsymbol{T}_{\nu,\rho}(\cdot, \cdot)$ denotes a bivariate t distribution function with degree of freedom ν and correlation ρ. $\Phi^{-1}(\cdot)$ the inverse function of the univariate standard Gaussian distribution function. $T_\nu^{-1}(\cdot)$ denotes the inverse function of univariate t distribution function.

[1] For differences in unconditional and conditional approaches, see [7, 14].

Table 2 Sample lower tail dependence and theoretical value given by each parametric copula between each pair of daily returns for three major stock indices

	Pair	$\hat{\lambda}_L(u)$	Gaussian	$t(6)$	$t(3)$	Rotated-Gumbel	Clayton
$\lambda_L(0.05)$	US/Euro	0.39	0.25	0.32	0.37	0.47	0.52
	US/JPN	0.35	0.20	0.30	0.36	0.40	0.43
	Euro/JPN	0.24	0.13	0.20	0.25	0.30	0.28
$\lambda_L(0.01)$	US/Euro	0.27	0.13	0.24	0.33	0.43	0.51
	US/JPN	0.29	0.09	0.23	0.32	0.36	0.41
	Euro/JPN	0.18	0.05	0.14	0.22	0.24	0.23

Table 2 shows the sample lower tail dependence $\hat{\lambda}_L(u)$ and the theoretical value by each parametric copula. The parameter for each parametric copula is obtained by equating it to Kendall's tau τ_K. We see that the sample lower tail dependence is much greater than the theoretical value given by the Gaussian copula. The sample lower tail dependence with a 5 % threshold approaches the theoretical value of the t copula with degree of freedom parameter $\nu = 3$. The lower tail dependence in this period is much stronger than that implied by the Gaussian copula. The data used for estimations for the US/JPN pair is one day lagged data for JPN. The parameter ν for t copula is fixed to 6 or 3 in this table.

Next, we estimate parameters for each parametric copula by maximizing the likelihood for the same data with empirical cumulative probabilities for each index return. Table 3 shows the maximum likelihood estimates and Schwarz's Bayesian information criterion (BIC) for each parametric copula. We adopt BIC to select the optimal parametric copula. The t copula with low degree of freedom parameters from 3 to 5 is selected by BIC. A Gaussian copula is worse than the t or rotated-Gumbel copula in terms of BIC. The result is consistent with the results of Tsafack [20], who selected a copula based on Akaike information criterion (AIC) and BIC for weekly stock return data for United States and Canada up to 2004.[2] We see that a Gaussian copula with weak lower tail dependence does not adequately capture the overall correlation structure in terms of likelihood.

Figure 1 depicts a joint density contour of the estimated copula with standard Gaussian margins for daily stock returns for the US and the Eurozone. Figure 1a depicts a Gaussian copula with standard Gaussian margins. Figure 1b depicts Student's t copula with standard Gaussian margins, adopted by BIC as the best-fit copula among several alternatives. We see that the lower tail dependence (the bottom-left corner) for Student's t copula is stronger than that for the Gaussian copula.

[2] Some information criteria, such as BIC and AIC, are used to select the optimal copula. Both criteria are calculated based on log-likelihood, with certain penalties applied by number of parameters. We adopt BIC for the criteria, which imposes more penalties in number of parameters than AIC. BIC is calculated by $-2l(\xi) + p \ln N$, where $l(\xi)$ is the maximum log-likelihood, p the number of parameters, and N the sample size. The model with the lowest BIC is selected.

Table 3 Maximum likelihood estimates and BIC for each parametric copula for each pair of daily returns for three major stock indices

Pair	Gaussian	t		Rotated-Gumbel	Clayton	Gumbel
	ρ	ρ	ν	γ	α	γ
US/Euro	0.519	0.520	3	1.539	0.831	1.533
	(−785)	(−1,055)		(−868)	(−695)	(−887)
US/JPN	0.443	0.427	3	1.389	0.640	1.366
	(−544)	(−637)		(−583)	(−488)	(−542)
Euro/JPN	0.270	0.273	5	1.220	0.387	1.180
	(−183)	(−253)		(−248)	(−217)	(−165)

The *upper values* in each cell are maximum likelihood estimates. The *lower values* in the parenthesis are the BICs. The parameter ν for the t copula is estimated with the restriction that ν is an integer value greater than 2

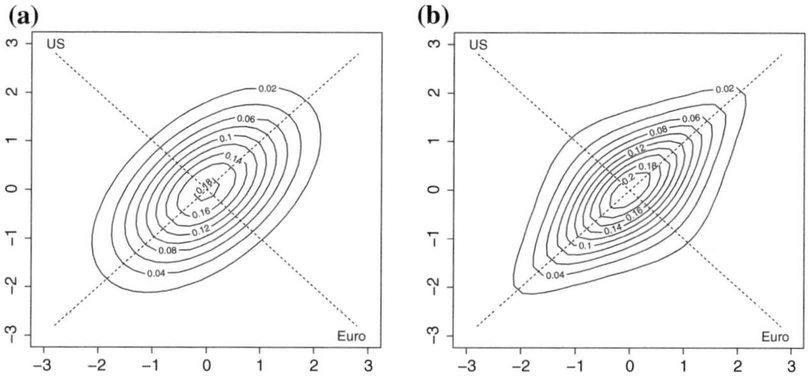

Fig. 1 Contour plot of copula density for two stock returns with standard Gaussian margins for **a** Gaussian copula and **b** Student's t copula

2.2 Correlation Structure Between Stock and Interest Rate

Japanese daily market data from 2007 to 2012 indicate a positive linear correlation between stock prices and interest rates, suggesting an increase in bond prices paired with a fall in stock prices. The measured risk of value-at-risk (VaR) or expected shortfall (ES) for the aggregated bond and stock portfolio becomes much smaller than the sum of the risk measures for those sub-portfolios. The reduction in risk measures for an aggregated portfolio is known as the diversification effect. Widely used aggregation methods that analyze recent Japanese data sometimes show up to a diversification effect of up to 60 %.

Yoshiba [22] has proposed using copulas for risk factors, focusing on stressed conditions. First, a bivariate copula with both positive and negative linear correlations is applied to the copula. Second, a copula estimated from stressed data such as

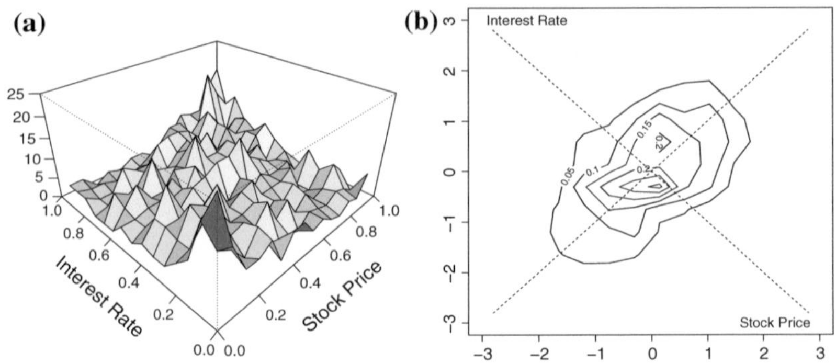

Fig. 2 Joint histogram and contour plot for the pseudo sample

Eurozone crisis data or post-Bubble data in Japan is applied to the copula. This paper gives an overview of the first method.

We select two risk factors; daily log return of Nikkei-225 index and daily changes in 5-year government interest rates.[3] The observation period is from October 1, 2007 to October 1, 2012. Marginal distributions are estimated by the skew-t distribution proposal by Azzalini and Capitanio [1]. Using the estimated marginal distribution functions $\hat{F}_1(x)$ and $\hat{F}_2(x)$ estimated, respectively, from stock returns and interest rate movement data, the pseudo sample $\{(u_{11}, u_{21}), \ldots, (u_{1N}, u_{2N})\}$ is obtained by $u_{ij} = \hat{F}_i(x_{ij})$ for $i = 1, 2;\ j = 1, \ldots, N$. Figure 2a depicts a joint histogram of the pseudo sample $\{(u_{11}, u_{21}), \ldots, (u_{1N}, u_{2N})\}$ during this period. The front side with zero for both axes indicates the largest drop in stock prices and interest rates. The figure indicates a relatively high frequency, suggesting that bond values will rise when stock prices fall, mitigating portfolio losses. Figure 2b plots the joint density contour after converting each u_{ij} to the quantiles of the standard Gaussian distribution, $\Phi^{-1}(u_{ij})$.[4] This contour rises diagonally up to the right with an elliptical shape, suggesting that this pseudo sample exhibits a positive linear correlation.

Table 4 the result of the maximum likelihood estimation. Here, we consider parametric copulas in Table 1 and the rotated ones for Gumbel and Clayton. We also consider a mixed-Gaussian copula implied in a mixed-Gaussian distribution. When we consider two state in the bivariate case, the mixed-Gaussian distribution is mixed with a negatively correlated Gaussian distribution and a positively correlated Gaussian distribution in the ratio of $\theta : (1 - \theta)$. The copula is expressed as follows:

$$C(u_1, u_2) = \theta \boldsymbol{\Phi}_{\rho_1}\left(\Phi^{-1}(u_1), \Phi^{-1}(u_2)\right) + (1-\theta)\boldsymbol{\Phi}_{\rho_2}\left(\Phi^{-1}(u_1), \Phi^{-1}(u_2)\right). \quad (3)$$

[3] We use generic interest rates calculated by Bloomberg for 5-year interest rates of government bonds.

[4] Joint density contours with standard Gaussian margins are visual representations of the various dependencies in the center and the tail area (see [8]). If the copula is Gaussian, the contour is elliptical (see Fig. 1a).

Table 4 MLE for the pseudo sample: The parameter ν for t copula is estimated without restriction on integer value

	Parameter	Est. Value	Std. Err.	BIC
Gumbel	γ	1.385	0.031	−239.8
Rotated-Gumbel	γ	1.416	0.031	−282.0
Clayton	α	0.662	0.050	−236.8
Rotated-Clayton	α	0.567	0.047	−184.5
Gaussian	ρ	0.436	0.021	−251.9
t	ρ	0.466	0.024	−307.2
	ν	5.481	0.918	
Mixed-Gaussian	ρ_1	−0.458	0.124	−313.0
	ρ_2	0.616	0.026	
	θ	0.145	0.036	

Table 5 VaR and ES for stocks and bonds, and their simple sum (billion yen)

	Stocks	Bonds	Simple sum
VaR (99 %)	2.61	2.47	5.08
ES (97.5 %)	2.82	2.77	5.59

The mixed Gaussian copula extracts a negatively correlated Gaussian component from positively correlated data. The mixed-Gaussian copula is selected by BIC.[5]

We consider a sample portfolio consisting of 50 billion yen in stocks and 700 billion yen in 5-year discount bonds, which is representative of the average portfolio held by Japanese regional banks. We adopt a daily 99 % VaR and 97.5 % expected shortfall (ES) estimates for risk measures.[6] For each category of stocks and bonds, we estimate the return distribution by applying a skew-t distribution. Table 5 shows the VaR and ES for each category, along with the simple sum of those risk measures.

Table 6 summarizes VaR and ES for the sample portfolio while accounting for the diversification effect. The diversification effect is given by the reduction rate of the aggregated VaR or ES from the simple sum of VaR or ES as follows:

$$\text{diversification effect} = \frac{\text{simple sum VaR/ES} - \text{aggregate VaR/ES}}{\text{simple sum VaR/ES}}. \tag{4}$$

[5]The mixed-Gaussian copula is not always selected by BIC. For example, Yoshiba [22] shows t copula is selected by BIC for data related to the Euro crisis and the post-bubble period in Japan. As for more complicated copulas, we can construct, as examples, mixed-t or mixed-Gaussian-$t$copulas. An examination of these copulas is left for the future.

[6]99 % VaR is the 99th percentile for a portfolio loss distribution. A 97.5 % ES is the average of the losses in the 2.5 % tail of the loss distribution. If the portfolio profit–loss distribution is Gaussian, a 97.5 % ES nearly equals 99 % VaR.

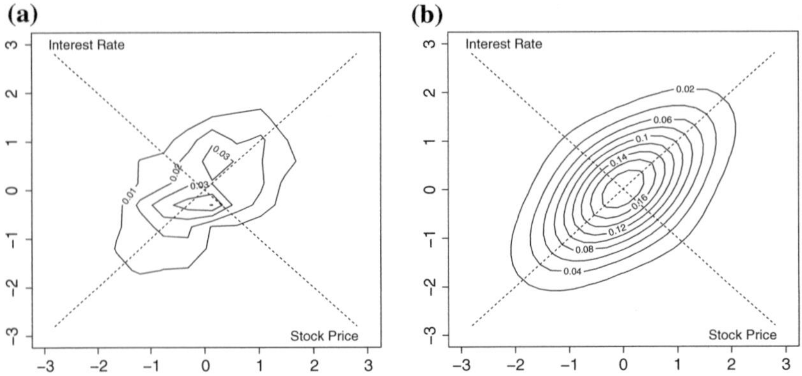

Fig. 3 Contour plot of joint density with Gaussian margins for **a** nonparametric copula and **b** mixed-Gaussian copula

The joint distribution for risk factors is constructed by each estimated copula with estimated marginal distributions. For the nonparametric copula, the pseudo sample is converted into a set of risk factors by taking quantiles for marginal skew-t distributions. For each parametric copula, we generate 100,000 random bivariate vectors and calculate VaR and ES with marginal skew-t distributions. Iterating the procedure 100 times, we obtain averages and standard deviations for VaR and ES.

The diversification effect for the negative correlation of a mixed-Gaussian copula (VaR:17 %; ES:19 %) selected by BIC is the smallest and is much smaller than that for other copulas. Unlike other parametric copulas, a mixed-Gaussian copula can capture both positive and negative linear correlations, allowing a better fit to the pseudo sample. The results indicate a negative linear correlation can be captured at a frequency of $\theta = 14.5\,\%$ (see Fig. 3). This negative correlation increases estimates for portfolio VaR and ES and reduces diversification effects.

Table 6 VaR and ES using the estimated copula: VaR and ES are given in units of billions of yen

Copula	VaR (99 %)	Std. dev.	Div. effect (%)	ES (97.5 %)	Std. dev.	Div. effect (%)
Nonparametric	3.01	–	41	3.24	–	42
Gumbel	2.66	0.03	48	2.90	0.04	48
Rotated-Gumbel	2.58	0.03	49	2.84	0.04	49
Clayton	2.68	0.03	47	2.96	0.04	47
Rotated-Clayton	2.81	0.03	45	3.05	0.04	45
Gaussian	2.65	0.03	48	2.95	0.04	47
t	2.60	0.03	49	2.85	0.04	49
Mixed-Gaussian (negative)	4.21	0.04	17	4.55	0.05	19

3 Risk Aggregation for Credit Portfolios

Correlation structures for firm asset values are a key element for credit portfolio risk aggregation. First, we will survey risk measurements for a banking industry credit portfolio by addressing prudent correlation structures. Second, we will provide an overview of applications to CDO valuation.

3.1 Credit Portfolio Risk Measurement in the Banking Industry

In the banking industry, the credit portfolio risk of lending exposure is assessed by VaR with confidence levels of up to 99.9 % in accordance with the Basel Accords [3]. In models for credit VaR, the asset log values for the lending firms follow a multivariate Gaussian distribution with a correlation matrix based on the structural model developed by Merton [15]. If the firm's asset value at maturity T is less than the firm's liability, the firm defaults. The maturity T is usually fixed to one year, which is the time interval required to cover the potential loss of the bank by economic capital. Bank losses are given by the loss given default (LGD) multiplied by the exposure at default (EAD). In most cases, EAD is capped at the lending amount. LGD is also fixed exogenously.

Banks estimate the default probability of each firm until maturity by applying a statistical model based on financial indicators, not by the above structural model. Even in the latter case, banks capture the same correlation structures for firm asset values as a multivariate Gaussian distribution, according to Merton [15]. The correlation structure is a Gaussian copula. Many credit VaR models, including J.P. Morgan's CreditMetrics, assume a Gaussian copula between firm asset values.

Frey et al. [6] have calculated the number of defaulting firms using a Gaussian copula and t copula with several degree of freedom parameters for a lending portfolio involving loans to N homogenous firms with 0.5 % default probability and 0.038 asset correlation. Following [6], Tozaka and Yoshiba [21] have applied Gaussian, t with degree of freedom parameter of 10, Clayton, and rotated-Gumbel copulas to asset correlation structures and calculated the number of defaulting firms for $N = 10{,}000$. Table 6 gives the distribution of the numbers obtained by a Monte Carlo simulation involving 100,000 iterations.

Table 7 indicates that a tail dependent copula like $t(10)$ and the rotated-Gumbel gives an extremely large number of defaulting firms with high confidence level (99.9 %).

This is a simple example of applying a copula to capture credit portfolio risk. In practice, the correlation structure may vary from industry to industry. A vine copula[7]

[7] For vine copulas, see the handbook [12].

Table 7 Number of defaulting firms given by each copula for specific confidence levels

	50 %	90 %	95 %	99 %	99.9 %
Gaussian	43	90	109	155	227
$t(10)$	9	133	240	586	1,305
Rotated-Gumbel	42	56	66	156	1,176
Clayton	26	122	179	343	643

or hierarchical Archimedean copula (HAC)[8] may be applied to capture flexible asset correlation structures. See [10] for a study of flexible asset correlation structures using vine copulas.

3.2 Application of Copulas to CDO Valuation

Li [13] has applied copulas to the valuation of CDOs, which consist of many CDS (credit default swap) to firms with 5-year maturity. To capture the number of defaulting firms at maturity, Li [13] uses a copula to represent the correlation structure for the underlying firm asset values at maturity. Credit ratings agencies, where Gaussian copulas are widely used for CDO valuations, have been criticized for the inability of their models to capture tail dependencies when the price of CDOs plunged around the time of the Lehman shock. The explanation for this shortcoming is that Gaussian copulas have no asymptotic tail dependence.

Burtschell et al. [5] have investigated various copulas with few parameters to fit iTraxx market data in August 2005. On the other hand, Shintani et al. [19] comparatively investigated the effects on credit spread by applying various copulas determined by historical asset value data.

Following standard settings for CDOs, we set a CDO with 100 homogenous underlying assets, 5-year maturity, and 40 % recovery rate for each asset. The default probability for each asset was 5 % per 5 years, somewhat higher than usual. The asset correlation ρ for the Gaussian copula is 0.15, a standard setting used by ratings agencies. In this case, the Kendall's tau τ_K of the Gaussian copula is 0.096. Equating this value to the theoretical value of each copula, we obtain the parameter for each copula as t's ρ: 0.15, Clayton's α: 0.21, and the rotated-Gumbel's γ: 1.11. We can divide the CDO portfolio into tranches of [0 %, 6 %] for equity, [6 %, 18 %] for mezzanine, [18 %, 36 %] for senior, and [36 %, 100 %] for super-senior, thereby maintaining ratings above AAA for super-senior, AA–AAA for senior, BBB–A for mezzanine by Gaussian copula's valuation.[9] Table 8 shows the credit spread for each

[8]For HAC, see [18].

[9]The tranche with [a %, b %] covers the portfolio loss, while the loss rate is in [a %, b %].

Table 8 Credit spread of each tranche (bp)

Copula	Equity	Mezzanine	Senior	Super senior
Gaussian	1147.43	63.38	0.65	0.000
$t(20)$	1061.07	86.94	2.33	0.002
$t(6)$	899.52	127.82	9.11	0.043
$t(3)$	735.55	165.40	21.81	0.196
Rotated-Gumbel	1018.34	59.01	19.04	2.685
Clayton	860.61	135.77	12.65	0.099

tranche derived from expected loss rates in this case.[10] The rows of $t(20)$, $t(6)$, and $t(3)$ are resulting from a t copula with fixed degree of freedom parameters $\nu = 20$, 6, and 3, respectively. The Gaussian copula evaluates the lowest credit spread for the upper tranches (senior, super-senior), while the rotated-Gumbel copula evaluates the highest credit spread for equity. For example, the credit spread for the senior tranche is 0.65bp by the Gaussian copula, which is much lower than 21.81bp by $t(3)$ copula or 19.04bp by rotated-Gumbel copula. We see that the recognition of tail dependencies yields large differences in risk sensitivity for the upper tranches.

4 Enterprise Risk Management

Banks measure risks within market, credit, and operational risk categories. Many major banks aggregate firm-wide risks for each risk category to calculate the economic capital needed out to a one-year horizon. The economic capital is allocated to each business unit and used as performance measures (see, for example, [11]). Rosenberg and Schuermann [17] have explored various risk aggregation methods empirically, and Brockmann and Kalkbrener [4] have introduced the Deutsche Bank model.

Since banks apply different methods to each risk category, the first problem in aggregating firm-wide risk is to determine which level of the correlation structure we should notice (see Fig. 4). Some advanced banks apply a tail dependent t copula or a high correlation of Gaussian copula to profit–loss for each risk category. This is called top level aggregation. Profit–loss is sometimes proxied by a time series of some index. Some other advanced banks apply a t copula to the joint distribution of risk factors (RFs) in each risk category. This is called base level or bottom up aggregation.

With top level aggregation, banks sometimes calibrate Gaussian or t copula parameters (including ν for t copula) by using each time series for market and credit profit–loss. They determine correlations with other risk categories a priori, without statistical estimates. In this context, intuitively identifiable correlation structures are desirable for actual practice.

[10] Credit spread s is calculated as $s = -\ln(1 - EL)/T$, where EL is the expected loss for the tranche and T is 5-year maturity.

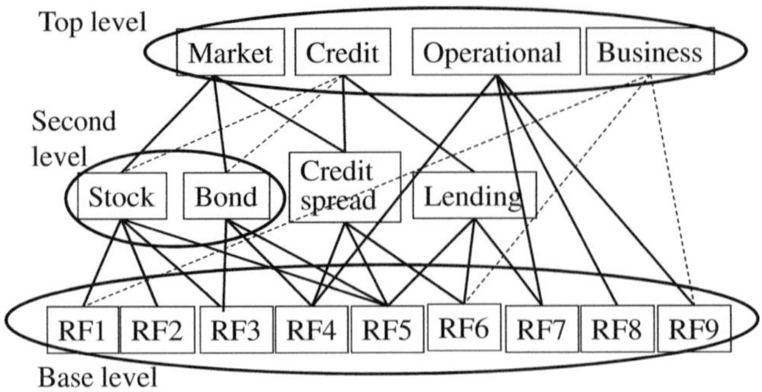

Fig. 4 Aggregation level

Differences in risk evaluation periods for each risk category represent a practical issue for enterprise risk aggregation. Although the risk horizon for calculating economic capital is one year, banks assume some positions may close in a short-term period of high market liquidity. One and ten days are standard risk evaluation periods for liquid market risks for trading positions. Three months and six months are typical evaluation periods for non-trading positions. Selecting the optimal time intervals for given correlation structures and methods for selecting ideal serial correlation structures when estimating economic capital remain outstanding issues.

5 Concluding Remarks

This paper surveys the application of copulas in the banking industry, focusing in particular on ways to incorporate stressed situations into risk measurements.

The application of copulas in the banking industry can differ from academic convention, incorporating drastic assumptions in certain cases. Research on applications that account for the needs of the banking industry is likely to prove beneficial.

Acknowledgments For their helpful comments, the author would like to thank Satoshi Yamashita and the participants of Forum "Math-for-Industry" 2014, held in Fukuoka from October 27 to 31, 2014. The views expressed in this paper are those of the author and do not necessarily reflect the official views of the Bank of Japan.

References

1. Azzalini, A., Capitanio, A.: Distributions generated by perturbation of symmetry with emphasis on a multivariate skew t-distribution. J. Roy. Stat. Soc. Ser. B **65**(2), 367–389 (2003)
2. Basel Committee on Banking Supervision: Joint Forum "Developments in Modelling Risk Aggregation" 21 Oct 2010

3. Basel Committee on Banking Supervision: Basel III: A Global Regulatory Framework for More Resilient Banks and Banking Systems. Revised version June 2011
4. Brockmann, M., Kalkbrener, M.: On the aggregation of risk. J. Risk **12**(3), 45–68 (2010)
5. Burtschell, X., Gregory, J., Laurent, J.P.: A comparative analysis of CDO pricing models under the factor copula framework. J. Deriv. **16**(4), 9–37 (2009)
6. Frey, R., McNeil, A., Nyfeler, M.: Copulas and credit models. Risk **14**(10), 111–114 (2001)
7. Isogai, T.: Benchmarking unconditional VaR and ES calculation methods: a numerical simulation with fitted truncated stable distribution. In: Bank of Japan Working Paper Series No. 14-E-1 (2014)
8. Joe, H.: Multivariate Models and Dependence Concepts. Chapman & Hall, London (1997)
9. Jorion, P.: Value at Risk: The New Benchmark for Managing Financial Risk, 3rd edn. McGraw-Hill, New York (2006)
10. Kawaguchi, M., Yamanaka, S., Tashiro, Y.: Improvement of credit concentration risk evaluation using copulas. Secur. Anal. J. **52**(3), 43–51 (2014)
11. Klaassen, P., van Eeghen, I.: Economic Capital: How It Works and What Every Manager Needs to Know. Elsevier Science, New York (2009)
12. Kurowicka, D., Joe, H.: Dependence Modeling: Vine Copula Handbook. World Scientific Publishing, Singapore (2010)
13. Li, D.: On default correlation: a copula function approach. J. Fixed Income **9**(4), 43–54 (2000)
14. McNeil, A.J., Frey, R., Embrechts, P.: Quantitative Risk Management: Concepts, Techniques, and Tools. Princeton University Press, Princeton (2005)
15. Merton, R.C.: On the pricing of corporate debt: the risk structure of interest rates. J. Financ. **29**(2), 449–470 (1974)
16. Nelsen, R.B.: An Introduction to Copulas, 2nd edn. Springer, New York (2006)
17. Rosenberg, J.V., Schuermann, T.: A general approach to integrated risk management with skewed, fat-tailed risks. J. Financ. Econ. **79**(3), 569–614 (2006)
18. Savu, C., Trede, M.: Hierarchies of Archimedean copulas. Quant. Financ. **10**(3), 295–304 (2010)
19. Shintani, K., Yamada, T., Yoshiba, T.: Correlation structures between assets' returns in financial crisis: evaluation using copulas. Kin'yu-kenkyu **20**(3), 89–122 (2010) (in Japanese)
20. Tsafack, G.: Asymmetric dependence implications for extreme risk management. J. Deriv. **17**(1), 7–20 (2009)
21. Tozaka, H., Yoshiba, T.: Commentary on concrete applications of copula in financial practices. Kin'yu-kenkyu **24**(S-2), 115–162 (2005) (in Japanese)
22. Yoshiba, T.: Risk aggregation by a copula with a stressed condition. In: Bank of Japan Working Paper Series No. 13-E-12 (2013)

Discrete Geometry for Reliable Surface Quad-Remeshing

Konrad Polthier and Faniry Razafindrazaka

Abstract In this overview paper we will glimpse how new concepts from discrete differential geometry help to provide a unifying vertical path through parts of the geometry processing pipeline towards a more reliable interaction. As an example, we will introduce some concepts from discrete differential geometry and the QuadCover algorithm for quadrilateral surface parametrization. QuadCover uses exact discrete differential geometric concepts to convert a pair (simplicial surface, guiding frame field) to a global quad-parametrization of the unstructured surface mesh. Reliability and robustness is an omnipresent issue in geometry processing and computer aided geometric design since its beginning. For example, the variety of incompatible data structures for geometric shapes severely limits a reliable exchange of geometric shapes among different CAD systems as well as a unifying mathematical theory. Here the integrable nature of the discrete differential geometric theory and its presented application to an effective remeshing algorithm may serve an example to envision an increased reliability along the geometry processing pipeline through a consistent processing theory.

Keywords Geometry processing · Simplicial surfaces · Surface parametrization · Branched covering · QuadCover algorithm · Hodge decomposition · Minimal surfaces

1 Calculus on Simplicial Surfaces

We begin with a 2-dimensional simplicial surface $M_h \subset \mathbb{R}^n$ where n is typically in $\{2, 3, 4\}$. On a simplicial surface M_h we consider two types of piecewise linear (PL) function spaces, the *conforming Lagrange space* $S_h(M_h)$ and the *non-conforming*

K. Polthier (✉) · F. Razafindrazaka (✉)
Freie Universität Berlin, Berlin, Germany
e-mail: konrad.polthier@fu-berlin.de

F. Razafindrazaka
e-mail: faniry.razafi@fu-berlin.de

R.S. Anderssen et al. (eds.), *Applications + Practical Conceptualization + Mathematics = fruitful Innovation*, Mathematics for Industry 11,
DOI 10.1007/978-4-431-55342-7_22

space $S_h^*(M_h)$. Both spaces are classic in finite element literature, but here we will see how both function spaces team up to mimic the concept of primality and duality of grids in the framework of function spaces on the *same* underlying geometry, thus allowing to use the simplicial surface M_h as single base geometry. For example, solutions of the discrete Cauchy-Riemann equations will consist of a pair of a discrete conforming and non-conforming harmonic map in S_h and in S_h^* resp. vice versa, all defined on M_h.

Definition 1 The *piecewise linear conforming* $S_h(M_h)$ *and non-conforming* $S_h^*(M_h)$ *function spaces* on a 2-dimensional simplicial surface $M_h \subset \mathbb{R}^n$ are given by:

$$S_h := \left\{ f : M_h \to \mathbb{R} \;\middle|\; f_{|T} \text{ is linear on each triangle } T, \text{ and } f \in C^0(M_h) \right\}$$

$$S_h^* := \left\{ f^* : M_h \to \mathbb{R} \;\middle|\; f^*_{|T} \text{ is linear, and continuous at edge midpoints} \right\}$$

On first sight, the missing global continuity of functions $f^* \in S_h^*$ sounds like a drawback but the space of non-conforming functions will turn out as a good match to S_h, see Fig. 3 for the graphs of two sample functions. Both function types have gradients from piecewise differentiation:

Definition 2 The *gradient field* ∇f of a function $f \in S_h$ or S_h^* is a constant tangent vector in each triangle. The *co-gradient field* $\delta f := J \nabla f$ is obtained by rotation J of the gradient ∇f, i.e. by $\frac{\pi}{2}$ in each triangle.

1.1 Discrete Vector Fields

Piecewise constant vector fields were introduced to geometry processing in [13] as a natural discretization of tangential vector fields on simplicial geometries. Among the useful properties of PC vector fields, say compared to Whitney type differential forms, are the formulation of the Hodge star operation in function spaces instead of introducing a pair of primary and dual meshes. After a short overview of PC vector fields and their integrability conditions we touch two sample problems, the discrete Cauchy-Riemann equation and the discrete Hodge decomposition, to highlight the efficiency of PC vector fields and to prepare some tools for the QuadCover application.

Definition 3 The space of *piecewise constant tangential vector fields* $\Lambda^1(M_h)$ on a 2-dimensional simplicial surface $M_h \subset \mathbb{R}^n$ is given by:

$$\Lambda^1(M_h) := \left\{ v : M_h \to TM_h \;\middle|\; v_{|\text{triangle } T} \text{ is a constant tangent vector in } T \right\}$$

The gradient and co-gradients fields of functions in S_h or S_h^* introduced above are examples of piecewise constant (PC) tangential vector fields, see Fig. 1.

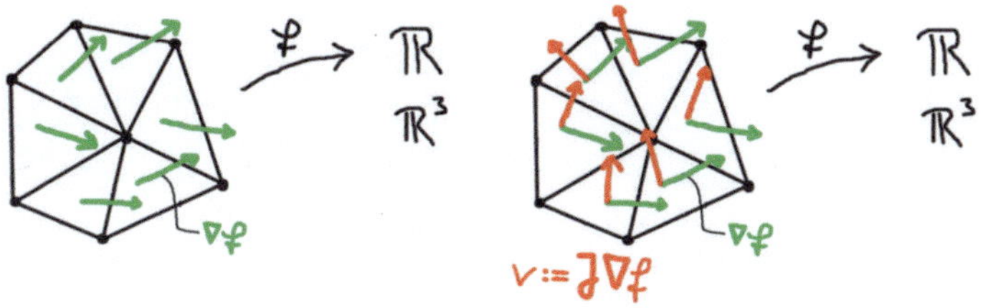

Fig. 1 Gradient field ∇f and co-gradient field $\delta f := J\nabla f$ are constant on each *triangle*

Definition 4 On a simplicial surface M_h let $v \in \Lambda^1(M_h)$, p a vertex and m an edge midpoint. Then the (total) *discrete curl* is given by

$$\operatorname{curl}_h\, v(p) := \frac{1}{2}\oint_{\partial\, \text{star}\, p} v = \frac{1}{2}\sum_{i=1}^{k}\langle v, c_i\rangle$$

$$\operatorname{curl}_h^*\, v(m) := \oint_{\partial\, \text{star}\, m} v = -\left\langle v_{|T_1}, c\right\rangle + \left\langle v_{|T_2}, c\right\rangle$$

where c_i are the edges of the oriented boundary of star p resp. c the edge with midpoint m.

Theorem 1 (Local integrability conditions) *Let M_h be a simply connected simplicial surface. Then a PC vector field $v \in \Lambda^1(M_h)$ can be characterized as gradient field in term of the discrete* curl *operator:*

1. *v is a gradient field of a function in S_h* $\iff$

$$\operatorname{curl}_h^*\, v(m) = 0 \textit{ at all edge midpoints } m.$$

2. *v is a gradient field of a function in S_h^** $\iff$

$$\operatorname{curl}_h\, v(p) = 0 \textit{ at all vertices } p.$$

Definition 5 On a simplicial surface M_h let $v \in \Lambda^1(M_h)$, p a vertex and m an edge midpoint. Then the (total) *discrete divergence* is given by

$$\operatorname{div}_h\, v(p) := \frac{1}{2}\oint_{\partial\, \text{star}\, p}\langle v, \nu\rangle\, ds = -\frac{1}{2}\sum_{i=1}^{k}\langle v, Jc_i\rangle$$

$$\operatorname{div}_h^*\, v(m) := \oint_{\partial\, \text{star}\, m}\langle v, \nu\rangle\, ds = \left\langle v_{|T_1}, J_{|T_1}c\right\rangle + \left\langle v_{|T_2}, J_{|T_2}c\right\rangle$$

where ν is the outer unit normal along ∂ star p resp. ∂ star m.

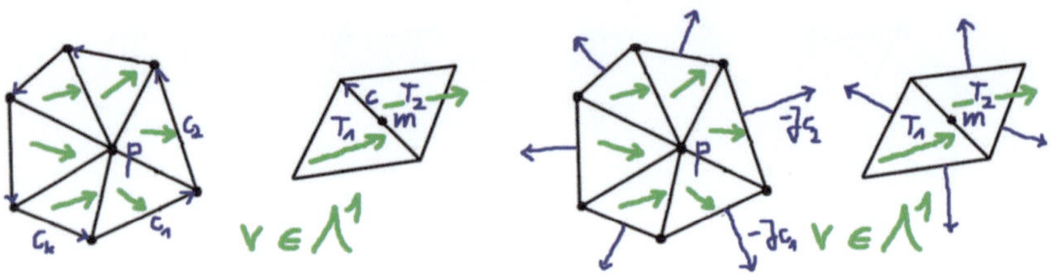

Fig. 2 Discrete curl and divergence operators

Remark 1 Discrete rotation and divergence are related by $\text{curl}_h\ Jv = \text{div}_h\ v$ and $\text{curl}_h^*\ Jv = \text{div}_h^*\ v$, compare Fig. 2.

1.2 Discrete Cauchy-Riemann Equation

With the first order operators ∇, curl and div at hand we deduce the discrete Laplace-Beltrami operators and a notion of conjugacy of harmonic vector fields (Fig. 3).

Definition 6 The *discrete Laplace-Beltrami operator* Δ of functions in S_h resp. in S_h^* on a simplicial surface M_h is given as divergence of the corresponding gradient functions, i.e. $\Delta_h f(p) := \text{div}_h \nabla f(p)$ for $f \in S_h$ and $\Delta_h^* f^*(m) := \text{div}_h^* \nabla f^*(m)$ for $f^* \in S_h^*$.

A natural question is the existence of solutions to the discrete Cauchy-Riemann equation which asks for pairs (f, g) of discrete maps f and g with $\nabla g = \delta f$, and thus for discrete holomorphic resp. conformal maps $z = f + ig$. In the smooth setting, a co-gradient field δf has a potential function g with $\nabla g = \delta f$ if and only if f and g are a pair of conjugate harmonic maps, i.e. real and imaginary parts of a

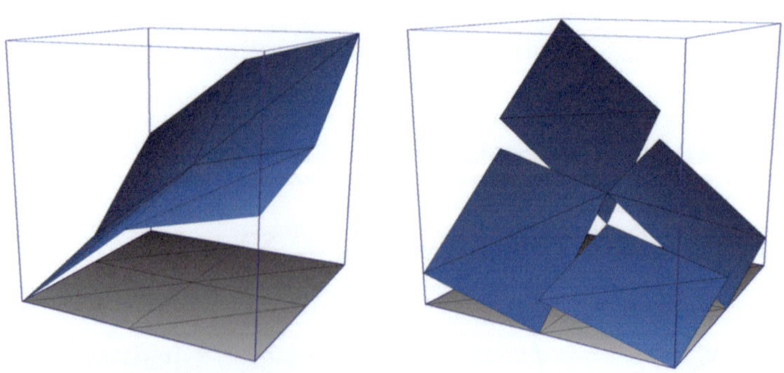

Fig. 3 Graph of a conforming harmonic function $f \in S_h$ and a non-conforming harmonic function $g \in S_h^*$, here satisfying the Cauchy-Riemann equation $\nabla g = \delta f$

holomorphic map. The discrete Cauchy-Riemann equation holds for a matching pair of a conforming and a non-conforming harmonic function:

Theorem 2 (Cauchy-Riemann equation) *On a simply connected simplicial surface $M_h \subset \mathbb{R}^n$ the harmonic maps in S_h and S_h^* come in conjugate harmonic pairs solving the* discrete Cauchy-Riemann equation $\nabla g = \delta f$ *[12]:*

1. *The co-gradient δf of $f \in S_h$ is a gradient field of a function $g^* \in S_h^* \iff f$ is discrete harmonic in S_h.*
   Furthermore, the conjugate map $g^* \in S_h^*$ is discrete harmonic.
2. *The co-gradient δf^* of $f^* \in S_h^*$ is a gradient field of a function $g \in S_h \iff f^*$ is discrete harmonic in S_h^*.*
   Furthermore, the conjugate map $g \in S_h$ is discrete harmonic.

1.3 Discrete Conjugate Minimal Surfaces

We extend the notion of piecewise linear functions to vector-valued piecewise linear functions, i.e. simplicial maps $F : M_h \to \mathbb{R}^3$ denoted by $F \in S_h(M_h)^3$ resp. for edge-based representations denoted by $F^* \in S_h^*(M_h)^3$. The discrete Laplace-Beltrami operator $\Delta_h F$ is defined by applying Definition 6 to the component functions of F. In the special case of $F = id_{M_h}$ its Laplace becomes the *discrete mean curvature vector* of M_h with versions at vertices

$$H(p) := \Delta_h id_{M_h}(p) \in T_p\mathbb{R}^3$$

and at edges

$$H(m) := \Delta_h^* id_{M_h}^*(m) \in T_m\mathbb{R}^3$$

which measures the variation of discrete surface area in the space of conforming $F(M_h)$ or non-conforming meshes $F^*(M_h)$.

Surfaces with mean curvature $H = 0$ are called *discrete minimal surfaces* since they are critical points of the discrete area functional. From $H = 0$ follows that the component functions of the identity map id_{M_h} are discrete harmonic, thus have conjugate harmonic pairs. Therefore, the conjugacy of harmonic maps extends to the conjugacy of discrete minimal surfaces composed of a pair of a conforming and a non-conforming discrete minimal surface. See Fig. 4 with a pair of a conforming catenoid and a non-conforming helicoid, resp. its reverse representations in the second row. See Fig. 12 with several more discrete minimal surfaces with a smooth quad-parametrization.

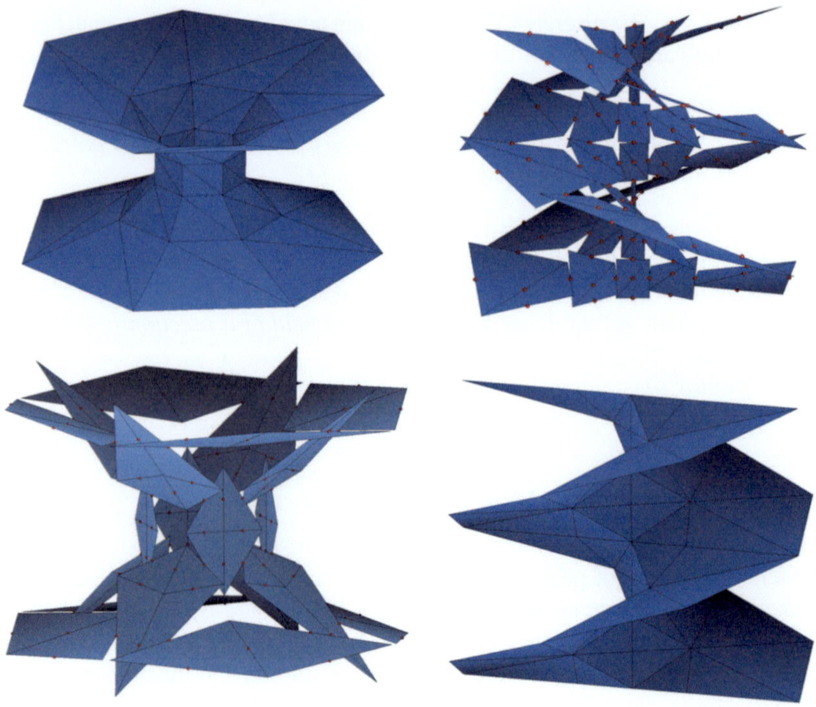

Fig. 4 Conjugate pairs of a discrete catenoid and a helicoid in each row

1.4 Discrete Hodge-Helmholtz Decomposition

The Hodge-Helmholtz decomposition of vector fields on surfaces provides a precise criterion for the local integrability properties of vector fields as well as their relation to globally defined harmonic vector fields. Later we will make flexible use of all three Hodge components of vector fields. As an example see Fig. 5 where a vector field on a simplicial torus is decomposed.

Theorem 3 (Hodge-Helmholtz decomposition) *The space of piecewise constant vector fields $\Lambda^1(M_h)$ on a simplicial surface M_h decomposes into an L^2-orthogonal sum of the spaces of gradient fields, co-gradient fields and harmonic fields:*

$$\begin{aligned} \Lambda^1 &= \nabla S_h \oplus \delta S_h^* \oplus \left(H := \operatorname{curl}_h^* \cap \ker \operatorname{div}_h\right) \\ v &= \underbrace{\nabla f}_{\operatorname{curl}_h^* \nabla f = 0} \oplus \underbrace{\delta g}_{\operatorname{div}_h \delta g = 0} \oplus \underbrace{w}_{\operatorname{curl}_h^* w = \operatorname{div}_h w = 0} \end{aligned}$$

This space of discrete harmonic fields H on a compact 2-surface of genus g has correct dimension $2g$. The roles of the conforming S_h and non-conforming S_h^* spaces may be exchanged [13].

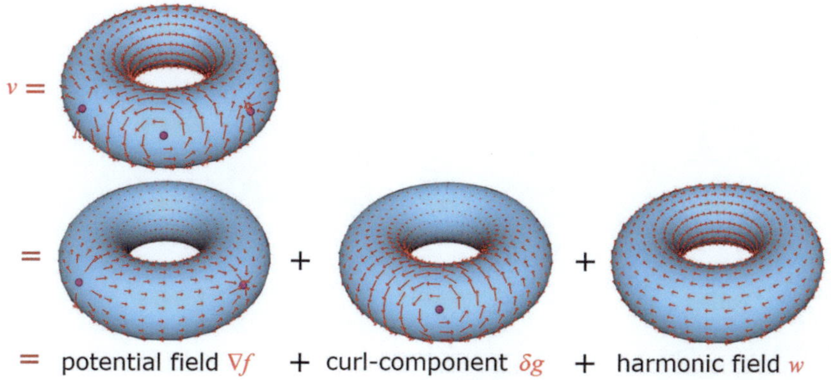

Fig. 5 Hodge decomposition of a piecewise constant tangent vector field on a simplicial torus to a gradient, co-gradient and harmonic field

Proof The gradient and co-gradient part of v can be computed by minimizing a quadratic energy, the harmonic part is then obtained as remainder. The gradient component of $v \in \Lambda^1$ is the unique minimizer of

$$\min_{f \in S_h} \int_{M_h} \|v - \nabla f\|^2$$

and the co-gradient component is the unique minimizer of

$$\min_{g \in S_h^*} \int_{M_h} \|v - \delta g\|^2$$

w.r.t. to appropriate boundary conditions. It is easy to show that curl and div components of the minimizer vanish.

Remark 2 Depending on applications, both spaces might be chosen of the same type when the correct dimension of H is negligible.

2 Mesh Parametrization

Triangle meshes are among the popular data structures for surface representations in computer graphics, geometry processing and finite element numerics. They provide a rich flexibility, for example, for adaptive mesh refinement according to local resolution needs, and they come along with a large set of efficient processing algorithms. On the other hand, sometimes more restrictive representations such as quad meshes are preferable. For example, when trying to convert a mesh into a hierarchical sub-

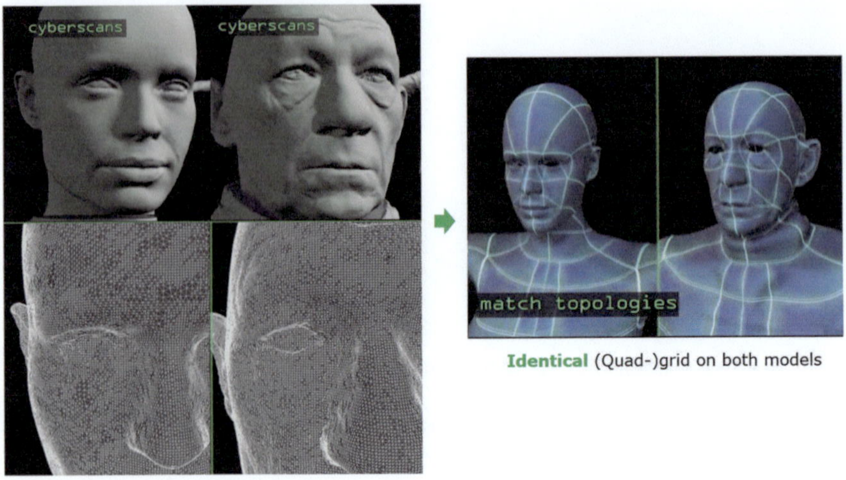

Fig. 6 Morphing a pair of 3d scans of two characters is currently performed by constructing two matching grids on both models by hand (*images* by Beau Janzen)

division surface, or when panelizing a roof construction in architecture with simple planar glass panels, or when trying to compute a morph between two scans of a pair of characters in computer animations.

Assume two characters have been scanned and each scan is given as a high-precision mesh with millions of triangles, see Fig. 6. A morphing between the two scans requires a bijective map between the two triangles meshes, certainly with additional restrictions of low distortion etc. Computing a homotopy automatically is typically very difficult and no perfect algorithm exists yet. In practice, computer animators generate a matching to a large extent by hand: on both meshes the same quad layout is drawn, thus generating a bijective correspondence between points and quads of the two meshes, which then extends to a smooth morphing of the two shapes. Here an automatic mapping algorithm would be more than appreciated and is currently the target of intensive research activities in geometry processing.

2.1 QuadCover Parameterization

We now use the discrete concepts derive in the previous sections to formulate the basic principles of the QuadCover algorithm [5] as an example and application of discrete differential geometric concepts to the solution of the intriguing problem of surface parametrization in geometry processing.

The QuadCover algorithm takes as input a triangle mesh plus a guiding frame field X on M_h, see Fig. 7. Formally, a *frame field* X on M_h consists of a pair of constant vectors $X_T = (X_1, X_2)_T$ in each triangle $T \in T_{M_h}$. In our application, it

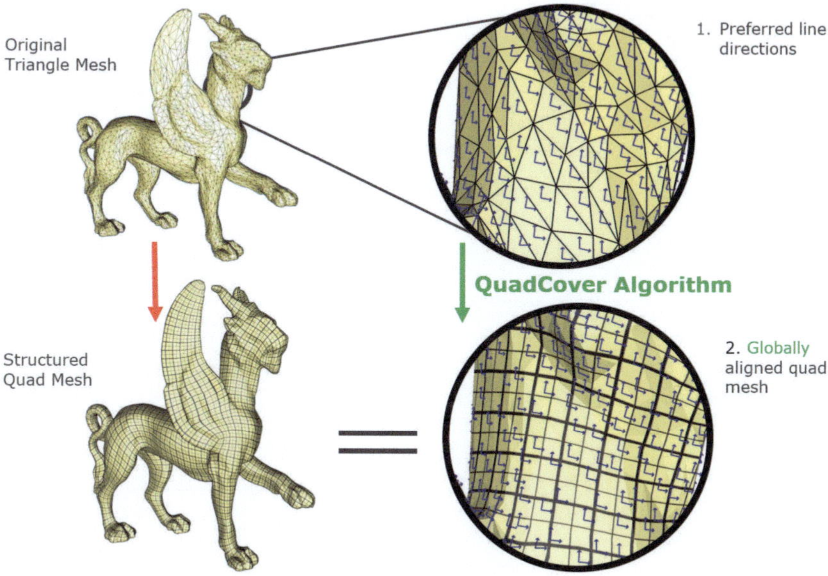

Fig. 7 QuadCover algorithm for quad-meshing guided by a frame field is described exactly in discrete differential geometry concepts

is convenient to extend a frame field X to a *cross field* $(X_1, X_2, -X_1, -X_2)$, also denoted by X. The continuity of a cross field across the common edge $e_{ij} = T_i \cap T_j$ of two adjacent triangles T_i and T_j is given by a periodic jump $r_{ij} \in \{0, 1, 2, 3\}$ denoting the pairing of the first vector X_1 on T_i with the r_{ij}-th vector in the adjacent triangle T_j; for simplicity we assume orthogonal frames in this description. Overall this edge pairing produces a well-defined continuous frame field resp. cross field on M_h.

The two tangent vectors of a frame field are intended to direct the parameter lines, namely, that optimally the parameter lines at each point are tangent to the two guiding vectors. The two additional directions of a cross field guide the same parameter lines, just in opposite direction. A frame field may be obtained either automatically as principal curvature directions of a discrete shape operator or a field may be carefully design to follow surface features or even aesthetic reasoning.

A central statement is, given a triangle mesh plus a frame field, then the QuadCover algorithm provides a merely exact and reliable computational algorithm for the generation of an *atlas of charts* which generates a globally consistent quad-layout on the surface. The inverse φ^{-1} of each chart φ maps the Z^2-grid of $\mathbb{R}^2$ onto the surface such that parameter lines are optimally aligned in direction of the guiding frame field, see Fig. 8. More precisely, a chart $\varphi \in S_h \times S_h : M_h \to \mathbb{R}^2$ is a simplicial map from the surface M_h to the texture domain $\mathbb{R}^2$. Using φ^{-1} the $\mathbb{Z}^2$-grid of the texture domain is mapped as texture onto the surface, such that the $\mathbb{Z}^2$-lines are tangential to the guiding frame field. An atlas of charts $\{\varphi_i\}$ is computed such that

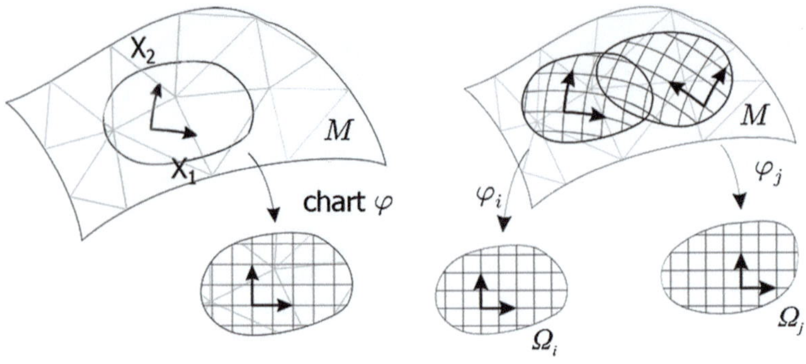

Fig. 8 A chart on a simplicial surface (*left*) with fulfilled compatibility condition for quad-parametrization (*right*)

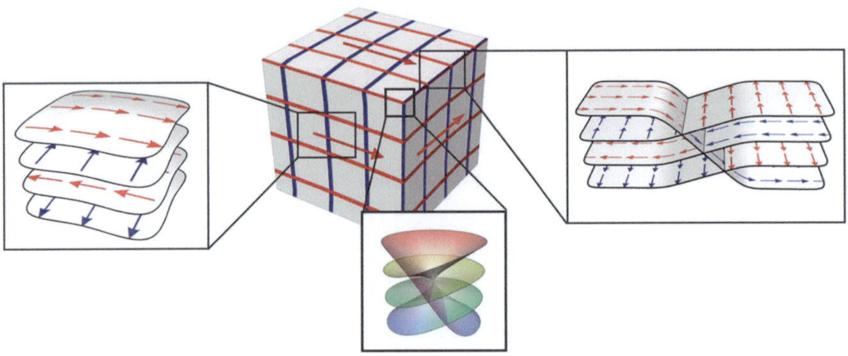

Fig. 9 Lifting a frame field to a vector field on a 4-fold branched simplicial covering surface

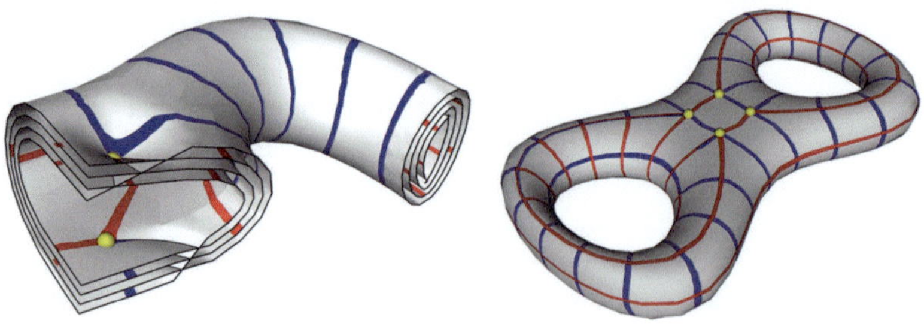

Fig. 10 The level lines of the generated potential function on M_h^* compose to the quad nets on M_h after projection

their inverses φ_i^{-1} generate a consistent quadrilateral grid $\left\{\varphi_i^{-1}\left(\Omega_i \cap \mathbb{Z}^2\right)\right\}$ on M_h. Note, that grid on M_h is given as texture map, that means at this stage as a functional representation on M_h (Figs. 9 and 10).

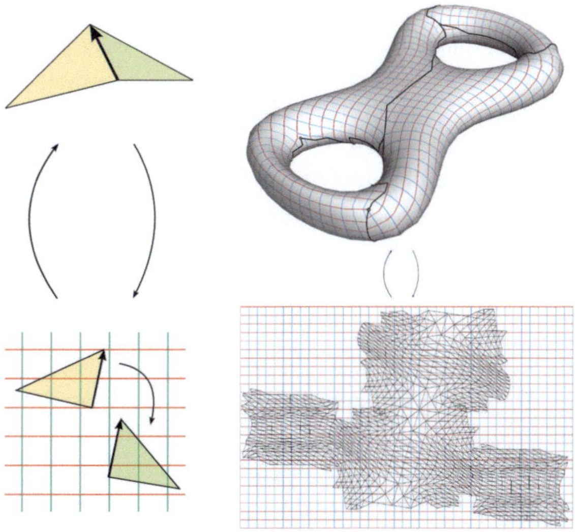

Fig. 11 Transition function between a pair of triangle based charts (*left*) and an atlas on a simplicial pretzel realized by a set of triangle based charts (*right*)

From a computer graphics point of view, the charts are nothing else than element based texture maps from the surface M_h to the Euclidean texture domain $\mathbb{R}^2$. That means that the images of triangles may not be connected to the images of adjacent triangles and there corresponds on chart φ_i to each triangle T_i. The *compatibility of condition* on the charts of a quad-based parametrization requires that the images of two triangles $\{T_i, T_j\} \subset M_h$ with non-empty common edge $e_{ij} = T_i \cap T_i$ are mapped to two triangles $\varphi_i(T_i)$ and $\varphi_j(T_j)$ such that the common edge is mapped to two edges $\varphi_i(e_{ij})$ and $\varphi_j(e_{ij})$ which are translated by an integer vector or rotated by a multiple of 90° or a combination of both. This compatibility condition on the atlas assures continuity of the grid lines on M_h, see Fig. 11.

2.1.1 QuadCover Algorithm

The main steps of the QuadCover algorithm are given below, relying on the discrete concepts introduced in the previous sections. For technical reasons we use a cross field instead of a frame field. Assume a simplicial surface M_h with cross field X is given. Then:

1. (Lift to 4-fold covering space) We lift the cross field to a 4-fold branched covering surface M_h^* of the mesh M_h, where each triangle is covered by four triangles of M_h^*. The four vectors of the cross field on M_h can be lifted to a single PC vector field X^* on M_h^* where each of the four vectors of the cross is lifted to a specific layer of M_h^*, such that maximal continuity is assured.

2. (Local integrability via Hodge) Compute the locally non-integrable curl-component $\delta g \in \delta S_h^*(M_h^*)$ of the vector field X^* on M_h^* and remove it $\overline{X^*} := X^* - \delta g$. Locally, $\overline{X^*}$ is now integrable and has a potential function.
3. (Global integrability via harmonic fields) The global matching of the grid spacing along all loops (so-called *global integrability*) requires that the path integrals of $\overline{X^*}$ along all homology loops $\gamma \in H_1(M_h^*)$ are integer valued. This integrability property is obtained by a correction of $\overline{X^*}$ by the L^2-smallest harmonic field $w \in H$ such that

$$\int_\gamma \overline{X^*} + w \in \mathbb{Z} \text{ for all homology loops } \gamma.$$

   Since the dimension of the space discrete harmonic fields is equal to the dimension of the first homology group $H_1(M_h^*)$, the minimizers exists. Note, harmonic fields are curl*-free, therefore the corrected field $\widetilde{X^*} := \overline{X^*} + w$ is still locally integrable.
4. (Integration) The vector field $\widetilde{X^*}$ is by 2. locally and by 3. globally integrable, thus on each chart on M_h^* we can solve the $1st$-order PDE $\widetilde{X^*} = \nabla\varphi^*$ with a function $\varphi^* \in S_h(M_h^*)$. On each chart, φ^* projects to four functions $(\varphi_1, \varphi_2, -\varphi_1, -\varphi_2)$ with $\varphi_i \in S_h(M_h)$ giving the component functions of the simplicial chart $\varphi = (\varphi_1, \varphi_2) : M_h \to \mathbb{R}^2$, the texture map we were looking for.

Note, step 1 is necessary since frame fields and cross fields typically do not globally decompose into vector fields, thus the Hodge theorem would not apply without the introduced 4-fold covering. Practically, the 4-fold covering surface is usually never created and all information stored otherwise.

The level lines of the chart functions φ^* on M_h^* obtained in the above process will project to the parameter lines on M_h we are looking for. Equivalently, the level lines of the two component functions of the projected chart function $\varphi = (\varphi_1, \varphi_2)$ on M_h directly yield the pair of parameter lines on M_h.

Some sample applications are shown on the two Figs. 12 and 13.

2.2 Implementation Issues

Smooth frame fields are often generated from principal curvature directions, if the surface is smooth. The shape operator on simplicial surfaces derived in [4] provides a reliable pair of principal curvature directions. Due to the instability of umbilical regions for principal curvature computations, often curvature directions are generated in a sparse set of trusted regions and then extended using parallel transport to unreliable regions to cover the whole surface. These extensions are smoothed either by a direct rounding approach [5], an iterative rounding approach [1] or using holomorphic energies [7]. Typically, orthogonal frame fields are used, but non-orthogonal frame fields are possible too, for recent work see also [11].

Fig. 12 QuadCover parametrizations of discrete minimal and constant mean curvature surfaces with boundary alignment

The principal idea behind the optimization is to find a texture map whose per triangle gradients align best to the input frame fields. The Hodge energy is either linearized by taking partial derivatives or put into a quadratic form. Solving the two linear systems is fast and widely used in frame field-driven parameterization methods [1, 3, 5, 8]. Unfortunately, it does not guarantee injectivity so that recent methods [2] rather minimize the quadratic energy using advanced non-linear solvers. The size of the system, in both methods, can be dramatically reduced by first introducing a dual spanning tree which defines a cut graph and then observing that the transition functions between triangle charts must be a grid automorphism [5] in $\mathbb{R}^2$. The cut graph is necessary to be able to flatten the mesh in a 2D domain. The grid automorphism condition assures the local integrability of the resulting parameterization.

Notice that the parameterization generated by the least square minimization has seams along the cut path of the geometry. This is due to the local integrability of the curl free frame field. To obtain a globally continuous parameterization, the patch integrals along all homotopy loops must be integers. Adding these constraints to the list square minimization makes the problem a mixed integer optimization which is an NP-hard problem. Several heuristics have been proposed to approximate the solution. The QuadCover algorithm [5] uses a simultaneous rounding via corrections with harmonic vector fields along the homology generators. The Mixed Integer Quadrangulation [1] approaches the problem by an iterative rounding which approximate a better solution at the cost of a slower running time. Recent methods consider the

Fig. 13 Extensions of QuadCover to stripe and hexagonal parametrizations, as well as its generalization to CubeCover, the cubification of 3d volumes

problem as a mixed integer quadratic problem [2] which aims at the global minimum of the energy.

The QuadCover algorithm is implemented based on JavaView available at www.javaview.de.

2.3 Examples of Parameterizations

We give some example of parameterizations generated by the QuadCover algorithm, among them are minimal and constant mean curvature surfaces as well as non-orientable surfaces such as the Klein bottle. A second slide shows the generalizations StripeCover [6] and HexCover [10] for stripe covering and hex-covering. The CubeCover algorithm [9] is an extension to 3d volumetric meshes which converts a tetrahedral mesh with 3D frame field to a uniform cubical mesh aligned with the surface boundary.

References

1. Bommes, D., Zimmer, H., Kobbelt, L.: Mixed-integer quadrangulation. ACM Trans. Graph. **28**(3), 77:1–77:10 (2009)
2. Bommes, D., Campen, M., Ebke, H.-C., Alliez, P., Kobbelt, L.: Integer-grid maps for reliable quad meshing. ACM Trans. Graph. **32**(4), 98:1–98:12 (2013)
3. Campen, M., Kobbelt, L.: Quad layout embedding via aligned parameterization. Comput. Graph. Forum **33**, 69–81 (2014)
4. Hildebrandt, K., Polthier, K.: Generalized shape operators on polyhedral surfaces. Comput. Aided Geom. Design **28**(5), 321–343 (2011)
5. Kälberer, F., Nieser, M., Polthier, K.: Quadcover—surface parameterization using branched coverings. Comput. Graph. Forum **26**(3), 375–384 (2007)
6. Kälberer, F., Nieser, M., Polthier, K.: Stripe parameterization of tubular surfaces. In: Pascucci, V., Hagen, H., Tierny, J., Tricoche, X. (eds). Topological Methods in Data Analysis and Visualization. Theory, Algorithms, and Applications. Mathematics and Visualization. Springer, New York (2010)
7. Knöppel, F., Crane, K., Pinkall, U., Schröder, P.: Globally optimal direction fields. ACM Trans. Graph. **32**(4) (2013)
8. Myles, A., Pietroni, N., Kovacs, D., Zorin, D.: Feature-aligned t-meshes. ACM Trans. Graph. **29**(4), 1–11 (2010)
9. Nieser, M., Reitebuch, U., Polthier, K.: CubeCover—parameterization of 3d volumes. Comput. Graph. Forum **30**(5), 1397–1406 (2011)
10. Nieser, M., Palacios, J., Polthier, K., Zhang, E.: Hexagonal global parameterization of arbitrary surfaces. IEEE Trans. Visual. Comput. Graph. **18**(6), 865–878 (2012)
11. Panozzo, D., Puppo, E., Tarini, M., Sorkine-Hornung, O.: Frame fields: anisotropic and non-orthogonal cross fields. ACM Trans. Graph **33**(4), 134:1–134:11 (2014)
12. Polthier, K.: Unstable periodic discrete minimal surfaces. In: Hildebrandt, S., Karcher, H. (eds). Geometric Analysis and Nonlinear Partial Differential Equations, pp. 127–143. Springer, New York (2002)
13. Polthier, K., Preuss, E.: Identifying vector field singularities using a discrete Hodge decomposition. In: Hege, H.-C., Polthier, K. (eds). Visualization and Mathematics III, pp. 113–134. Springer, New York (2003)

Erratum to: Age Structures in Mathematical Models for Infectious Diseases, with a Case Study of Respiratory Syncytial Virus

Alexandra B. Hogan, Kathryn Glass, Hannah C. Moore and Robert S. Anderssen

Erratum to:
R.S. Anderssen et al. (eds.), *Applications + Practical Conceptualization + Mathematics = fruitful Innovation*, Mathematics for Industry 11, DOI 10.1007/978-4-431-55342-7_9

Erratum DOI 10.1007/978-4-431-55342-7_23

The book was unintentionally published with an incorrect placement of equations in chapter 'Age Structures in Mathematical Models for Infectious Diseases, with a Case Study of Respiratory Syncytial Virus', the chapter and the book has been updated by deleting the equation from the respective chapter.

The updated original online version for this chapter can be found at
DOI 10.1007/978-4-431-55342-7_9

A.B. Hogan (✉) · K. Glass
The Australian National University, Canberra, Australia
e-mail: alexandra.hogan@anu.edu.au

K. Glass
e-mail: kathryn.glass@anu.edu.au

H.C. Moore
The University of Western Australia, Perth, Australia
e-mail: hannah.moore@telethonkids.org.au

R.S. Anderssen
The Commonwealth Scientific and Industrial Research Organisation (CSIRO), Acton, Australia
e-mail: bob.anderssen@csiro.au

R.S. Anderssen et al. (eds.), *Applications + Practical Conceptualization + Mathematics = fruitful Innovation*, Mathematics for Industry 11,
DOI 10.1007/978-4-431-55342-7_23

Index

R.S. Anderssen et al. (eds.), *Applications + Practical Conceptualization + Mathematics = fruitful Innovation*, Mathematics for Industry 11,
DOI 10.1007/978-4-431-55342-7

MIX
Papier aus verantwortungsvollen Quellen
Paper from responsible sources
FSC® C105338

If you have any concerns about our products,
you can contact us on
ProductSafety@springernature.com

In case Publisher is established outside the EU,
the EU authorized representative is:
Springer Nature Customer Service Center GmbH
Europaplatz 3, 69115 Heidelberg, Germany

Printed by Libri Plureos GmbH
in Hamburg, Germany